SPACEFLIGHT LIFE SUPPORT AND BIOSPHERICS

Author

PETER ECKART
Research Fellow at the Institute of Astronautics
Technical University of Munich, Germany
Graduate of the International Space University
Summer Session, Kitakyushu, Japan, 1992
Research Fellow at the NASA Johnson Space Center
Crew and Thermal Systems Division, Houston, TX 1994/95

Drawings: Herbert Utz, Munich, Germany   Layout: Thomas R. Neff, Munich, Germany

SPACE TECHNOLOGY LIBRARY

Editorial Board

Also in the Space Technology Library

An Introduction to Mission Design for Geostationary Satellites, J. J. Pocha
Space Mission Analysis and Design, James R. Wertz and Wiley J. Larson
Handbook of Geostationary Orbits, E. M. Soop
Spacecraft Structures and Mechanisms, Thomas P. Sarafin

Spaceflight Life Support and Biospherics

by

Peter Eckart

Revised and Expanded from *Life Support & Biospherics*,
Herbert Utz Publishers, Munich, Germany, 1994

Space Technology Library

Published Jointly by

Microcosm Press
Torrance, California

Kluwer Academic Publishers
Dordrecht / Boston / London

Library of Congress Cataloging-in-Publication Data

A C.I.P. Catalogue record for this book is available from
the Library of Congress

ISBN 1-881883-04-3 (pb) (acid-free paper)
ISBN 0-7923-3889-8 (hb) (acid-free paper)

Published jointly by
Microcosm Press
2601 Airport Drive, Suite 230, Torrance, CA 90505 USA
and
Kluwer Academic Publishers,
P.O. Box 17, 3300 AA Dordrecht, The Netherlands.

Kluwer Academic Publishers incorporates
the publishing programmes of
D. Reidel, Martinus Nijhoff, Dr. W. Junk and MTP Press.

Sold and distributed in the USA and Canada
by Microcosm, Inc.
2601 Airport Drive, Suite 230, Torrance, CA 90505 USA
and Kluwer Academic Publishers,
101 Philip Drive, Norwell, MA 02061 USA

In all other countries, sold and distributed
by Kluwer Academic Publishers Group,
P.O. Box 322, 3300 AA Dordrecht, The Netherlands.

Printed on acid-free paper

Printed in the United States of America

To Nuschin

CONTENTS

PREFACE

Arthur C Clarke, CBE

LIFE SUPPORT SYSTEMS BY PETER ECKART

I would like to congratulate Peter Eckart on his monumental study 'Life Support & Biospherics' on human life in extraterrestrial environments. (In fact many of his conclusions are applicable to terrestrial environments as well!)

Although there is now a certain hiatus in manned exploration of space, beyond close-to-earth-orbits, this is only a temporary state of affairs. Early in the next century, the developement of more efficient and completely reusable space transportation systems will once again open up the path to the Moon - and then, to Mars, and the worlds beyond.....

It is not generally realized that the cost in energy of sending a man into space is less than $100: the fact that we now have to pay millions of times as much is merely a measure of our present incompetence. The time will come when space flight will be as affordable as jet transportation is today.

I hope that Mr Eckart - who I am happy to note is a graduate of the International Space University - will live to see that day. When it arrives his book may well become a standard textbook, for the pioneers of our first homes beyond the Earth.

Arthur C Clarke, CBE

Fellow of King´s College, London
Chancellor: International Space University
Chancellor: University of Moratuwa

23 May 1994, Colombo, Sri Lanka

FOREWORD

I am not an expert in anything. I am the person who keeps my eyes open, watches from the sides and becomes aware of where things might come together that otherwise might not come together.

Jacques Cousteau

About two years ago I was looking for a book summarizing the basic knowledge on life support systems - their history, requirements, components. Furthermore, I was interested in a book that discusses the links between life support systems and biosphere research. I wanted to find out what kind of life support systems for application in space could be expected in the future. Also, I was interested to learn to which extent the research on ecological questions and on the development of future life support systems for spaceflight are related. After visits and discussions at ESA-ESTEC, NASA-MSFC, Boeing and other institutions I knew that such a book did not exist.

Thus, I decided to write "Life Support & Biospherics", which was published in late 1994 with Herbert Utz Publishers. Only one year later and thanks to the success of the first edition and Kluwer Academic Publishers / Microcosm Inc., you hold "Spaceflight Life Support & Biospherics" in your hands, now - the revised and updated edition of "Life Support & Biospherics". The book is intended to be the "Introduction to Life Support Systems" that has been lacking. The structure of this book is such that it, step by step, answers the basic questions concerning life support systems of any scale - from small microbial systems to the Earth's biosphere. The questions relate to the chapters as follows:

Chapter 1 *Why life support system development and biosphere research?*

Chapter 2 *How does our natural life support system, the biosphere, work?*

Chapter 3 *What are the environmental conditions for life support systems in space?*

Chapter 4 *What are the fundamental terms and requirements of life support?*

Chapter 5 *Which physico-chemical life support subsystems do currently exist?*

Chapter 6 *Which are the potential bioregenerative life support technologies of the future?*

Chapter 7 *What are the experiences of the largest artificial ecosystem - Biosphere 2?*

Chapter 8 *What are life support systems of future planetary habitats going to look like?*

Chapter 9 *What are the potential terrestrial benefits of life support development?*

"Spaceflight Life Support & Biospherics" is intended to be a source of information for everyone involved in the life support system design and development process, and for all those who are simply interested in this exciting disciplin. I hope, that it will not only be a useful tool, but that it may also be enjoyed by everyone who is interested in the preservation of planet Earth and the future exploration of our extraterrestrial environment.

Like everything created by a human, this volume cannot be perfect. I am convinced that somewhere on our planet very exciting research projects are underway that should be mentioned here. I want to apologize at this point to have not been aware of those. Also, I want to encourage everyone to submit his / her comments and suggestions how any future edition of this book could be improved.

Of course, I cannot end this foreword without saying 'Thank you !' - it would have been impossible to complete this work without the invaluable help and support of several institutions, acquaintances, and friends all over the world. I am very much indebted to all of them and, thus, I want to express my deep gratitude to:

Prof. Dr.-Ing. Harry O. Ruppe - my teacher and promoter, Head of the Chair of Astronautics of the Technische Universität München, Germany, for his continuous support and for giving me maximum freedom in my work.

Herbert Utz - Managing Director, Herbert Utz Publishers, and Research Fellow at the Chair of Astronautics of the Technische Universität München, Germany, for his endless patience in drawing all the figures, reading and rereading the script over and over again, and for being my friend in every respect.

Thomas Neff - Director of Finances, Herbert Utz Publishers, and Research Fellow at the Chair of Lightweight Construction of the Technische Universität München, Germany, for doing all the layout work for this book plus an endless amount of further major and minor jobs, helping to give birth to our "baby".

Ursula Kiening - our wonderful secretary at the Institute of Astronautics, a special thanks for all her support.

Susan Doll - Life Support Engineer, Boeing, Huntsville, AL, for being a wonderful host, supporter, and friend - our "Woman in the U.S.".

Dr. James Wertz - Microcosm Inc., CA, for the chance to publish "Spaceflight Life Support & Biospherics", and for being a very cooperative publisher.

Dr. Wiley Larson - U.S. Air Force Academy, Colorado Springs, CO, for initiating the contact with Jim Wertz that led to the publication of this book.

Bernd Zabel - "Biospherian" and General Manager of Construction of Biosphere 2 for also being my host and showing me around the Biosphere 2 complex.

Dr. Christian Tamponnet - Life Support Division, ESA-ESTEC, Noordwijk, The Netherlands, for inspiring my work and taking his time to support.

Dr. Arthur C. Clarke, CBE - for encouraging our work and submitting the preface.

All of my friends of the **International Space University** Community, at the **Lehrstuhl für Raumfahrttechnik,** and the **Johnson Space Center** for their inspiration.

And to my **Nuschin**.

Peter Eckart
München, Germany, August 1995

I RATIONALE

Pictures of Earth taken from space show us how unique and beautiful our planet is and how fragile and alone it looks in space. Today, our global life support system that provides air, water, food, and power is being stressed by pollution, poor management, and population pressure. There are numerous early warning signs that are beginning to appear at various places, e.g., excessive erosion of best agricultural soils, and dying trees in industrial regions.

At 10:13 p.m., April 13, 1970, as the spacecraft of the APOLLO 13 mission neared Moon, an explosion occurred in the service module, and alarm lights flashed on the command module control panel. It was quickly evident that the explosion had ruptured one or both oxygen tanks. Computers and staff at Mission Control sprang into feverish activity to plot rescue options that would use the lunar module, which had its own life support apparatus, as a lifeboat. It was not exactly known how long the command module would be inhabitable, because information on "consumables" was in pieces, with no total picture on which to base an estimate of how much time was available to get the astronauts safely back to Earth. Precious time was wasted putting all the pieces of information together. One is reminded that a similar situation exists here on Earth. We do not have the total picture of our life support "consumables" or understand how they interact.

The life support systems so far used in manned spaceflight are mechanically controlled "storage systems". For the most part, vital necessities such as oxygen and food are produced on Earth and stored onboard, not regenerated as they are here on Earth. Likewise, waste products such as carbon dioxide and urine are chemically stored, not recycled. In contrast, Earth is bioregenerative, i.e., plants, animals, and especially microorganisms regenerate, recycle, and control life's necessities.

Biospherics - A New Science

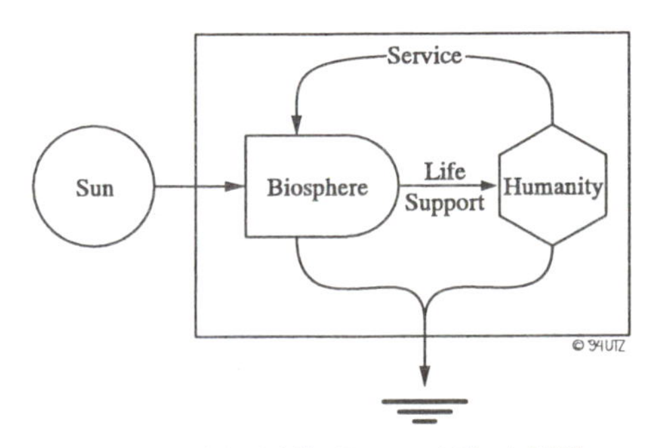

Figure I.1: **A Life Support Model [8]**

The diagram in figure I.1 shows that the biosphere, nourished by the Sun, provides life support for humanity, including all the artifacts that make our lives rich and comfortable. We have to pay for the servicing of this wonderful biomachine if we expect to continue to receive its goods and services. In this case, "service" involves protection of vital parts, maintenance of vital functions, and repair when we overtax the biosphere's selfrepair capacity. Mankind is now entering a new age where it is going to have to pay for the "free" goods and services, because billions of people are begetting more billions, whose expectations and demands on our life support system are growing. Since man did not build Earth's life support systems, and since they involve a complex array of subsystems, there is not yet a clear understanding of how the whole thing works. But the necessity of comprehending the laws of development and functioning of the biosphere as a single whole is becoming more and more obvious and urgent.

Because it is so precious and fragile, the biosphere cannot be subjected to any direct study which may harm it. Therefore, this science depends on the study of analogs and small models, i.e., artificial ecological systems with differing degrees of complexity and closure. On such model ecosystems both the development of individual elements and components of the ecosystems, and the general principles of turnover of the entire biospheric system can be studied. The principal objects of study are closed ecological systems, from simple microsystems to more sophisticated human life support systems, under extreme conditions on Earth and in space.

The name for this new discipline, "Biospherics", was first discussed at the First International Conference on Closed Life Systems, held under the auspices of the Royal Society in London in July 1987. Biospherics was seen as an integrating science drawing on the achievements of many individual sciences. The term Biospherics was finally adopted by unanimous vote of delegates from Russia, ESA, UK, and USA during the Second International Conference on Closed Life Systems in Krasnoyarsk, Siberia, in September 1989. A resolution from the same conference defined the goals of this new discipline :

- To create working models of the Earth's biosphere and its ecosystems and thus to understand better the regularities and laws that control its life.

- To create biospheres for human life support beyond the limits of the Earth's biosphere.

- To create ground-based life support systems that provide a high quality of life under the extreme conditions of the Earth's biosphere, as at polar latitudes, deserts, mountains, and under water.

- To use closed ecological systems to develop technologies for the solution of pollution problems in urban areas and for developing high yield sustainable agriculture. [1]

Since so far all attempts to build a large bioregenerative life support system that would support a large number of people in space, without supply of a "umbilical cord" to Earth, have failed, the stay of man in space is still limited by the amount of life supporting consumables that can be carried onboard. Human presence on Moon or Mars will be characterized by long mission duration and by very limited chances of resupply from Earth. This leads to the need for a high degree of self-sufficiency which cannot be achieved by techniques presently applied in the life support subsystems of the current manned projects. Future life support in space will involve the progressive integration of biology-based components into life support systems, gradually replacing physico-chemical, i.e., first generation subsystems, and leading to a scenario of a fully closed, so called Controlled Ecological Life Support System (CELSS). The development of CELSS technology is a complex and long-term activity. Due to this complexity, theoretical predictions for the evolution of such closed ecological systems are very difficult and need to be supported and corrected by experiments and observations both on ground and in space. To date, at least on the ground, many experiments to study those systems have been set up, e.g., BIOS-1 to 3 and Biosphere 2, to mention only the biggest and most prominent ones. [13]

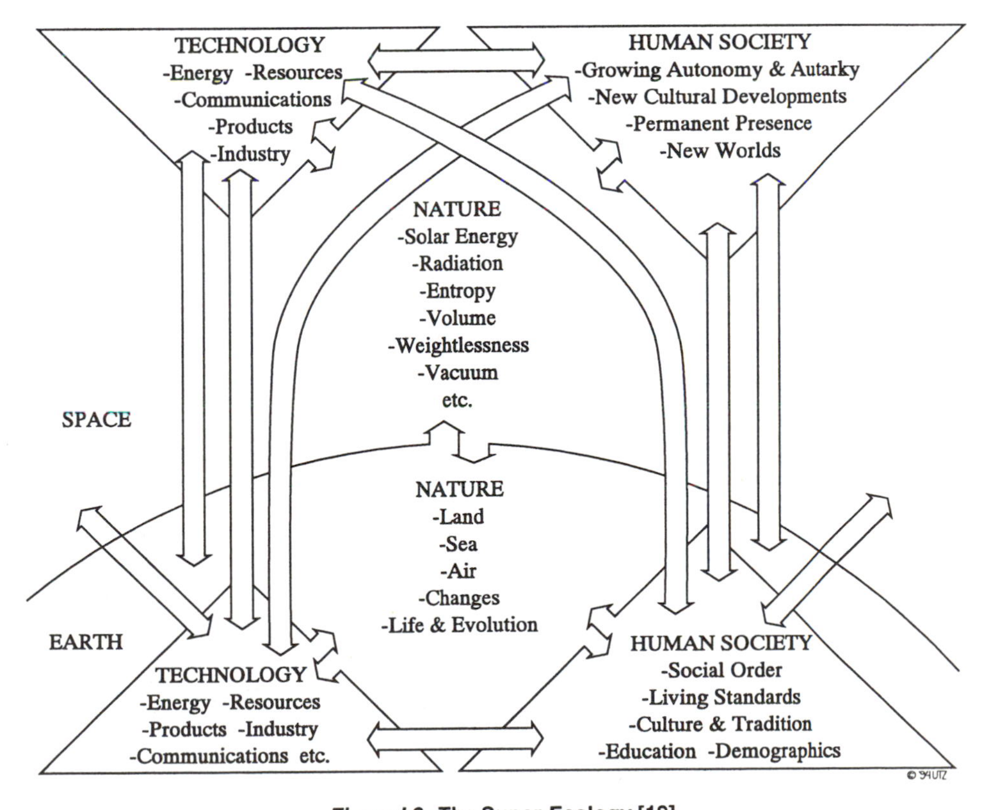

Figure I.2: **The Super-Ecology [19]**

What is mentioned above indicates how challenging and important the work on any kind of life support system is. At this point, the work of two disciplines that, at least at first sight, do not seem to have much in common, mingles. The science of ecology provides the background for the understanding. Anyway, the concept of ecology, as introduced by Ernst Haeckel in 1866, meant the interactions in the environment of land, sea, and air. It was a geocentric view which even today continues to govern the thinking of environmentalists. It totally ignores the fact that Earth is not at all alone and not the center of universe, but that in reality it is embedded in space. By adding space to land, sea, and air as a fourth regime, a new type of super-ecology is being created which introduces many more interactive loops, as shown in figure I.2. That Earth's biosphere and outer space are not separated should be kept in mind when reading through this book. At this point traditional ecology may learn from those dealing with the development of life support systems for application in space, since an ecology neglecting the influence from outer space on the biosphere is incomplete.

The Future of Manned Spaceflight

According to the development plan of the U.S. Space Exploration Initiative (SEI), given in figure I.3, the first level concerning the presence of man in space will be to achieve easy access into and return from space. The second level would be the accomplishment of permanent presence in space by means of the space station. It will serve as a laboratory platform and test bed on which to develop the technologies leading to the next level - the capability to go beyond Earth orbit, to Moon and Mars.

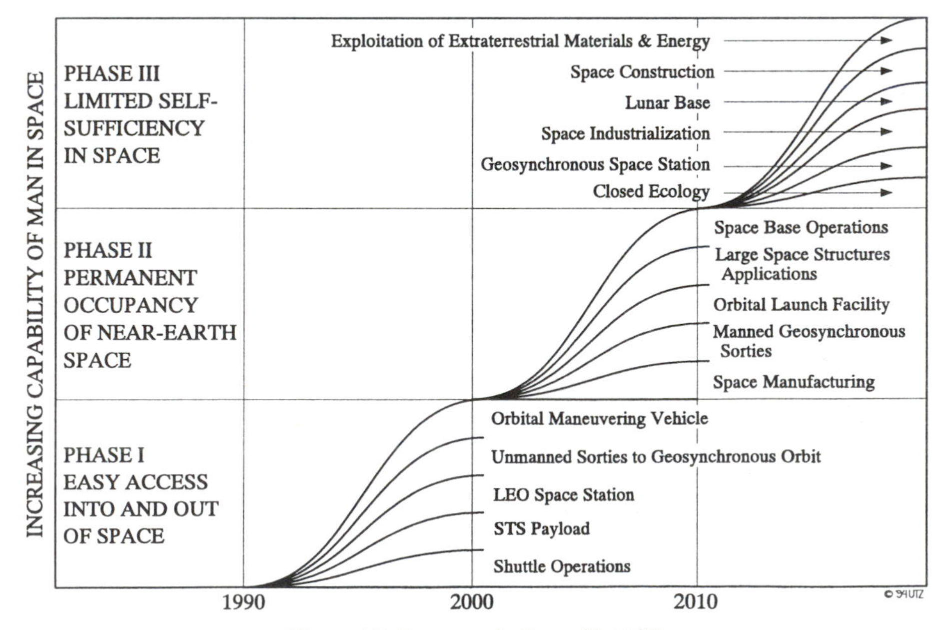

Figure I.3: **Progress in Spaceflight [6]**

The international space station, to be built in the coming years, is, first of all, designed as a laboratory for scientific research, for studying life sciences and materials processing. The evolutionary extension of the space station after the year 2000 will lead to an orbital launch pad, i.e., a kind of spaceport. From this spaceport, humans may then be able to assemble platforms, antennas and other application facilities in LEO and GEO. It will be at that time that mankind may also establish a permanently manned base on Moon and humans may fly to Mars. By then, bioregenerative life support systems must be available, as also indicated in the figure.

Although the above mentioned explains that there are many common goals of ecologists and space life support system engineers, one big difference still remains. While the need to preserve mankind's basis of existence, Biosphere 1, is absolutely obvious, the goal of a permanent human presence in space always raises one big question: *Why?* Why spend money and man- power on such a dubious adventure while facing much more urgent problems to be solved here on planet Earth? The author is well aware that it is not easy to find a really satisfying answer to this question. Especially, if it shall not be an answer involving doubtful economical justifications and extrapolations. These kinds of justifications have, unfortunately, been given too often in the past. After space enthusiasts have learned not to make such uncertain promises, it is currently in fashion to come up with always new "reason-why reports" that, in general, give no, or at best only doubtful, evidence for the necessity of manned spaceflight.

In principle, there are three classes of potential benefits of spaceflight as listed in table I.1. It is unpredictable to which extent all of these, or even only a few benefits, may become reality in the near future, since there are too many technological, economical, and political problems that may occur on the way, and these cannot be estimated or extrapolated. In fact, the manned exploration of the solar system is a cultural challenge - not more and not less. One day mankind will go to Moon, Mars, and beyond. Does it matter whether it will be today or tomorrow? It's on us. At least, the fields described in the following chapters constitute a lot of potential for interdisciplinary and also international cooperation focusing on the preservation of our natural environment and also the manned exploration of space - not more but not less. [6]

Classes of Potential Benefits	**Comments**
Intellectual	Derived from science - new knowledge and new technologies
Utilitarian / Materialistic	Industrial products, terrestrial applications, commercialization
Humanistic	Communications and informations from space, e.g., social and health services, international cooperation, cultural and spiritual effects

Table I.1: **Potential Benefits from Spaceflight**

I References

[1] Blüm V.
Ecosystems on Earth and in Space
European International Space Year Conference, ESA, ISY-4, p. 393-398, 1992

[2] Brack A.
Exobiology and Terrestrial Life
European International Space Year Conference, ESA, ISY-4, p. 387-391, 1992

[3] Brown L. et al
State of the World 1992
Earthscan Publications Ltd., London, 1992

[4] Gethmann C. et al
Bemannte Raumfahrt im Widerstreit
DLR-Nachrichten, Vol. 68, p. 10-14, 1992

[5] Hallmann W.; Ley W.
Handbuch der Raumfahrttechnik
Carl Hanser Verlag, München, 1988

[6] Harris P.
Living and Working in Space
Ellis Horwood Ltd., Chichester, 1992

[7] Horneck G.
European Activities in Exobiological Research in Space
Life Sciences Research in Space, ESA, SP-271, p. 185-192, 1987

[8] Horneck G.
Life In and From Space
European International Space Year Conference, ESA, ISY-4, p. 207-210, 1992

[9] Kiefer J.
Potentials, Message and Challenges of Life Science Research in Space
Life Sciences Research in Space, ESA, SP-307, p. 639-642, 1990

[10] Lovelock J.
The Ages of Gaia
Bantam Books, New York, 1990

[11] Lubchenko J. et al
The Sustainable Biosphere Initiative: An Ecological Research Agenda
Ecology, 72(2), p. 371-412, 1991

[12] Ockels W.
Why Life in Space? The Step to Go Beyond
Life Sciences Research in Space, ESA, SP-271, p. 315-316, 1987

[13] Odum E.P.
Ecology and Our Endangered Life Support Systems
Sinauer Associates Inc., Sunderland, 1990

[14] Perchurkin N.
Biospherics: A New Science
Life Support & Biosphere Science, Vol. 1, No. 2, p. 85-87, 1994

[15] Ruppe H.O.
Die grenzenlose Dimension Raumfahrt, Vol. 1+2
ECON Verlag, Düsseldorf, 1980

[16] Stafford T.
America at the Threshold
NASA, 1991

[17] Vester F.
Neuland des Denkens
Deutscher Taschenbuch Verlag, München, 1984

[18] von Puttkamer, J.
Der Mensch im Weltraum - Eine Notwendigkeit
Umschau Verlag, Frankfurt, 1987

[19] von Puttkamer, J.
Handout of Presentation, Symposium on Aerospace and Environment, München, 1992

II BIOSPHERE 1 - THE LIFE SUPPORT SYSTEM OF EARTH

II.1 THE TERRESTRIAL ENVIRONMENT

In this chapter the basic features of the terrestrial environment are summarized. A lot of numbers concerning Earth's geology and atmosphere are given. Also, the radiation from outer space that penetrates Earth, magnetic fields, and gravity are discussed.

II.1.1 Geology

Planet Earth is not an exact sphere. While the average radius of Earth can be given as 6370 km, in fact, the radius at the equator is 6378 km and at the poles it is 6356 km. The overall surface is $5.1 \cdot 10^8$ km². The mass of Earth is $5.97 \cdot 10^{24}$ kg, the volume is $1.083 \cdot 10^{21}$ m³ and, thus, the average density is 5514 kg/m³. Earth basically consists of three different layers: the core ($0 < r < 3600$ km), the mantle (3600 km $< r <$ 6330 km), and the crust ($6330 < r < 6370$ km, i.e., Earth's surface). Some characteristics of these layers are summarized in table II.1.

	Crust	Mantle	Core
Temperature	300-1500 K	1500-4000 K	4000-7000 K
Pressure	$1 - 0.2 \cdot 10^6$ bar	$0.2 \cdot 10^6 - 1.5 \cdot 10^6$ bar	$1.5 \cdot 10^6 - 4 \cdot 10^6$ bar
Main Components	O, Si, Al, Fe	O, Si, Mg, Fe	Fe, Ni, Co

Table II.1: **Characteristics of the Layers of Earth [13]**

Chemically, Earth mainly consists of Oxygen (49.4 %), Silicium (25.8 %), Aluminum (7.5 %), Iron (4.7 %), Calcium (3.4 %), Sodium (2.6 %), Potassium (2.4 %), Magnesium (1.9 %), Hydrogen (0.9 %), and Titan (0.6 %). While 70.7 % of Earth's surface are covered with water ($3.61 \cdot 10^8$ km²), only 29.3 % are covered with land ($1.49 \cdot 10^8$ km²). All oceans together contain about $1.37 \cdot 10^9$ km³ of water. It may be interesting to know that this is about 97.2 % of all the water on Earth. Another 2.8 % of all the water are underground or frozen in glaciers. Only 0.01 % of all Earth's water can be found in rivers, lakes, and the atmosphere ($1.4 \cdot 10^7$ km³). [13]

II.1.2 Atmosphere

The atmosphere of Earth can be divided into four different layers:

- Troposphere 0-11 km
- Stratosphere 11-50 km
- Mesosphere 50-80 km
- Thermosphere > 80 km

The distribution of temperature, pressure, and density in these layers is outlined in figure II.1.

At ground level Earth atmosphere is composed of 78.08 % N_2, 20.95 % O_2, 0.93 % Argon, 0.03 % CO_2, and minimum quantities of other gases. Additional components include trace contaminants, dust, and smoke particles. Water vapor is also present in amounts up to the saturation level. The degree of saturation, i.e., the relative humidity, depends on temperature and pressure, e.g., at 15° C the saturation pressure of water vapor is 1.7 kPa and the absolute humidity is 12.8 g/m³. [5,11]

Atmospheric pressure is induced by the weight of the atmosphere itself. Thus, it decreases with increasing distance from Earth's surface. The nominal atmospheric pressure at sea level is p_0 = 101.3 kPa. For altitudes a of up to 11000 m the atmospheric pressure can be determined with the following formula:

$$p = p_0 \left(1 - \frac{0.0065a}{288}\right)^{5.255} kPa$$

Further characteristics of Earth's atmosphere are summarized in table II.2. [5]

Characteristics		Value
Density	r	1.29 kg/m³
Dynamic Viscosity	h	0.0181 mPa·s
Gas Constant	R	287 J/kgK
Specific Heat Capacity	c_p/c_v	1.009 / 0.72 kJ/kgK
Conductivity	l	0.02 W/mK

Table II.2: **Atmosphere Characteristics at Sea Level (Nominal, i.e., at 0° C and p_0) [4]**

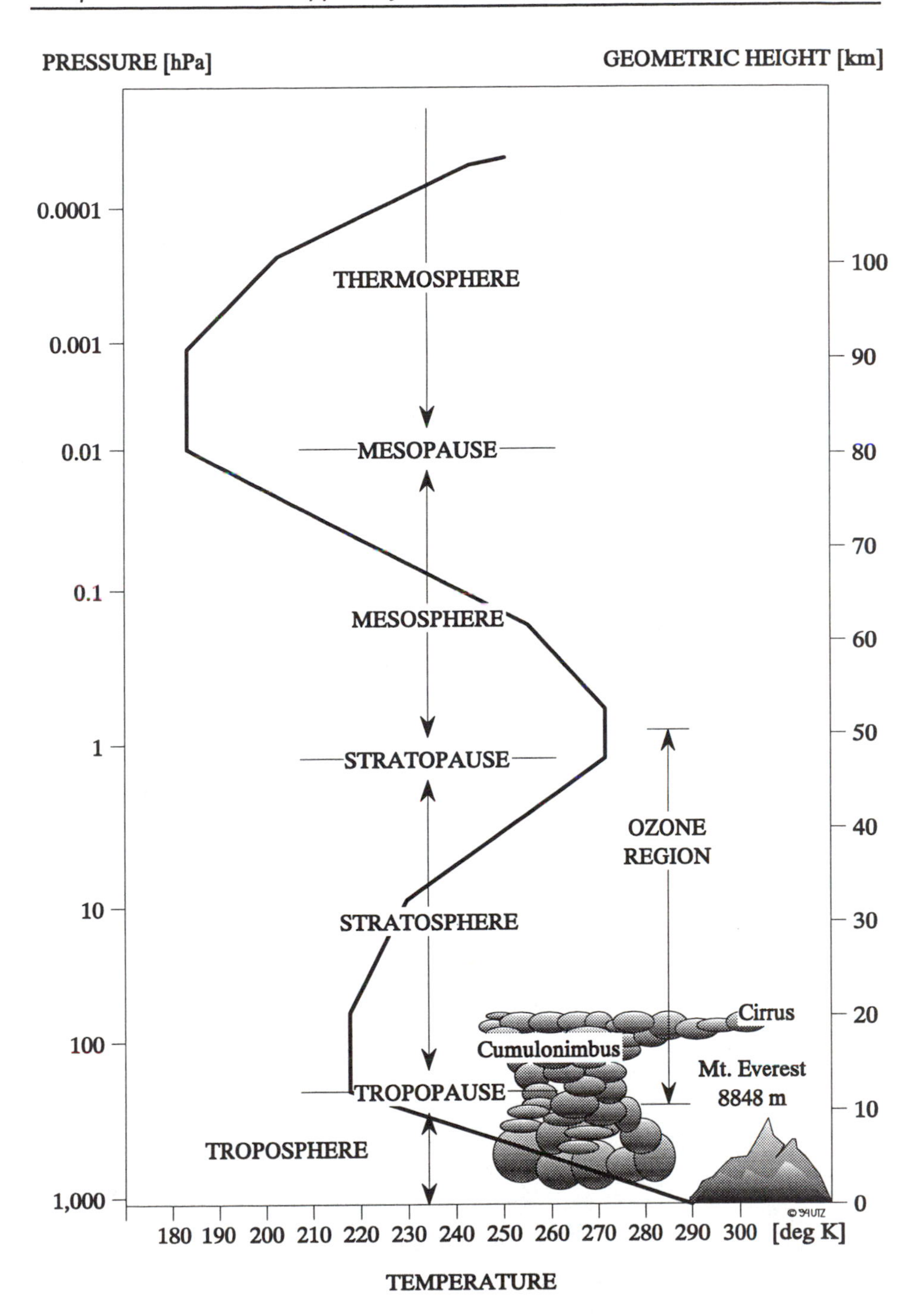

Figure II.1: **Temperature, Pressure, and Density Distribution of Earth's Atmosphere [12]**

II.1.3 Gravity

Not only many physical, but also most biological processes on planet Earth are very much dependent on the existence of the terrestrial gravity field. This includes, of course, also the life and health of all human beings. The general formula for the gravity force *F* acting on any mass *m* in Earth's gravitational field is:

$$F = mg$$

The average value of the gravity constant on Earth's surface (r_0 = 6370 m) is g_0 = 9.81 m/s². Thus, for objects at higher altitudes the following formula may be applied:

$$g = g_0 \left(\frac{r_0}{r}\right)^2$$

Distance from Earth's Center: r

The role of gravity in free space, and especially the effects of zero gravity, are further dealt with in section III.2.

II.1.4 Radiation

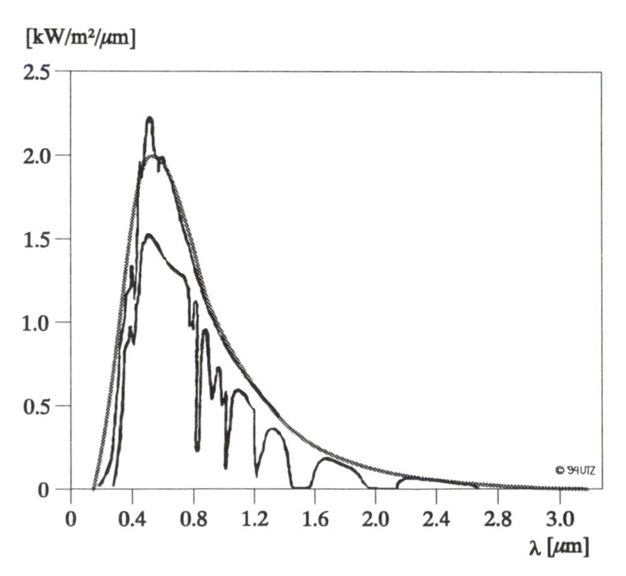

Figure II.2: **Solar Spectra above and below the Atmosphere**

Earth is constantly hit by all kinds of radiation. Basically, one can distinguish electromagnetic and ionizing radiation. Earth's atmosphere and the magnetic field block most of their harmful components, e.g., most of the ionizing component that is dealt with in great detail in section III.1. Parts of the electromagnetic radiation, which is mainly emitted by the Sun, are filtered by the atmosphere. This can be seen in figure II.2, where the two curves for the electromagnetic spectrum in outer space and on Earth's surface are compared. Note that especially most of the UV (0.01-0.4 μm) is filtered by the atmosphere, i.e., mainly by the ozone in the stratosphere. Further details on electromagnetic radiation may be found in section III.1.1.

II.1.5 Magnetic Field

Although Earth's magnetic field has no direct impact on the terrestrial environment its existence is essential for the survival of life on Earth. This is because the magnetic field shields most of the ionizing radiation coming from outer space. The fact that this protection is missing in outer space poses great problems to manned spaceflight (see section III.1).

II.2 FUNDAMENTALS OF ECOLOGY

According to the best understanding of geological history, Earth did not in the beginning support life. The first tiny microorganisms that appeared more than two billion years ago had to survive in an environment with no oxygen, lethal ultraviolet radiation, poisonous gases, and extreme temperature variation, i.e., conditions that would be lethal to much of today's life. Over millions of years, organisms interacting with geological and chemical processes gradually changed the environment by putting oxygen into the atmosphere and forming a green mantle over the surface of Earth, where sunlight could be converted into all kinds of food, supporting an increasing number and variety of creatures, including, eventually, humans. These facts are discussed by James Lovelock who states in his Gaia Hypothesis (see section II.2.6.) that "the biosphere is a self-regulating entity with the capacity to keep our planet healthy by controlling the physical and chemical environment". In fact, man is able to breathe, drink, and eat in comfort, because millions of organisms and hundreds of processes are operating in a coordinated manner out there in the environment. Life support is provided by a vast, diffuse network of processes operating on different time scales. Unfortunately, there is a tendency to take nature's services for granted because no money has to be paid for most of them. [6,10]

The ecological systems and processes that provide life support can be identified. To do this one must think about the environment as a whole and partition the landscape into functional units in some systematic manner. What can be seen in an aerial view from an airplane can be listed under three categories: fabricated, domesticated, and natural environments. In less formal language, the landscape can be divided into developed sites, cultivated sites, and natural sites. The developed environment includes cities, industrial parks, and transportation corridors. From the standpoint of energy use the fabricated environment is comprising fuel-powered systems. The domestic environment includes agricultural lands, managed woodlands and forests, and artificial ponds and lakes. This part of the landscape is made up of what ecologists call subsidized solar-powered systems. The Sun provides the basic energy, but this source is augmented by human-controlled work energy in form of human labor, machines, fertilizers, etc. "Self-supporting" and "self-maintaining" are the key-words characterizing the natural environment. Natural areas operate without energetic or economic flows directly controlled by humans. These are basic solar-powered sources dependent on sun-light and other natural forces that are indirect forms of solar energy, e.g., wind and rainfall. Natural environments include natural streams, rivers, woodlands, prairies, mountains, lakes and the oceans.

The three basic media that support life are air (atmosphere), water (hydrosphere), and soil (pedosphere). Soil is the product of physical weathering of Earth's crust and activities of organisms, especially vegetation and microorganisms. Soil is composed of distinct layers, which often differ in color. These layers are called soil horizons, and the sequence of horizons is called soil profile. The *A horizon* (topsoil) is composed of bodies of plants and animals reduced to finely divided organic matter, mixed with clay, sand, silt, and other mineral matter. The *B horizon* is composed of mineral soil in which organic matter is mineralized. The *C horizon* is more or less unmodified parent material, i.e., original geological material that is disintegrating in place.

The life support environment can be defined as that part of Earth that provides the physiological necessities of life, namely, food and other energy, mineral nutrients, air, and water. The functional term "life support system" is used to describe the environment, organisms, processes, and resources interacting to provide these physical necessities. Here, processes mean such operations as food production, water recycling, waste assimilation, air purification, etc. Some of these processes are organized and controlled by humans, but many are natural and driven by solar or other natural energies. All life supporting processes involve the activities of organisms other than human, i.e., plants, animals, and microbes. In terms of landscape, agricultural systems + natural systems = life support systems. Earth's life support environment exhibits considerable ability to recover from periodic, short-duration disturbances such as storms, fires, pollution episodes, or harvest removal, because organisms and ecosystem processes are adapted to natural disturbances. What is new in recent times is the increase in intensity and geographical extent of anthropogenic disturbance, and the large scale introduction into the environment of new chemical poisons.

Since Earth's life support systems have not been built by man, and since they involve a complex array of subsystems, a clear understanding of how the whole thing works could not be achieved, yet. This is the main reason why so far all attempts to build a large bioregenerative life support system that would support a large number of people in space, without a supply "umbilical cord" to Earth, have failed. At this point, the stay of man in space is limited by the amount of life supporting "consumables" that can be carried onboard the spacecraft.

However, in 1987, construction was started on an experimental earthbound capsule that is designed to be bioregenerative: Biosphere 2. In addition to setting up experiments such as Biosphere 2, which is described in more detail in chapter VII, a lot more needs to be learned about how the current real-world life support systems of Biosphere 1, our Earth, function, so not only the quality of these systems can be preserved and maintained, but so also someday self-maintaining spacecraft might be built and the establishment of space colonies on a large scale might be considered. Even more important may be the need to understand how the life supporting non-market, i.e., unprized and taken-for-granted, goods and services provided by the natural environment interact with and support economic, social, cultural, and most other human endeavors. In a very broad sense, the science of ecology provides the background for this understanding. This chapter will give an overview about the basic expressions and concepts of ecology. Moreover, the terms and facts outlined here are of importance for any engineer who wants to design bioregenerative life support systems for future spaceflight. [8]

II.2.1 Levels of Ecological Hierarchy

For a better understanding of the complex ecological world at ground level, it is helpful to think in terms of levels of organizational hierarchies. A hierarchy is defined as an arrangement into a graded series of compartments. The levels of ecological hierarchy are shown in table II.3.

↓	Biosphere
↓	Biogeographic Region
↓	Biome
↓	Landscape
↓	Ecosystem
↓	Biotic community
↓	Population (species)
↓	Organism

Table II.3: **Levels of Ecological Hierarchy [8]**

In ecology, the term *population* denotes groups of individuals of any species that live together in some designated area. *Community* is used in the sense of biotic community to include all of the populations living in a designated area. The community and the non-living environment function together as an ecological system or *ecosystem.* Groups of ecosystems along with human artifacts make up *landscapes* which in turn are part of large regional units called *biomes,* e.g., an ocean or a grassland region. The major continents and oceans are *biogeographic regions*, each with its own special flora and fauna. *Biosphere* is the widely used term for all of Earth's ecosystems functioning together on a global scale. The biosphere merges imperceptibly into the *lithosphere,* i.e., rocks, mantle, and core of Earth, the *hydrosphere,* i.e., surface and ground water, and the *atmosphere.* Each level in a hierarchy influences what goes on in adjacent levels. Processes at lower levels are often constrained in some way by those at higher levels. Accordingly, study or management of any one level is never complete until relevant aspects of adjacent levels are also studied or managed. In this respect it may be interesting to know that the word *economics* is derived from the same root *oikos* ("household") as is the word *ecology.* Since *nomics* means "management", economics translates as "the management of the household". Thus, at least in theory, ecology and economics should be companion disciplines.

An important consequence of hierarchical organization is that as components are combined to produce larger functional wholes, new properties emerge that were not present or not evident at the level below. Accordingly, an emergent property of an ecological level is one that results from the functional interaction of the components, and therefore is a property that cannot be predicted from the study of components that are isolated or decoupled from the whole unit. This principle is a more formal statement of the fact that the "whole is more than the sum of the parts".

Ecology is a discipline that emphasizes a study of both, parts and wholes. While the concept of the whole being greater than the sum of the parts is widely recognized, it tends to be overlooked by modern science and technology, which emphasize the detailed study of smaller and smaller units on the theory that specialization is the way to deal with complex matters. In the real world, the truth is that although findings at any one level aid the study of another level, they cannot fully explain the phenomena occurring at that level, which must also be studied to get the complete picture, e.g., to understand and manage a forest, one must not only have a certain knowledge about trees, but also has to know about the unique characteristics of the forest as it functions in its entirety.

The phenomenon of hierarchical organization, functional integration, and homeostasis suggests that one can begin the study of ecology at any one of the various levels without having to learn everything there is to be known about adjacent levels. The challenge is to recognize the unique characteristics of the level selected and then devise appropriate methods of study and/or action. Different tools are required for the study at different levels. To get useful answers the right questions must be asked. Otherwise, the attempts to solve an environmental problem may fail, because the wrong question is asked, or the wrong level is focused on. [8]

II.2.2 Ecosystems and Ecosystem Models

In order to describe something as complex as an ecological system one has to begin by defining simplified versions of the real world, i.e., models, which encompass only the more important or basic properties or functions. A model is a simplified formulation that mimics real-world phenomena so that complex situations can be comprehended and predictions made. In its formal version, a working model of an ecological situation may consist of five components as described in table II.4.

Component		Description
Properties	P	State variables
Forces	E	Forcing functions, e.g., outside energy sources or causal forces that drive the system
Flow pathways	F	Showing energy or material transfer connections with each other and forces
Interactions	I	Interaction functions, showing the interaction of forces and properties to modify, amplify, or control flows
Feedback loops	L	Output loops back to influence an "upstream" component or flow

Table II.4: **Main Components for the Modelling of Ecological Systems [8]**

Modelling usually begins with the construction of a diagram that may take the form of a compartment diagram as illustrated in figure II.3. Shown are two properties P_1 and P_2 which interact at I to produce or affect a third property P_3 when the system is driven by a forcing function E. Five flow pathways are shown, with F_1 representing the input and F_6 the output for the system as whole. Also shown is a feedback loop, signifying that a downstream output, or some part of it, is fed back, i.e., recycled, to affect or control an upstream component or process. To use and experiment with models for any theoretical or practical purpose, the chart models must be converted to mathematical models by quantifying properties and drawing up equations for flows and their interactions.

Like all kinds and levels of biological systems, ecosystems are open systems, that means things are constantly entering and leaving, even though the general appearance and basic functions may remain constant for long periods of time. Inputs and outputs are an important part of the concept. A graphical model of an ecosystem can consist of a box that can be labelled the system, representing

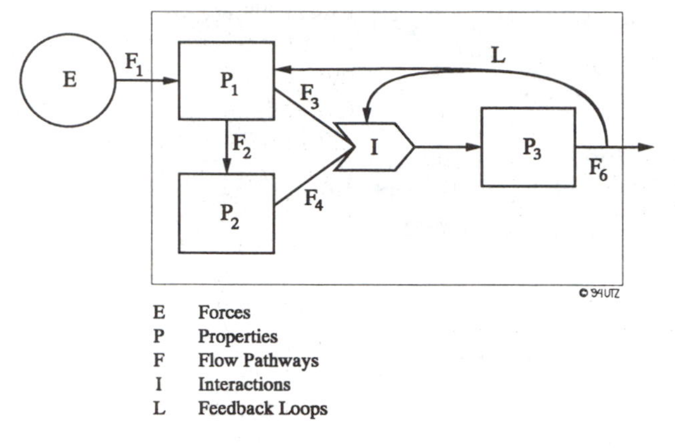

Figure II.3: **Main Components of an Ecological System [8]**

the area of interest, and two large funnels that can be labelled input environment and output environment. The boundary for the system can be arbitrary or natural. Energy is a necessary input. The Sun is the ultimate source of energy for the biosphere, and directly supports most natural ecosystems within the biosphere, but there are other energy sources that may be important for many ecosystems, e.g., wind, rain, or water flow. Energy flows out in the form of heat and other transformed or processed forms such as organic matter and pollutants. Water, air, and nutrients necessary for life, along with all kinds of other materials, constantly enter and leave the ecosystem and, of course, organisms and their propagules, e.g., seeds, enter or leave. [8]

There are two major biotic components. First is an *autotrophic*, *i.e.*, self-nourishing component. Autotrophic organisms are capable of synthesizing the organic compounds which they contain from inorganic compounds, e.g., water, carbon dioxide, nitrites. Based on the source of energy used for organic synthesis, they may be divided into *phototrophic,* using electromagnetic emission in the visible spectral region, and *chemotrophic,* which receive energy as a result of oxidation of different substances, e.g., iron, sulfur, hydrogen, nitrates.

Generally, the green plants, i.e., vegetation on land, algae, and water plants in aquatic habitats, constitute the autotrophic component. These organisms may be thought of as producers. They form an upper "green belt" where the solar energy is greatest. The second major unit is the *heterotrophic*, i.e., other-nourishing, component, which utilizes, rearranges, and decomposes the complex materials synthesized by the autotrophs. Fungi, non-photosynthetic bacteria and other microorganisms, and animals, including humans, constitute the heterotrophs, which concentrate their activities in and around the "brown belt" of soil and sediment below the green canopy. These organisms may be thought of as consumers, since they are unable to produce their own food, and must obtain it by consuming other organisms. Heterotrophes may be further subdivided according to the source of their food energy. Thus, there are the herbivores or the grazers which feed on plants, the carnivores or predators which feed on other animals, the omnivores that feed on both plants and animals, and the saprovores (largely microorganisms) which feed on decaying organic

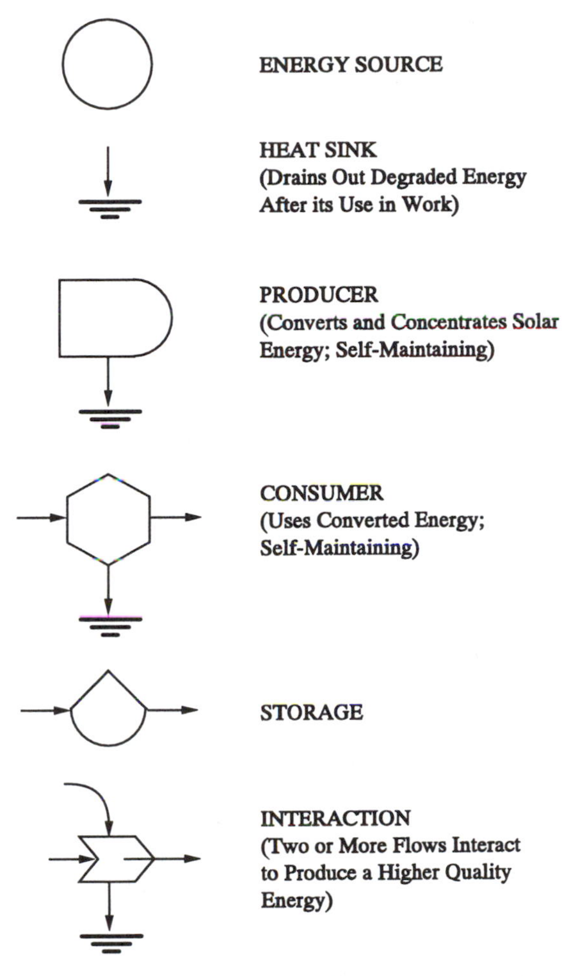

Figure II.4: **H.T. Odum's "Energy Language" Symbols [8]**

materials. Knowing this and using the "energy language" symbols developed by H.T. Odum (see figure II.4), having different shapes according to their basic functions, functional diagrams of ecosystems may be drawn. Such a functional diagram of an ecosystem is given in figure II.5. Here, circles represent renewable energy sources, bullet-shaped modules are autotrophs, hexagons are heterotrophs, tank-shaped boxes are storages, and arrow-into-ground are heat sinks. In the diagram, the autotrophic (A) and heterotrophic (H) components are shown linked together in a network of energy transfers, called the *food web*.

The terrestrial and aquatic ecosystems are contrasting types. Land ecosystems and water ecosystems typically are populated by different kinds of organisms. Despite wide differences in species composition, the same basic ecological components are present and function in the same manner in both ecosystems. On land, the predominant autotrophs are usually rooted plants, ranging in size from grasses to large forest trees. In shallow water situations rooted aquatic plants occur, but in vast open waters the autotrophs are microscopic suspended plants called phytoplankton (*phyto:* plant; *plankton:* floating) that include various kinds of algae, green bacteria, and green protozoa. Because of size differences in plants, the biomass of terrestrial systems may be very different from that of aquatic systems. Plant biomass may be 10000 or more grams of dry matter per m² in a forest, in contrast to 5 grams or less in open waters. Despite this biomass discrepancy, 5 g of phytoplankton are capable of manufacturing as much food in a given amount of time as are 10000 g of large plants, given the same input of light energy and nutrients. This is because the rate of metabolism of small organisms is much greater per unit of weight than

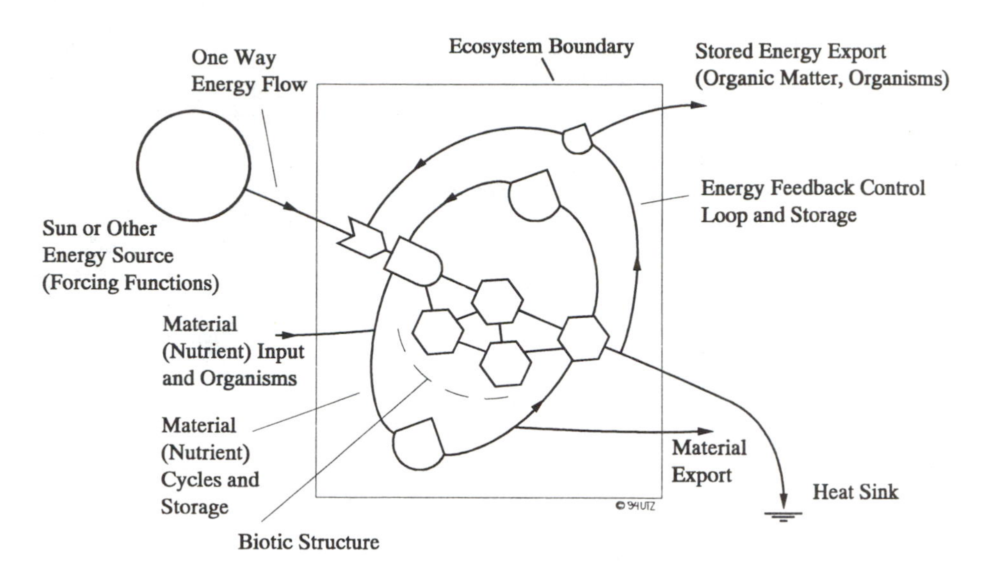

Figure II.5: **A Functional Diagram of an Ecosystem [8]**

that of large organisms. Furthermore, large land plants such as trees are composed mostly of woody tissues that are relatively inactive photosynthetically. Only the leaves photosynthesize, and in a forest, leaves comprise only one to 5 % of the total plant biomass. Thus, the concept of turnover may be introduced. The turnover is the ratio of the standing stock of biotic or abiotic components to the rate of replacement of the standing stock. The reciprocal is the turnover rate. On land, plant biomass tends to accumulate over time so it can be conveniently harvested when a large or perhaps a maximum standing crop has accumulated. So, basic human food produced on land is in plant form. In contrast, the turnover at the autotrophic level in the sea is so fast that very little biomass accumulates. What accumulates in the sea is animal biomass.

In natural and semi-natural landscapes that contain a variety of ecosystems, autotrophic and heterotrophic activity taken as a whole tends to balance. The organic matter produced is utilized in growth and maintenance over the annual cycle. Sometimes production exceeds use, in which case organic matter may be stored or exported to another ecosystem or landscape. In contrast, cities consume much more food and organic matter than they produce, and are accordingly heterotrophic ecosystems. Nature's capacity to support the ever more expanding and demanding cities is being stretched to the limit in many places. Thus, recycling water and wastes, growing food on rooftops, and using solar energy directly to heat buildings and produce electricity are some of the things that need to be done on a larger scale than they are at present.

The two basic abiotic functions that make the ecosystem operational are energy flow and materials cycles. Energy flows from the Sun or other external sources, through the biotic community and its food web, and out of the ecosystem as the heat, organic matter, and organisms produced in the system. Although energy may be stored and used later, energy flow is one-way in the sense that once energy has been utilized that is converted from one form to another, it cannot be used again. Sunlight must continue to flow in if food production is to continue. In contrast, chemical elements can be used over and over without loss of utility. In a well-ordered ecosystem, many of these materials cycle back and forth between abiotic and biotic components. These are the *biogeochemical cycles.* Of the large number of elements and simple inorganic compounds found at or near the surface of Earth, certain ones are essential for life. These are called biogenic substances or nutrients. These tend to be retained and recycled within living systems to a greater extent than nonessential ones. Carbon, hydrogen, nitrogen, phosphorus, and calcium, among others, are required in relatively large amounts and hence are designated as *macronutrients.* These occur abundantly in simple compounds such as carbon dioxide, water, and nitrates that are readily available to organisms. They also occur in chemical forms that are not readily available. Nitrogen in gaseous form in the air, e.g., is not available to plants until it is converted to inorganic salt forms, e.g., nitrate and ammonia, by specialized microorganisms or by other means. Phosphorus in the soil may also occur in chemical forms unavailable to roots of plants. Other elements, no less vital than the macronutrients but required only in small amounts by organisms, are known as *micronutrients* or trace elements. These include a number of metal ions such as iron, magnesium, manganese, zinc, cobalt, or molybdenum. The carbohydrates, the proteins, and the lipids, which make up the bodies of living organisms, are also dispersed widely in nonliving forms in the environment. These and hundreds of other complex compounds make up the organic component of the abiotic environment. As the bodies of organisms decay they become dispersed into fragments and dissolving materials, collectively called organic detritus. Organic detritus not only is a food source for saprovores, it improves soil texture and enhances the retention of water and minerals. Unfortunately, the byproducts of industry, including the petrochemicals, have become increasingly voluminous and increasingly poisonous in recent decades, and the technology of waste management has lagged far behind the ability to produce toxic substances.

Other important terms used by ecologists are the term *habitat,* which describes the place where a species can be found, and the term *ecological niche* to mean the ecological role of an organism in the community. The habitat is the "address" and the niche is the "profession", i.e., how it lives, including how it interacts with and is constrained by other species. Nature, just like well-ordered human societies, has its specialists and its generalists when it comes to niches or professions. In general, specialists are efficient in the use of their resources. Therefore, they often become abundant when their resources are in ample supply.

But the specialist is vulnerable to changes or perturbations that adversely affect its narrow niche. Since the niche of nonspecialized species tends to be broader, they are more adaptable to changing or fluctuating environments, even though they are never so locally abundant as the specialist. We see the same pattern in agriculture. The best solution is diversity of cultivars and crop species, so no matter what the conditions, there won't be a total crop failure. This is "nature's plan".

The few common species in a particular community are called *ecological dominants.* Although dominants may account for most of the standing crop and community metabolism, this does not mean that the rare species are unimportant. Species that exert some kind of controlling influence, whether or not they are dominants, are called *keystone species.* In the aggregate, rare species have an appreciable impact, and they determine the *diversity* of the community as a whole. Should conditions become unfavorable for the dominants, rarer species adapted to or tolerant of the changes may increase in abundance and take over vital functions. Redundancy in the biotic community thus contributes to the resilience of the ecosystem. Basically, two components of diversity have to be recognized: (1) the species richness or variety component, which can be expressed as the number of kinds per unit of space or as a ratio of kinds to numbers, and (2) the relative abundance component, or the apportionment of individuals among the kinds. Thus, two communities could have the same number of species but be very different in terms of relative abundance or dominance of each species. Diversity tends to be lowest where physical conditions are limiting to life and highest in benign environments where conditions are favorable for a large variety of life. Variety among living organisms certainly enriches humans' lives, but it also has a very practical value. It is much safer to have more than one kind of organism that can carry out vital functions. Currently, there is much concern not only about loss of species diversity, but also about loss of genetic diversity due to human activity. Resilience stability is enhanced by the presence of many different species in a landscape. Whether a high species diversity increases resistance stability, i.e., the ability of the ecosystem to remain the same in the face of disturbances, is a question much debated by ecologists and most important when designing a bioregenerative life support system for spaceflight. [8,9]

II.2.3 Energetics and Photosynthesis

Something that is absolutely essential and involved in every action of life on Earth is energy. The primary energy source for heterotrophs is food. For autotrophs it is light and the indirect solar energies, i.e., wind and rain, required for photosynthesis. In addition, human societies require large amounts of concentrated energy in the form of fuels.

The behavior of energy is governed by two laws, known as the laws of thermodynamics. The first law states that energy may be transformed from one form, such as light, into another type, such as food, but is never created or destroyed. The second law states that no process involving an energy transformation will occur unless there is a degradation of energy from a concentrated form, such as fuel, into a dispersed form, such as heat. Because some energy is always dispersed into unavailable heat energy, no spontaneous transformation can be 100 % efficient. The second law is also known as the entropy law. Entropy (*en:* in; *trope:* transformation) is a measure of disorder in terms of the amount of unavailable energy in a closed thermodynamic system. Thus, although energy is neither created nor destroyed during transformation, some of it is degraded to an unavailable or less available form when used. Organisms and ecosystems maintain their highly organized, low-entropy, i.e., low-disorder, state by transforming energy from high to low utility states. Anyway, entropy is not at all negative. Since the quantity of energy declines in successive transfers the quality of the remainder may be greatly enhanced. As mentioned before, the primary source of energy of all processes on Earth is the Sun. It emits radiation energy, i.e., electromagnetic radiation. The entire spectrum of electromagnetic radiation is shown in figure II.6.

Solar radiation is in the middle range of this spectrum, with wavelengths largely between 0.1 and 10 µm. It consists of visible light and two invisible components, ultraviolet (UV) and infrared (IR) light. The long-wave infrared radiation is the "heating" part of sunlight. The visible range is the energy used in photosynthesis. Much of the UV-light reaching the upper atmosphere is kept out by the ozone layer, which is fortunate, since UV is lethal to exposed protoplasm. The absorption of solar radiation by the atmosphere as it enters the biosphere greatly reduces UV, broadly reduces visible light, and irregularly reduces IR light. Radiant energy reaching the surface of Earth on a clear day is about 10 % UV, 45 % visible, and 45 % IR. Vegetation absorbs the blue and red visible wavelengths, which are the most useful for photosynthesis, and the far IR strongly, the green less strongly, and the near IR very weakly. By rejecting the near IR, in which the bulk of the Sun's heat energy is located, leaves of terrestrial plants avoid lethal temperatures (they are also cooled by the evaporation of water). Because the green light and the near IR light are reflected by vegetation, these spectral bands are used in aerial and satellite remote sensing and photography to reveal patterns of natural vegetation (see figure II.7).

On a square meter basis, solar energy comes in at a rate of about 21 million kJ per year. This flow is reduced exponentially as it passes through the atmosphere, so the annual amount actually reaching the autotrophic layer of the ecosystem is only 4-8 million kJ/m². Of this amount, about half is absorbed by a well-stocked green layer, and about 1 % of this on average is converted to organic matter by photosynthesis. About a quarter of the solar energy flow is used to recycle water, one of the biosphere's most vital nonmarket services. It is the

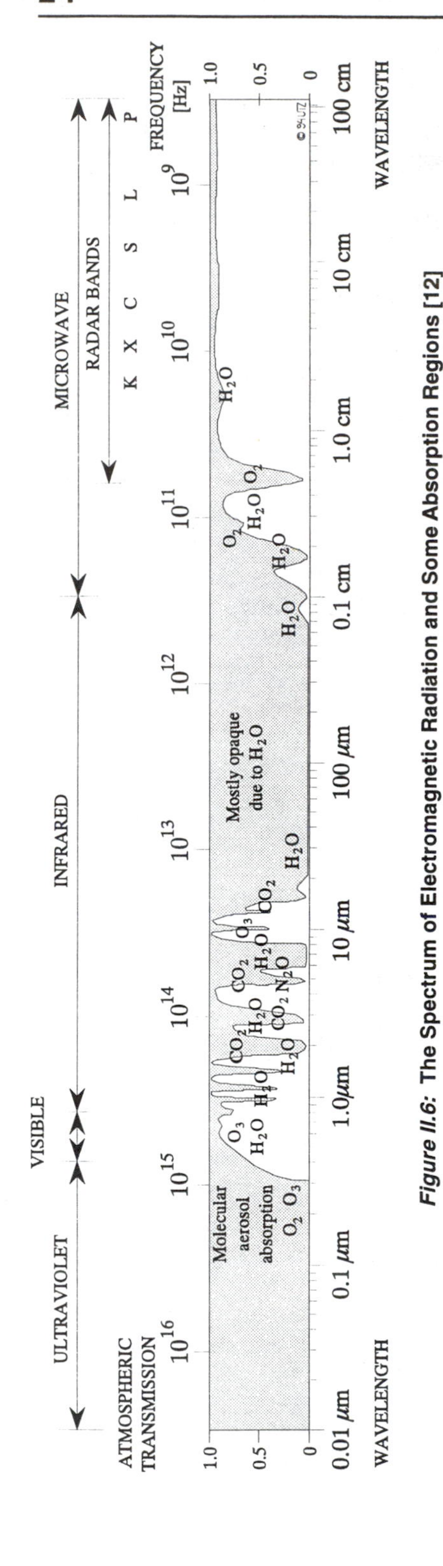

Figure II.6: **The Spectrum of Electromagnetic Radiation and Some Absorption Regions [12]**

flow of energy that drives the cycles of materials. To recycle water and nutrients requires an amount of energy that is not recyclable. It's because of this fact that artificial recycling of resources such as water, metals, or paper is not an instant and free solution to shortages. A more detailed description of the distribution of incoming solar energy is given in figure II.8.

Nature basically makes good use of solar power. One advantage of solar energy compared to fossil fuels is that it is renewable. On the other hand, solar energy is much more dilute, i.e., it has to be concentrated and for this kind of transformation there is an entropy cost. As mentioned earlier, as energy quantity decreases, its "quality" increases. In the natural food chain, the amount of energy declines with each step, but concentration in terms of the number of solar kJ dissipated increases. Thus, the energy of the Sun is concentrated by green plants, further concentrated by the fossilization process that produces coal, and finally enhanced some more in the production of electricity, which accordingly, is a form of energy derived from solar energy but concentrated many thousands of times. This energy concentration process is outlined figure II.9.

Photosynthesis is the most efficient process for tapping the small portion of solar radiation that can be upgraded to high-utility organic matter. The basic photosynthetic process is chemically an oxidation-reduction reaction:

$$6CO_2 + 6H_2O \xrightarrow{light} C_6H_{12}O_2 + 6H_2O$$

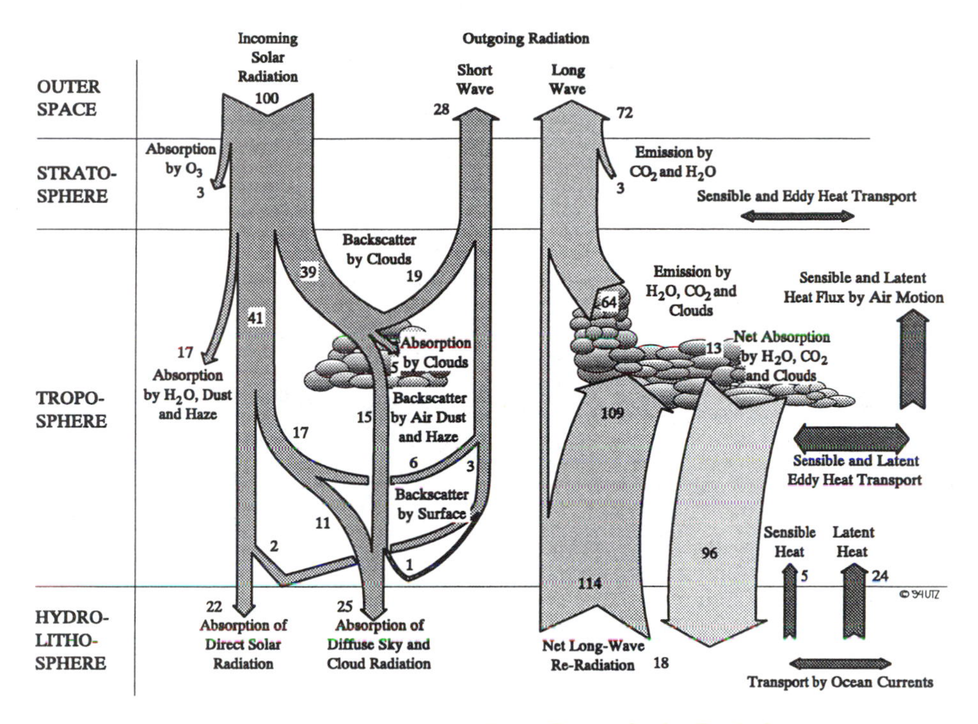

Figure II.7: **Distribution of Absorbed Solar Energy in the Earth System [12]**

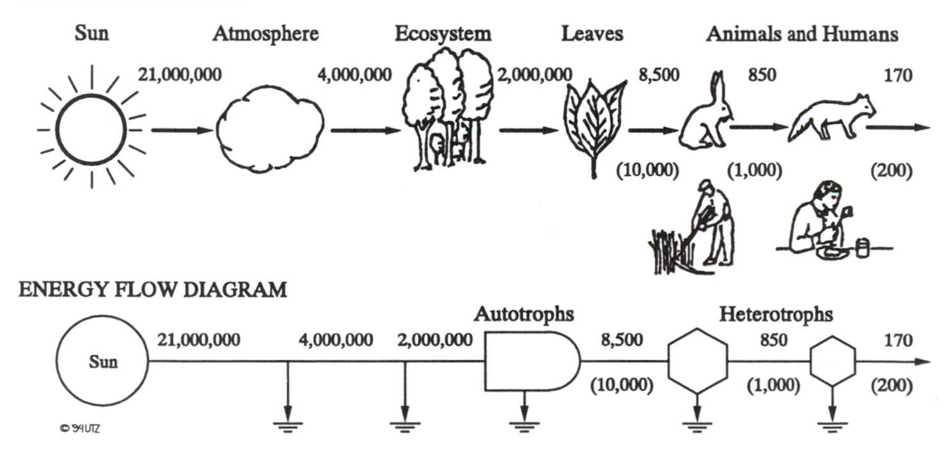

Figure II.8: **Solar Energy Flow through the Biological Food Chain (in kJ/m²) [8]**

FOOD CHAIN

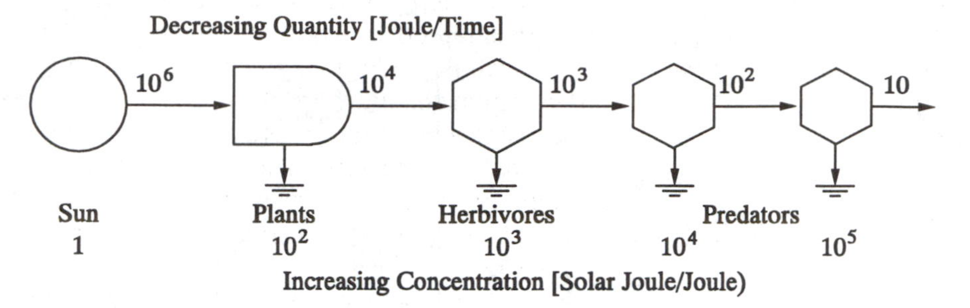

ELECTRIC ENERGY CHAIN

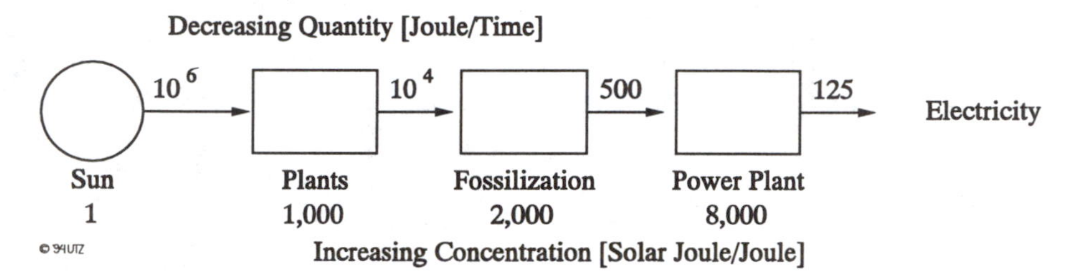

Figure II.9: **Food Chain and Electric Energy Chain [8]**

The amount of organic matter fixed by *autotrophs* in a given area over a given period of time is called primary production. *Gross primary production* (P_g) is the total amount, including that used by the plant for its own needs, while *net primary production* (P_n) is the amount stored in a plant in excess of its respiratory needs and therefore, potentially available to heterotrophs. The amount left after the biotic community, i.e., autotrophs and heterotrophs, have taken all the food they need is called net community production. Finally, energy storage at consumer levels, e.g., in cows or fish, is referred to as secondary production. High rates of primary production in both natural and cultured ecosystems occur when physical factors, e.g., water, nutrients, etc., are favorable, and especially when auxiliary energy from outside the system reduces maintenance costs. Secondary energy that supplements the Sun and allows a plant to store and pass on more of its photosynthate can be thought of as energy subsidy.

In most plants, CO_2 fixation starts with the formation of three carbon-compounds, but recently it was discovered that certain plants reduce CO_2 in a different manner, starting with four carbon carboxylic acids. Theses two types of plants are designated as C_3 and C_4 plants. C_4 plants have a different arrangement of chloroplasts within their leaves and respond differently to light, temperature and water, as shown in figure II.10.

C_3 plants tend to peak in photosynthetic rate at moderate light intensities and temperatures, and to be inhibited by high temperatures and the intensity of full sunlight. In contrast, C_4 plants are adapted to high light and temperature conditions, and use water more efficiently under these conditions. C_4 plants dominate desert and grassland communities in warm temperate and tropical climates and are rare in forests and in the cloudy north. C_3 plants account for most of the world's primary production, presumably because they are more competitive in communities of mixed species where light, temperature, and so on are rather average than extreme. Major human food plants such as wheat, rice, and potatoes are C_3 plants, as are most vegetables. Crops of tropical origin such as corn, sorghum, and sugarcane are C_4 plants.

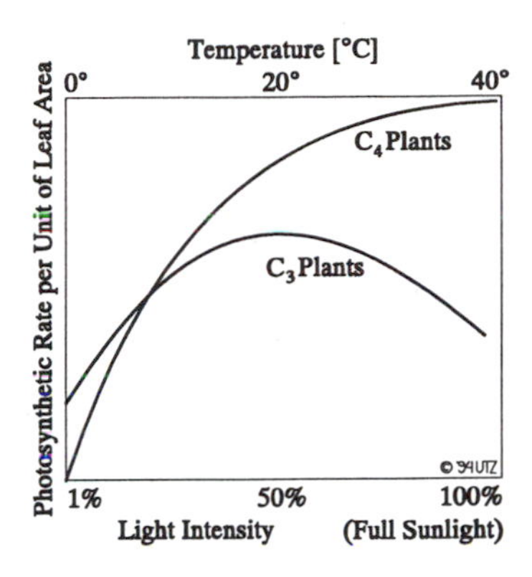

Figure II.10: **Photosynthetic Responses of C_3 and C_4 Plants [8]**

In this context, it might be interesting to know the general pattern of world distribution of primary production. As it is also shown in figure II.11, large parts of open ocean and land deserts have an annual production rate of 4000 kJ/m² or less. The ocean is nutrient limited, and deserts are water limited. Grasslands, coastal seas, shallow lakes, and ordinary cultural communities range between 4000 and 40000 kJ/m². Estuaries, coral reefs, moist forests, wetlands, and intensive agricultural and natural communities on fertile plains have annual production rates of 40000-100000 kJ/m².

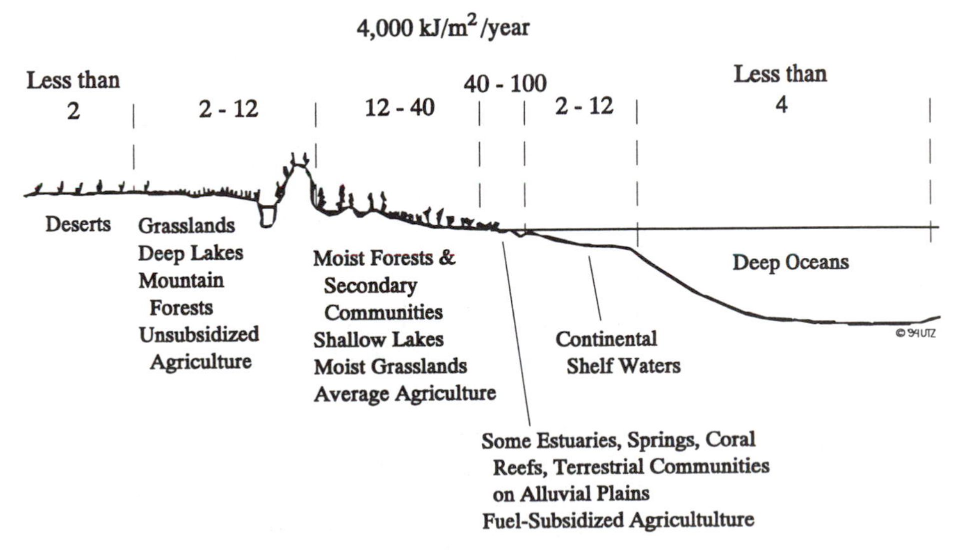

Figure II.11: **World Distribution of Primary Production (in [kJ/m²/year]) [8]**

Human food production per unit of area has been greatly increased in recent decades by increasing mechanization, fertilizers, irrigation, and pesticides. Genetic selection for increased harvest index, i.e., the ratio of edible to inedible part of the crop plant, is another way yields have been increased. On Earth, good land on which to crow crops is in short supply. At best 24 % of the land area is arable. Most of this is already in use for crops and pastures. [8]

II.2.4 Materials Cycles

The more or less circular paths of the chemical elements passing back and forth between organisms and environment are called *biogeochemical cycles.* In this context, *bio* refers to living organisms and *geo* to rocks, soil, air, and the water of Earth. Biogeochemistry is, thus, the study of exchange of materials between the living and nonliving components of the biosphere. The scheme of a typical biogeochemical cycle is shown in figure II.12. The most important materials cycles on Earth are outlined below.

Natural recycling is mostly driven by natural energy such as sunlight. For artificial recycling to have a net benefit, work energy has to be available at cost not exceeding the value of the recycled product. Like water, the vital nutrient elements, e.g., carbon, nitrogen, phosphorus, etc., are not homogeneously distributed or present in the same chemical form throughout an ecosystem. Rather, materials exist in compartments or pools, with varying rates of exchange between them. In figure II.12, the large reservoir is the box labelled "nutrient pool", and the rapidly cycling material is represented by the shaded circle going from autotrophs to heterotrophs and back again. Decomposition releases not only minerals but also organic by-products which may affect the availability of minerals to autotrophs. One way this occurs is by a process known as chelation in which organic molecules "grasp" or form complexes with calcium, magnesium, iron, etc. Chelated minerals are more soluble and often less toxic than some of the inorganic salts of the same element. From the standpoint of the biosphere as a whole, biogeochemical cycles fall into two groups: gaseous types with a large reservoir in the atmosphere, and sedimentary types with a reservoir in the oils and sediments of Earth's crust. These major cycles of the biosphere are indicated in a general way in figure II.13.

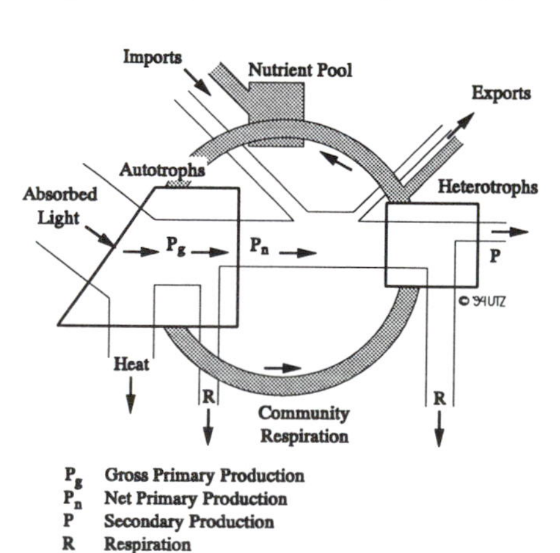

Figure II.12: **A Biogeochemical Cycle [8]**

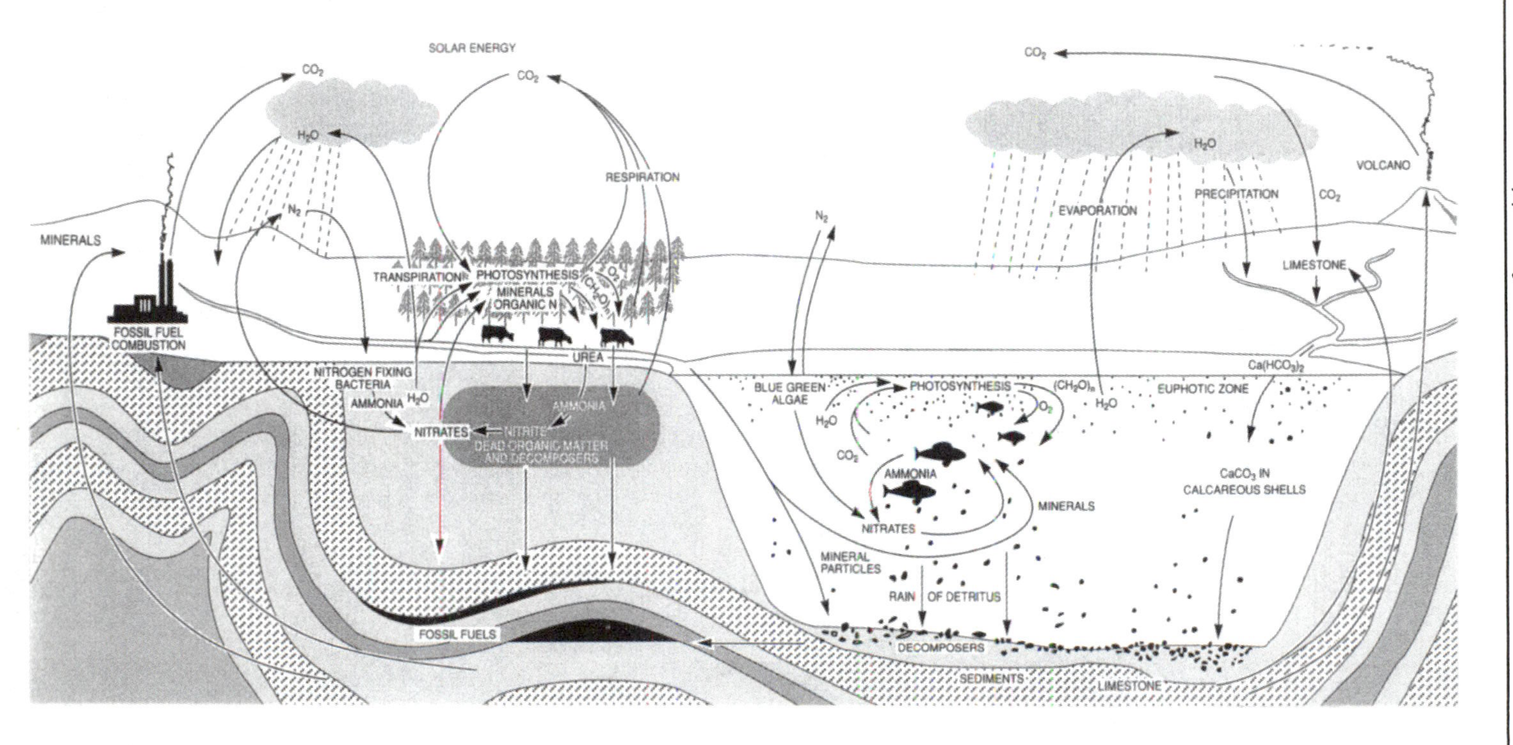

Figure II.13: **The Major Cycles of the Biosphere [1]**

The Water Cycle

Water, a major component of our weather, is a vital life supporting material that cycles back and forth between living organisms and the abiotic environment. As water is used, it evaporates from vegetation, lakes, and other surfaces, percolates through soil into ground water, and runs off in streams and rivers to the sea. No matter how water leaves the ecosystem, it must be eventually replaced by rain. The hydrological or water cycle is divided in two phases: the upstream phase, driven by solar energy, and the downstream phase, which provides goods and services that humans and the environment require (see figure II.14). More water evaporates from the sea than returns there as rainfall, and vice versa for the land. Thus, a considerable part of the rainfall that supports land ecosystems and most human food production comes from water evaporated from the sea. About one third of the solar-energy reaching Earth's surface is dissipating in driving the hydrological cycle.

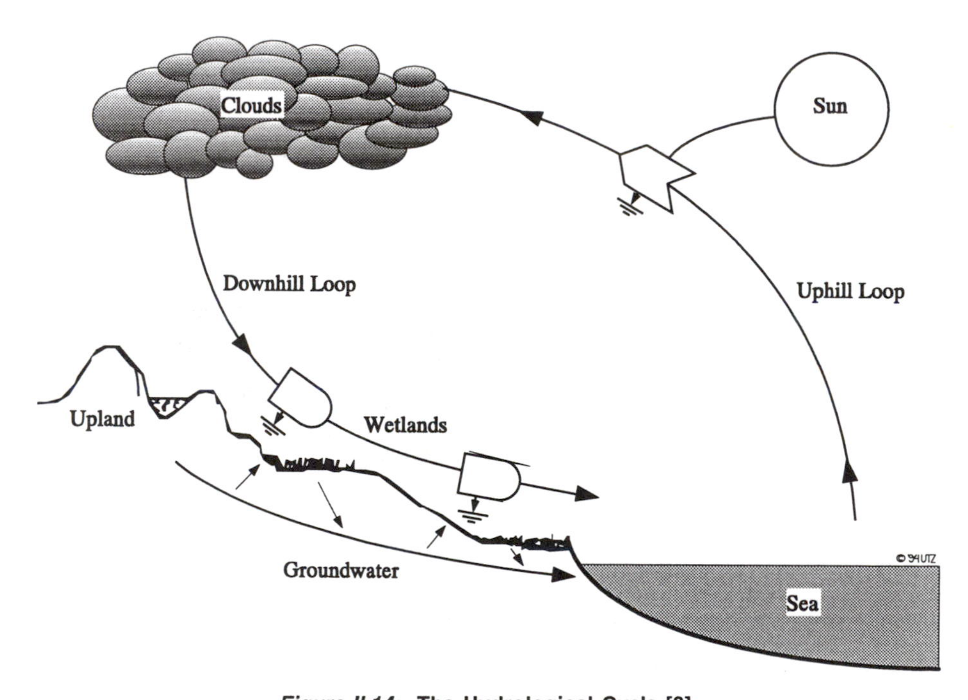

Figure II.14: **The Hydrological Cycle [8]**

The Nitrogen Cycle

Nitrogen is constantly feeding into and out of the atmospheric reservoir and the rapidly recycling pool associated with the organisms. Both biological and nonbiological mechanisms are involved in the denitrification, which puts nitrogen into the air, and nitrogen fixation, the conversion of gaseous nitrogen, which is

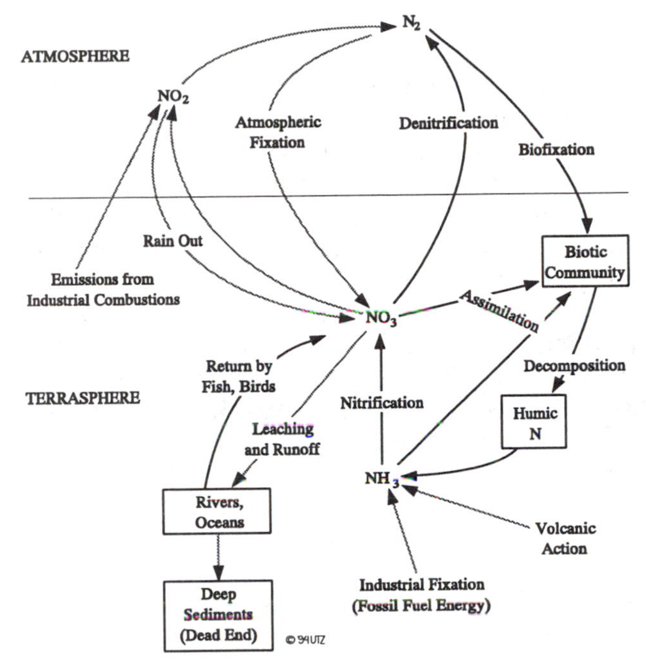

Figure II.15: **The Nitrogen Cycle [8]**

not usable directly by autotrophs, into ammonia, nitrite, and nitrate, which are usable. Specialized microbes play key roles in most of the steps in the nitrogen cycle, e.g., only a few primitive bacteria, i.e., prokaryotes, including the blue-green algae, can fix nitrogen. Legumes and some other higher plants fix nitrogen only through the prokaryotic bacteria that live in special nodules on their roots. A scheme of the nitrogen cycle is given in figure II.15.

The Carbon Cycle

A model of the carbon cycle is given in figure II.16. The CO_2 is mainly distributed in four major compartments: atmosphere, oceans, terrestrial biomass, and soils and fossil fuels. The flux between compartments is shown by the arrows. The atmospheric pool is small in comparison to the amounts in the other compartments, but is a very active pool which is being increased by the burning of fossil fuels and the clearing and plowing of land for agriculture. While more CO_2 in the air could have the positive effect of increasing primary production, it raises concern about possible undesirable changes in climate due to a greenhouse effect. In addition to CO_2, two other forms of carbon are present in the atmosphere in smaller amount: carbon monoxide (CO) and methane (CH_4). Both arise from incomplete or anaerobic decomposition of organic matter.

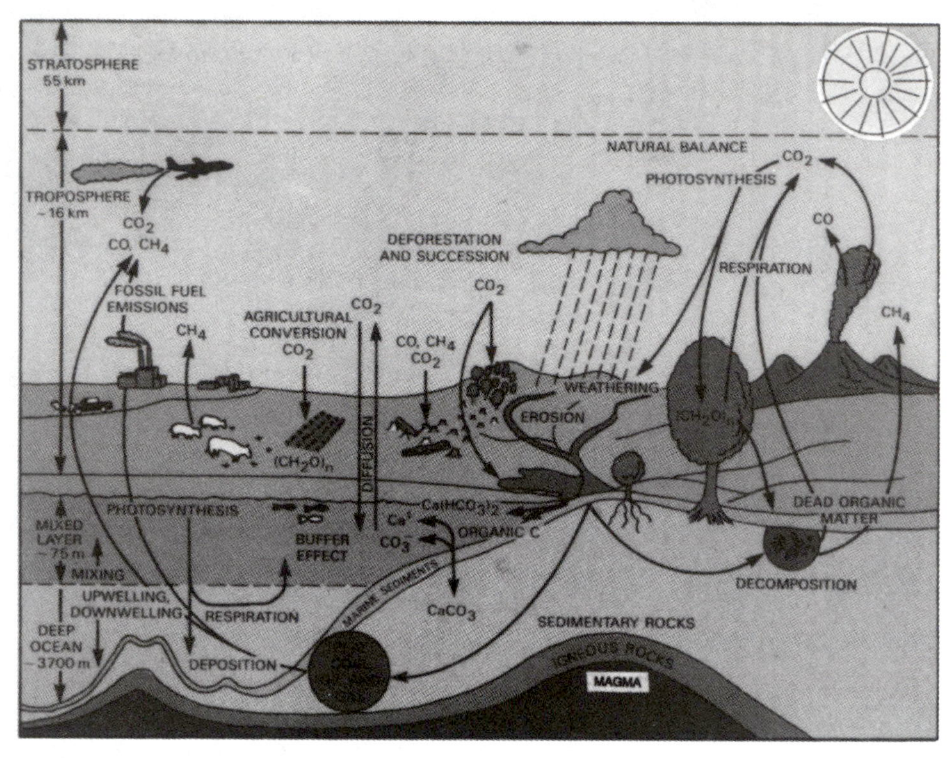

Figure II.16: **The Carbon Cycle [12]**

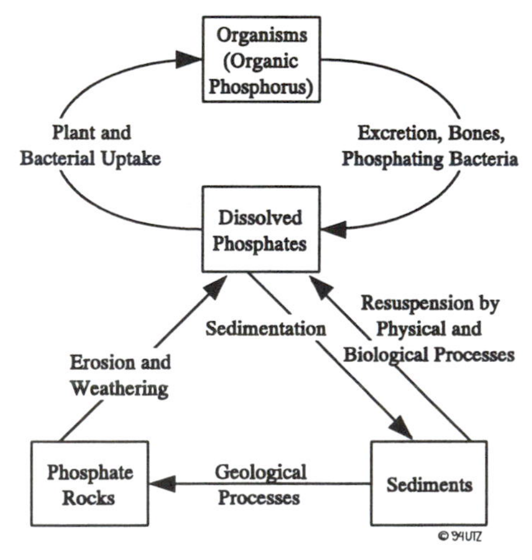

Figure II.17: **The Phosphorous Cycle [8]**

The Phosphorous Cycle

The phosphorous cycle is an example of a sedimentary cycle of utmost importance. Phosphorus is required for the energy transformation that distinguishes living protoplasm from nonliving material. Organisms have devised any mechanisms for hoarding this element. Hence, the concentration of phosphorus in a gram of biomass is usually many times that in a gram of surrounding environment. [8]

II.2.5 Recycling Pathways and Limiting Factors

A number of ways by which resources are recycled are shown in figure II.18. Recycling of many vital nutrients involves microorganisms and energy derived from the decomposition of organic matter (pathway 1 - M = microorganisms, D = detritus consumers). Where small plants such as grass or phytoplankton are heavily grazed, recycling by way of animal (A) excretion may be important (pathway 2). In nutrient-poor situations, a direct return (path 3) is accomplished by symbiotic microorganisms (S) that become a part of autotrophs (plants). Many substances are recycled by using physical means involving physical energy (path 4). Finally, fuel energy is used by humans to recycle water, fertilizers, metals, and paper (path 5). Note that each recycling method requires energy dissipation from some source, such as organic matter (paths 1, 2, 3), solar energy (path 4), or fuel (path 5).

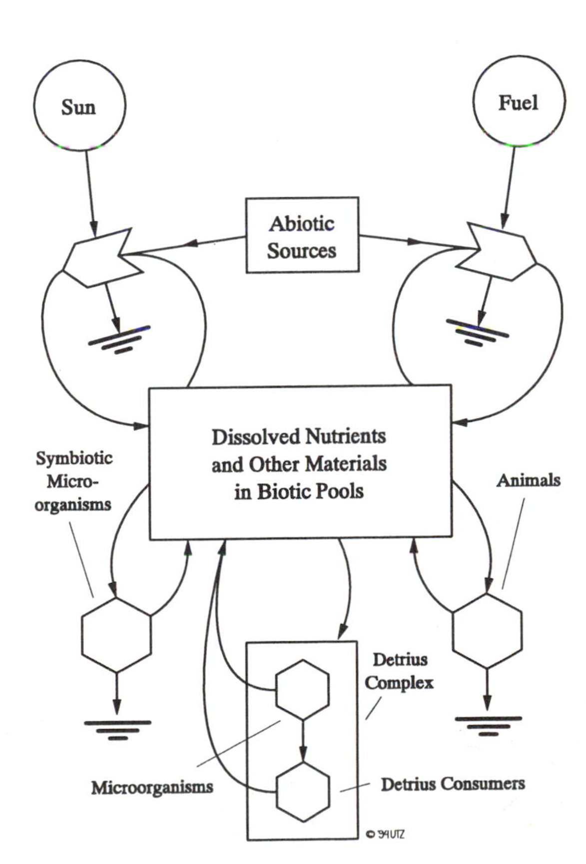

Figure II.18: **Recycling Pathways [8]**

The idea that organisms may be controlled by the weakest link in an ecological chain of requirements goes back to Justus Liebig. Liebig's law of the minimum has come to mean that growth is limited by that nutrient that is least available in terms of need. An extended *concept of limiting factors* may be restated as follows: The success of an organism, population, or community depends on a complex of conditions. Any condition that approaches or exceeds the limit of tolerance for the organism or group in question may be said to be a limiting factor. Liebig's law is most applicable to steady-state conditions where inflows balance outflows, and least applicable under transient state conditions, where flows are unbalanced and where rates of function will likely depend on rapidly changing concentrations and the interactions of many factors.

Species with wide geographic ranges often develop locally adapted genetic races or subpopulations, called *ecotypes*, with different growth forms or different limits of tolerance for temperature, light, nutrients, etc. Many species have narrow ranges of tolerance and are, accordingly, sensitive to change. Such species can be useful ecological indicators of changes in environmental conditions. Organisms not only adapt to physical environments, but also use natural periodicities in the physical environment to time their activities and to "program" their life histories so they can benefit from favorable conditions. They accomplish this by means of biological clocks, physiological mechanisms for measuring time. The most common and perhaps basic manifestation is the cir-cadian rhythm (*circa:* about, *dies:* day), or the ability to time and repeat functions at about 24-hour intervals even in the absence of conspicuous environmental clues such as daylight. The biological clock couples environmental and physical rhythms and enables organisms to anticipate daily, seasonal, tidal, and other periodicities. A dependable clue used by organisms to time their seasonal activities in the temperate zone is the length of day or photoperiod. [8]

II.2.6 The Gaia Hypothesis

The *Gaia Hypothesis* (*Gaia* is the Earth goddess in Greek mythology) of James Lovelock states that "the biosphere is a self-regulating entity with the capacity to keep our planet healthy by controlling the physical and chemical environment". In other words, Earth is a superecosystem with numerous interacting functions and feedback loops that moderate extremes of temperature and keep the chemical composition of the atmosphere and the oceans relatively constant. Also, and this is the most controversial part of his hypothesis, Lovelock says that the biotic community plays the major role in biospheric homeostasis, and organisms began to establish control soon after the first life appeared on Earth more than three billion years ago. The contrary hypothesis is that purely geological (abiotic) processes produced conditions favorable for life, which then merely adapted to these conditions.

In the evolutionary process of planet Earth, the first or primary atmosphere was formed from gases rising from the hot core of Earth, a process geologists call outgassing. In contrast, the present atmosphere, called secondary atmosphere, is a biological product, according to the Gaia Hypothesis. This reconstruction began with the first life, the primitive microbes that do not require oxygen, i.e., the anaerobes. When the green anaerobic microbes began to put oxygen into the air, the plants and animals that require gaseous oxygen, i.e., the aerobes, evolved, and the anaerobes retreated to the oxygenless depths of soils and sediments, where they continue to thrive and play a major role in various ecosystems. Comparison of the atmosphere of Earth with that of the planets Mars and Venus provides strong indirect evidence for the Gaia hypothesis (see table II.5). Also, this hypothesis is the basis for thoughts about *terraforming* of other planets, especially Mars.

	Mars	Venus	Earth	Hypothetical Earth
Carbon Dioxide	95 %	98 %	0.03 %	98 %
Nitrogen	2.7 %	1.9 %	79 %	1.9 %
Oxygen	0.13 %	Trace	21 %	Trace
Temperature	- 53° C	477° C	13° C	290±50° C

Table II.5: **Atmospheres of Earth, Venus, Mars, and a Hypothetical Earth without Life [6]**

Without the critical buffering activities of early life form and the continued coordinated activities of plants and microbes that dampen fluctuations in physical factors, conditions on Earth would be similar to current conditions on Venus. In summary, according to the Gaia Hypothesis, the biosphere is a highly integrated and self-organized cybernetic (*kybernetes:* pilot) or controlled system. But control at the biosphere level is not accomplished by external, goal-oriented thermostats, chemostats, or other mechanical feedback devices. Rather, control is internal and diffuse, involving hundreds and thousands of feedback loops and synergetic interactions in subsystems like the microbial network that controls the nitrogen cycle. Since humans did not build this system, it is not yet fully understood and it has not been possible to construct even a simplified biologically controlled life support system for space travel. Much has to be learned about what really goes on in the impenetrable networks in the oceans and the "brown belts" of soils and sediments that determine when, where and at what rate nutrients are recycled and gases exchanged. [6,8]

II.3 EARTH'S ENDANGERED LIFE SUPPORT SYSTEM

On Earth, the time has come to view and manage entire landscapes as a whole. This is where the science of ecology can help, since it deals with the interconnectedness of humans and nature. As mentioned before, the word *ecology* is derived from the Greek, literally meaning "the study of households". This includes the plants, animals, microbes, and people that live together as independent beings on Spaceship Earth. The environmental house within which mankind has placed its human-made structures and operates its machines provides most vital biological necessities. Hence, ecology can be thought of as the study of Earth's life support system.

On planet Earth good farmland is limited. Worldwide, only about a quarter of the land area has the soils, water, and climate to sustain the high level of food production needed to feed Earth's billions of inhabitants. It is important to recognize that the city is a parasite on the natural and domestic environments, since it makes no food, cleans no air, and cleans very little water to a point where it could be reused. The larger the city, the greater the need for undeveloped or lightly developed countryside to provide the necessary host for the urban parasite. In a host-parasite relationship a parasite does not live for very long if it kills or damages its host. At present, mankind does not do an adequate job of caring for its life support environment. Wastes produced are discharged into rivers, millions of tons of solid wastes are dumped each year. Because the large body of water is biologically and physically active, it has been able, so far, to "digest" all or most of this enormous discharge. But natural waste treatment systems have become overloaded in recent years. Thus, there are but two options: (1) increase costly artificial treatment, or (2) reduce the amount of waste that requires disposal or treatment. Soil erosion and runoff of toxic chemicals increased with increasing agricultural production. The dilemma is that efforts to enhance one part of the life support environment, i.e., agriculture, have degraded other equally vital components, i.e., the natural systems.

Passage and enforcement of laws regulating discharges from industry, power plants, and sewage treatment plants have reduced point-source pollutions in many streams and rivers as well as in the air. However, nonpoint-source pollution, such as soil and pesticide runoff from agricultural lands, or auto exhaust, has increased, so there has been little or no overall improvement in water and air quality. The accumulation of greenhouse gases, like CO_2, CH_4, CFC, NO_x, in the atmosphere that originate from anthropogenic sources have probably begun to change the global climate. In the pre-industrial atmosphere the CO2 level was about 280 ppm, increasing to about 355 ppm in the early 1990's. If this rise is continuing the global mean temperature may increase by 1.3-2.5 °C by the year 2020. Greenhouse gases and the likely resulting climate change would affect the terrestrial ecosystems. Feedbacks from these ecosystems to global climat change processes may significantly affect the rate and magnitude of climate change. Thus ideas on how to reduce the stress on Earth's life support environment are emerging. Nonpoint-source can be controlled by input management, e.g., by reducing the amount and toxicity of agricultural chemicals applied to cropland, removing sulfur and other pollutants from coal before it is burned in power plants before it is burned, or recycling paper instead of dumping it in a landfill. It is also necessary to give more attention to increasing the efficiency of resource use, thereby reducing the deleterious impacts of the fabricated and domesticated environment on the life support environment. [3,8]

II References

[1] Allen J.
Biosphere 2 - The Human Experiment
Penguin Press, New York, 1991

[2] Brack A.
Extraterrestrial Organic Molecules and the Emerge of Life on Earth
Life Sciences Research in Space, ESA, SP-307, p. 565-569, 1990

[3] Dixon R.
Editorials
International Journal of Life Support and Biospheric Sciences,
Vol.1, No.1, p. 27-28, 1994

[4] Hallmann W.; Ley W.
Handbuch der Raumfahrttechnik
Carl Hanser Verlag, München, 1988

[5] Kuchling H.
Physik - Formeln und Gesetze
VEB Fachbuchverlag, Leipzig, 1977

[6] Lovelock J.
The Ages of Gaia
Bantam Books, New York, 1990

[7] Lubchenco J. et al
The Sustainable Biosphere Initiative: An Ecological Research Agenda
Ecology 72(2), p. 371-412, 1991

[8] Odum E.P.
Ecology and Our Endangered Life Support Systems
Sinauer Associates Inc., Sunderland, 1990

[9] Shepelev Y.
Biological Life Support Systems
Translation of Biologicheskiye Sistemy Zhizneobespecheniya, Academy of Sciences, Moscow, 1972

[10] Voitkevich G.
Origin and Development of Life on Earth
Mir Publishers, Moscow, 1990

[11] Wieland P.
Designing for Human Presence in Space: An Introduction to Environmental Control and Life Support Systems
NASA Marshall Spaceflight Center, 1992 (to be published)

[12] From Pattern to Process: The Strategy of the Earth Observing System
NASA

[13] Grosser Atlas der Erde
Lingen Verlag, Köln, 1975

III THE EXTRATERRESTRIAL ENVIRONMENT

The extraterrestrial environment differs in many ways from the familiar environment here on Earth:

- *Radiation*
  All kinds of particle and wave radiation with a wide spectrum, different from those on Earth, exist in outer space and on the other planets of the solar system.

- *Gravity*
  In free space basically no gravitational loads act upon free-flying space vehicles. The gravitational acceleration on other planets varies by a great margin.

- *Atmosphere*
  In free space there is a vacuum, i.e., practically no atmosphere. The structure and behavior of the atmosphere of other planets are physically and chemically different from Earth's atmosphere.

- *Magnetic Fields*
  The magnetosphere of every planet in the solar system is different in orientation and strength. Also there exists an interplanetary magnetic field (IMF).

These different and mostly extreme conditions of the extraterrestrial environment have a varying, but mostly strong impact on the design and technological development of any kind of space vehicle, especially the life support systems that are required on manned missions. In the following chapter the several characteristics of the different environmental effects of outer space are described. [22]

III.1 RADIATION IN FREE SPACE

The study of space radiation began early in this century. Nevertheless, most knowledge has been acquired since the launch of the first satellites in the late 1950's. Also, the advent of manned spaceflight provided an immense number of observations as a basis for analytical treatment of space radiation hazards. Although research efforts have provided a basic understanding of these hazards, a great deal remains unknown. Earth is constantly bombarded by many types of radiation. One can basically distinguish two kinds of radiation:

- *Electromagnetic Radiation*
- *Ionizing Radiation*

Earth's atmosphere blocks most of their harmful components, but in space this protective shield is not available. Radiation found in space poses a serious problem for future long-duration manned missions, and any kind of culturing of microbes, algae, or plants in space. [27]

III.1.1 Electromagnetic Radiation in Space

Basically all of the electromagnetic radiation in the solar system is emitted by the Sun. While the solar electromagnetic radiation has an energy density of about 1390 W/m^2 near Earth, the energy density of starlight is hardly more than 10^{-9} W/m^2. In the visible region, the solar radiation corresponds to that of a black body of 5700 K surface temperature. In general, the solar radiation density may be written as a function of the distance from the Sun: [22]

$$I = \frac{I_0}{r^2}$$

Distance from the Sun in [AU]: r Radiation energy flux at 1 AU: $I_0 = 1390$ W/m^2

Thus, for the solar intensity on planets and moons in the solar system relative to the solar intensity at the top of Earth's atmosphere the values that are shown in table III.1 can be calculated. The wavelength areas and the electromagnetic spectrum of solar radiation are shown in table III.2 and also in figure III.1.

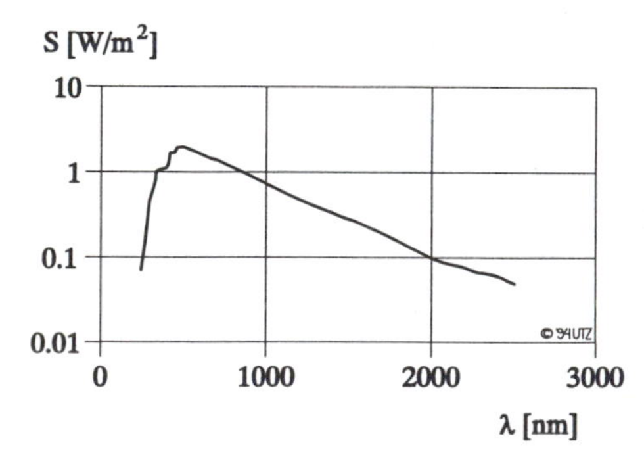

Figure III.1: **Electromagnetic Spectrum of the Sun**

For any kind of spacecraft in planetary orbits the electromagnetic radiation emitted by the planet also has to be taken into account. The mean radiation energy of Earth is about 225 W/m^2. Earth's radiation is in infrared, corresponding to a black body at 290 K. A method for calculating the equilibrium temperature of an orbiting spacecraft exposed to electromagnetic radiation is given in section III 3.1. [17]

Planet	Solar Intensity Relative to Earth's Value
Mercury	666
Venus	1.91
Earth	1
Mars	0.43
Jupiter	0.04
Saturn	0.01
Uranus	0.003
Neptune	0.001
Pluto	0.0006
Earth's Moon	1
Mars' Moon	0.43
Asteroids	0.28

Table III.1: **Solar Intensity on Planets of the Solar System [17]**

Radiation	γ-rays	X-rays	Ultraviolet	Visible light	Infrared	Radio waves
Wave length	10^{-14}- 10^{-12}m	10^{-12}- 10^{-8}m	10^{-8} - 10^{-7}m	10^{-7}m	10^{-6} - 10^{-3}m	10^{-3} - 10^{4} m

Table III.2: **Wave Length Areas of the Electromagnetic Spectrum**

III.1.2 Ionizing Radiation Sources in Space

There are three different kinds of relevant ionizing radiation in space:

- *Solar Cosmic Rays (SCR)*
  The Sun's radiation may be subdivided in a regular and an irregular portion. The regular portion, the solar wind, is a proton-electron gas that blows away from the Sun in radial direction. The solar wind exists because the Sun's corona is very hot ($2 \cdot 10^6$ K), thus, literally "boiling off" its outer

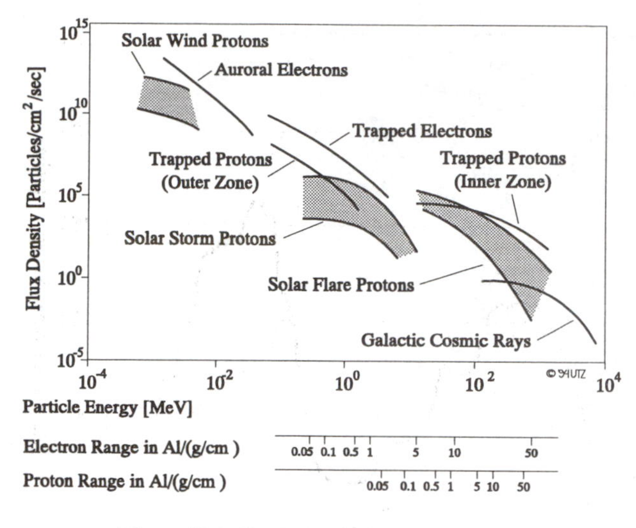

Figure III.2: **Radiation in Free Space [20]**

atmosphere. The irregular portion, solar flares, are produced by "storms" in the solar magnetosphere. These eruptions yield very high radiation doses within very short periods of time (hours to days). Solar flares show a correlation with the 11-year solar cycle. The largest events normally occur in the months following sunspot maximum. Individual events vary considerably in particle constituents, energy spectra, and particle flux. The occurrence of solar flares is basically not predictable and, thus, the warning period is only a few minutes to hours.

- *Galactic Cosmic Radiation (GCR)*
  GCR is emitted by distant stars and even more distant galaxies. They diffuse through space and arrive at the solar system from all directions.

- *The Van Allen Belts*
  The Van Allen Belts consist of particles, i.e., protons and electrons, trapped by Earth's magnetic field.

Ionizing radiation is a general term referring to high-energy particles and photons which detach electrons from molecules upon close passage. It consists of numerous particles with various energy spectra (see figure III.2). GCR are the most penetrating because of their high energy. The large fluxes associated with major solar flare events are potentially lethal. The same is true for very long stays in the inner Van Allen belt. [10, 25, 30]

III.1.2.1 Solar Wind and Solar Cosmic Rays (SCR)

The solar wind is a plasma wind that basically consists of a proton-electron gas with an average flow velocity of 400-500 km/s and a mean proton and electron density of about 5/cm³. The streaming solar plasma interacts with Earth's intrinsic magnetic field and the resulting interaction (Van Allen belts) has important repercussions on space systems operating in the near Earth environment.

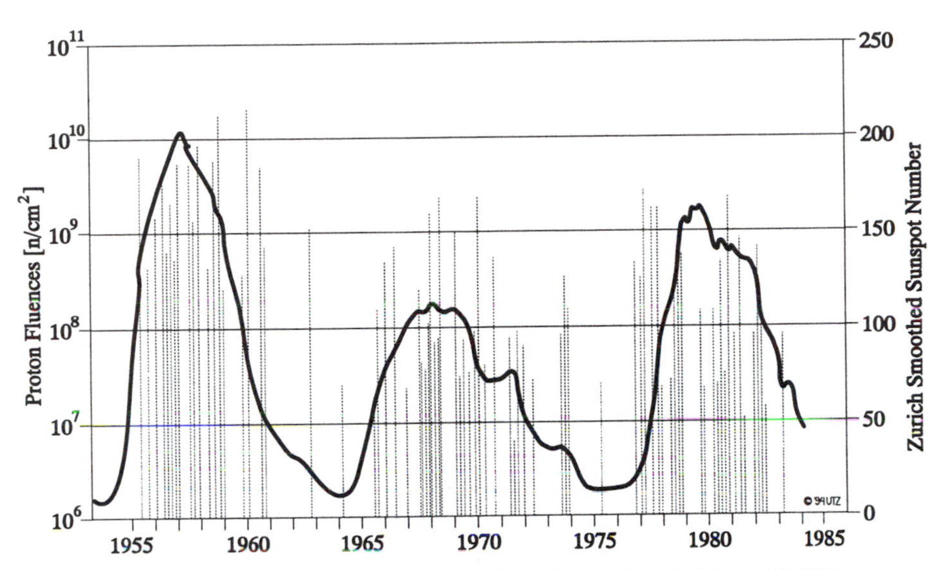

Figure III.3: **Solar-Flare Proton Events for Solar Cycles 19, 20, and 21 [20]**

Solar cosmic rays (SCR) represent the most variable component of natural space radiation. SCR are composed of protons, electrons, or other heavy nuclei accelerated to energies between 10^7 and 10^9 eV during very large solar flares that are occurring once or twice in a solar cycle of about 11 years. These particles can be responsible for a thousandfold increase in the radiation dose over a short period of time. Due to the "individuality" among events, the radiation dose accumulated due to solar protons may vary from negligible to well above lethal. The occurrence of solar flares is basically not predictable and, thus, the warning period is only a few minutes to hours. Solar flares are differentiated according to their total energy released because, ultimately, the total energy emitted is the deciding factor in the severity of a flare's effects. The radiation from a solar flare extends from radio to X-ray frequencies. The brilliance of a flare is measured in two frequency bands: optical and X-ray. While the visible, i.e., optical, emission from a flare increases by, at most, a few percent, the X-ray emission may be enhanced by as much as four orders of magnitude. The total energy released during a flare may range from 10^{21} to 10^{25} Joules integrated over the three phases of a flare: precursor , flash, and main phase. The precursor phase can last from minutes to hours. The flash phase begins with an increase in optical and X-ray emissions by at least 50 % above background within one to five minutes. Within the flash phase there may be impulsive burst of microwaves and X-rays. The main phase can last for hours and is characterized by the slow decay to preflare levels. The values for the solar-flare proton events since the early 1950's are given in figure III.3. [27] It should also be noted that the intensity of solar activity influences the levels of trapped particles and of galactic cosmic radiation (see section III 1.2.2). [2]

III.1.2.2 Galactic Cosmic Radiation (GCR)

The galactic cosmic radiation (GCR) is a permanent radiation that consists of particles originating from outside the solar system. GCR consists of particles with atomic numbers Z greater or equal 1, namely high energy (> 0.1 GeV) protons (85%), α-particles (=^{4}He-nuclei - 14%), and heavy nuclei (1%). For modelling purposes one can define heavy nuclei as nuclei of all atoms with atomic numbers of 2 < Z < 29, i.e., from lithium (Li, Z=3) to nickel (Ni, Z=28). In figure III.4 the relative abundances of even-numbered galactic cosmic ray nuclei are given. For comparison, the open bars give these abundances when weighted with the square of the particles charge to give a measure of the ionizing power of each element.

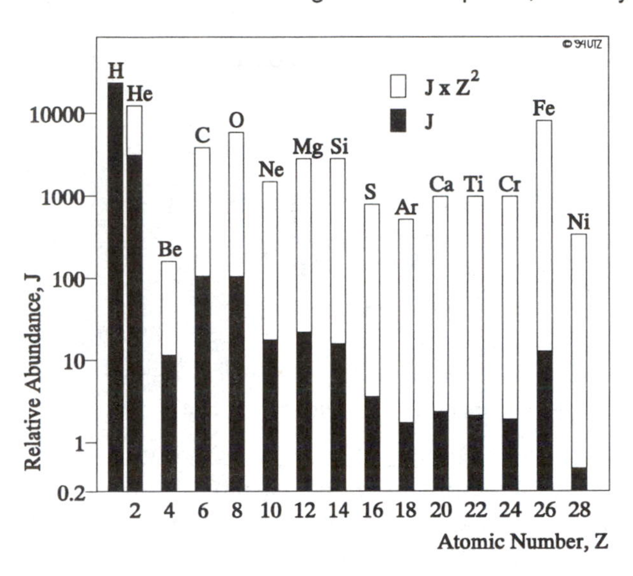

Figure III.4: **Relative Abundances of Even-Numbered GCR Rays [20]**

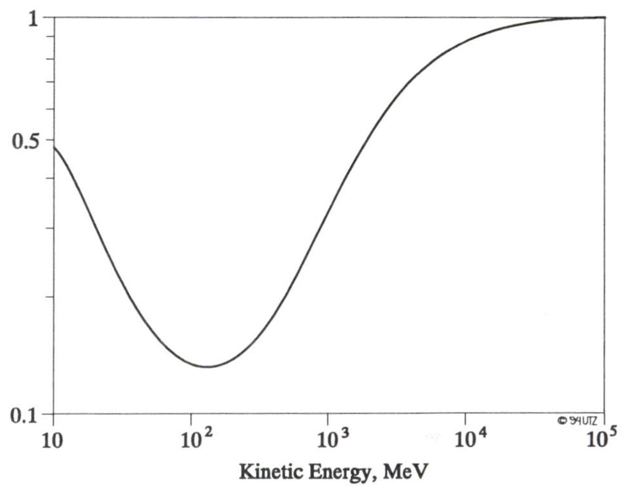

Figure III.5: **GCR Proton Flux Reduction Factors at Solar Maximum Conditions [25]**

Emitted by distant stars and even more distant galaxies, GCR diffuses through space and arrives at Earth from all directions. The most important temporal variation in flux is associated with the 11-year solar cycle. During solar maximum, when the interplanetary magnetic field strength is greatest, cosmic ray particles are scattered away from the Earth. This produces a GCR flux minimum.

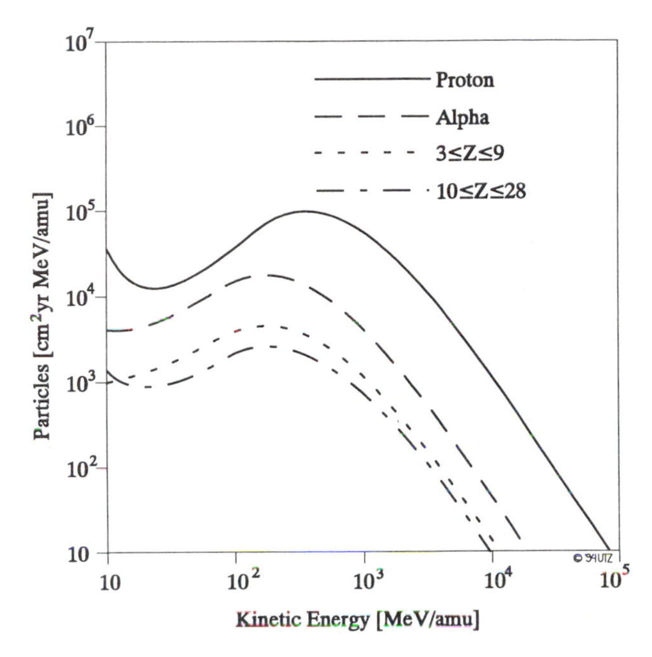

Figure III.6: **Flux vs. Energy Distribution for GCR Ions During Solar Minimum [24]**

Conversely, GCR flux is largest during solar minimum. Figure III.6 shows the GCR particle spectra at solar minimum conditions. The 11-year solar cycle produces a factor of two variations in the cosmic ray dose at a geosynchronous orbit. Low-altitude, low-inclination orbits would experience almost no dose variations due to the strong shielding produced by the combined effects of the atmosphere and geo-magnetic field.

Although not very numerous, these particles constitute a deeply penetrating radiation due to their extremely high energy. Thus, spacecraft shielding is not very effective in reducing the variation dose. Fortunately, GCR flux is comparatively low, so it does not pose a serious threat to humans. For example, several particles have probably passed through your body since you started reading this section. In all orbits, approximately 5-10 % of the total effective radiation dose is due to GCR. This small amount is sometimes referred to as background radiation. As mentioned earlier, the level of GCR is influenced by the intensity of solar activity. A reduction factor for solar maximum conditions, i.e., minimum GCR flux, in comparison to solar minimum conditions, i.e., maximum GCR flux, is given in figure III.5. [2, 14, 19, 20, 27]

III.1.2.3 The Van Allen Belts

The Van Allen radiation belts are doughnut-shaped regions which surround Earth. They consist of energetic (keV to MeV) particles, i.e., electrons and protons that were caught by Earth's magnetic field. They oscillate around Earth along the magnetic field lines. The radiation belts are divided into two concentric zones, an inner belt and an outer belt as illustrated in figure III.7.

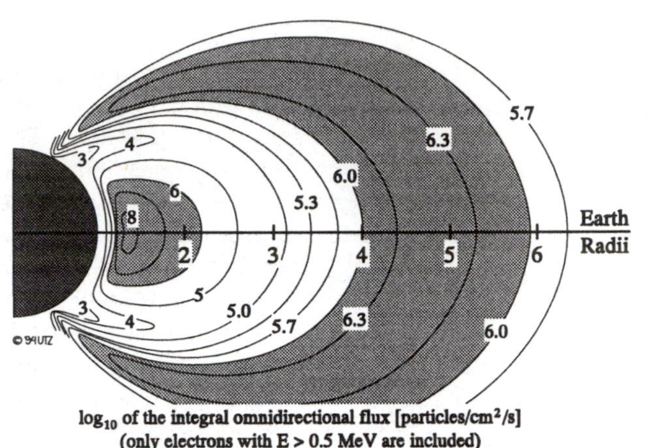

Figure III.7: **The Van Allen Belts [30]**

The outer belt is fed by solar particle inputs, the source for the inner belt are mainly neutron albedo and decay, the neutrons being a result of interaction between cosmic radiation and the upper atmosphere. The trapped particles oscillate about field lines at a cyclotron speed, are reflected at a point called mirror point, when their speed component parallel to the field line is null, and drift to the East or the West, depending on their charge. The belts present an anomaly in which an important concentration of particles can be found. This anomaly is due to the offset of the dipole field of Earth and it takes place at the vertical of the South Atlantic. Thus, it is called South Atlantic Anomaly (SAA). The position of the SAA is shown in figure III.8.

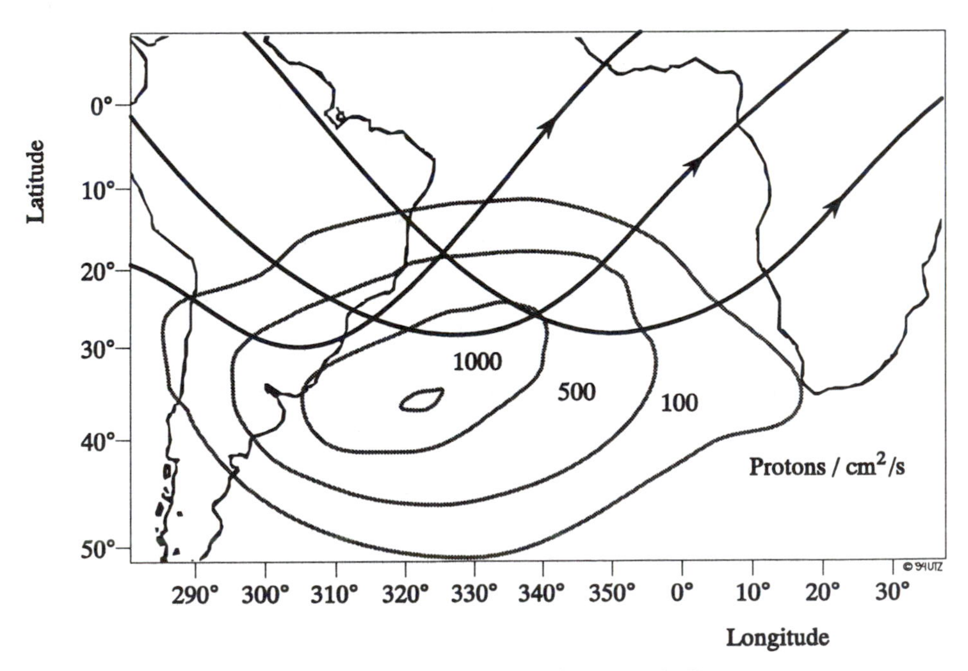

Figure III.8: **The South Atlantic Anomaly [11]**

The Inner Belt

Energetic protons trapped in the inner belt are the major source of radiation for Earth orbiting spacecraft above 500 km. The amount of radiation varies with latitude and longitude (the inner belt extends to about 45° latitude). The prospective orbit of the space station will be at 520 km altitude and 28° inclination and, thus, intercepts the fringes of the inner radiation belt in the SAA. The predominant part of radiation exposure is due to the protons trapped in the SAA. The inner belt proton population is also susceptible to solar-induced variations. Population density varies out of phase with the 11-year solar cycle, so that the inner belt is most inflated during solar minimum. This change in particle population causes a variation in radiation dose rate of a factor of two during the solar cycle for low orbiting spacecraft.

The Outer Belt

The outer Van Allen belt contains both electrons and protons. However, the electrons have much higher number densities and are responsible for most of the radiation dose within this region. The outer belt is asymmetric, with the nightside being elongated and the dayside flattened. Generally, particle energy and outer boundary location vary with the 11-year cycle. During solar maximum, the outer boundary of the electron belt is closer to Earth and contains higher energy particles. At solar minimum, the outer boundary moves outward and contains fewer energetic electrons. Outer belt electron densities undergo order of magnitude changes over scales of weeks. These short-term variations can produce significant radiation dose variations and are related to the level of geophysical activity. [2, 3, 11, 22, 27, 30]

III.1.3 The Concept of Radiation Dose

The quantity called absorbed dose D provides a measure which quantifies the absorbed radiation and is correlated with radiation effects. It is a measure of the energy deposited in tissue by radiation. The standard (SI) unit of absorbed dose is called Gray [Gy]:

$$1\ Gray = 1\ Joule/kg = 100\ rad = 10000\ ergs/g$$

One Gray is the amount of ionizing radiation corresponding to 1 Joule absorbed by 1 kilogram of material. Note that a dose of 1 Gy from high energy protons is the same as 1 Gy from X-rays. The Gray represents an amount of absorbed radiation energy and not the source or type of radiation.

The concept of relative biological effectiveness (RBE) has been introduced to quantify the effects of radiation on biological tissue. To define this unit, each type of radiation was given a RBE compared to a beam of 200 keV X-rays. According to these RBE values, the complex field of GCR may be partitioned into three categories:

- The sparsely ionizing component, called background component, consisting of mainly electromagnetic radiation, electrons and fast penetrating protons (RBE = 1).
- Neutrons and stopping protons with intermediate or moderately high ionizing power, called neutron component (RBE = 1 to 20).
- The densely ionizing component, called HZE (= High atomic number Z and high Energy) component, which consists of the heavy ions, or HZE particles, of the cosmic radiation which induce extremely high ionization densities. The concept of absorbed dose as a measure of radiation exposure breaks down for this component, and hence RBE loses its meaning. [3, 8, 10, 27, 30]

Radiation	RBE	Occurrence
X-rays	1	Radiation belts, Solar radiation, Bremsstrahlung
5 MeV γ-rays	0.5	Radiation belts, Solar radiation, Bremsstrahlung
1 MeV γ-rays	0.7	Radiation belts, Solar radiation, Bremsstrahlung
200 keV γ-rays	1.0	Radiation belts, Solar radiation, Bremsstrahlung
Electrons	1.0	Radiation belts
Protons	2.0 - 10.0	Cosmic radiation, Inner radiation belt
Neutrons	2-10	Close to the Earth, the Sun and any matter
α-particles	10-20	Cosmic radiation
Heavy particles		Cosmic radiation

Table III.3: **RBE and Occurrence of Different Kinds of Radiation [27, 30]**

In table III.3, one can see that protons can be twice as damaging as 200 keV rays, and therefore, a 1 Gy proton dose will be twice as damaging as a 1 Gy dose from X-rays. For human exposure, the dose equivalent H is defined by introducing the quality factor Q which is essentially the same as RBE:

$$H = D \cdot Q$$

The SI unit of the dose equivalent is the Sievert [Sv]:

$$1\ Sievert = 1\ J/kg = 100\ rem = 10000\ ergs/g$$

For example: A 1 Gy dose of 200 keV X-rays gives a biological equivalent dose of 1 Sv, but a 1 Gy dose from protons gives a biological equivalent dose of 2 Sv. The large Sv value for protons accounts for the increased biological damage. [13]

In general, Q is a function of linear energy transfer (LET), which in turn is a function of both particle type and energy. LET, usually given in [keV/µm] of tissue, expresses the spatial density of ionization generated in the irradiated material by ionizing radiations of different qualities. Radiation with different LET causes different biological effects and defects in matter. [25]

There is some uncertainty associated with the quality factors. The International Commission on Radiation Protection (ICRP) 26 adopted the conservative posture of assigning a Q of 20 to all radiations with LET greater than 175 keV/µm uniformly absorbed in tissue, in order to provide a standard for radiation protection purposes. The total dose of, e.g., an interplanetary mission, however, is mainly due to GCR and is calculated based on the Q of 20 for heavy ions with LET greater than 170 keV/µm. Experimental results have cast serious doubts about the validity of this assumption.

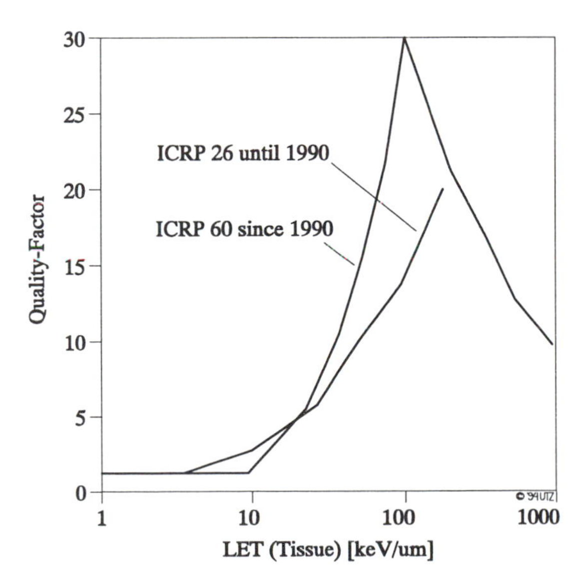

Figure III.9: **Quality Factor vs. LET [29]**

The relationship between Q (or RBE) and LET is not a simple one. Q increases with LET up to about 100-200 keV/µm and then decreases steadily to a value less than 1.0 at very high LET for both cell inactivation and neoplastic transformation. The region of the Q vs. LET curve up to about 100 keV/µm was fairly well supported by the experimental data. However, for high LET heavy ions found in GCR, the assumption of a plateau where Q=20 was not supported by the experimental results. Thus, in 1990 a new relationship between Q and LET, as shown in figure III.9, has been introduced by ICRP 60.

For the characterization of radiation damage in plants, e.g., gene mutation, chromosome aberration, or cell lethality, induced in space, a parameter D, for summarized damage, has been introduced. It classifies and summarizes the sensitive biological effects and is a biophysical approach to the preliminary estimation of the quality factors of densely ionizing radiation. For calculating D the following formula, including five empirical weight factors for certain damages, has been defined [1]:

$$D = \sum_{i=1}^{5} d_i \cdot (n_i / n_0)$$

Weight factors for certain damages:	d_i
Number of damaged plants:	n_i
Number of examined plants:	n_0

III.1.4 Radiation Monitoring and Dosimetry

The complex radiation environment in spaceflight rises a number of problems in dosimetry, mainly due to the great number of radiation sources including GCR, SCR, and the Van Allen belts, and radiation induced in on-board materials by high-energy charged particles such as protons, α-particles and heavier nuclei. The fluences and energy spectra of these radiations depend on altitude, orbit inclination, solar activities, spacecraft positions and orientations in orbit and types of shielding materials used in the spacecraft. [16]

Past and present spaceflight dosimetry measurements, both in the United States and the former Soviet Union, have mainly been made using passive detectors that yielded the integrated dose and integral LET spectra. Information about the LET spectrum is required for the biologically relevant equivalent dose. For determination of the absorbed dose, thermoluminiscence dosemeters (TL) are commonly used in space dosimetry. Heavy charged particles are usually detected with nuclear track detectors such as track etch foils, emulsions, and combinations of fission and track etch foils. Active detectors have been used only on several occasions and gave data on solar flares and photon burst from electrons precipitation from the Van Allen belts. Due to high LET particles existing in the space environment, the knowledge of the quality factor is one of the priorities in radiation protection purpose. In this aim, the experiment CIRCE (Compteur Intégrateur de Rayonnement Complex dans l´Espace) has recorded the dose rate and quality factor values inside the MIR station. CIRCE is able to measure the absorbed dose D in the 1 μGy/h to 5 mGy/h range, the dose equivalent rate H in the 2 μSv/h to 20 mSv range, and the quality factors on a range from 1 to 20, with an accuracy of 10 %. The Soviet-French mission in December 1988 found that the dose equivalent inside the MIR station was equal to 0.6 mSv/day and the absorbed dose was equal to 0.3 mGy/day. Long-term measurements onboard MIR from December 1988 to April 1989 gave an average quality factor value of 1.9, and an average high LET factor of 7.7. Through SAA, the dose equivalent rapidly increased to 1.2 mSv/h, the quality factor was equal to 1.4. Even if crossing durations through the SAA vary only from 8 to 10 minutes, more than 30 % of total dose per day are due to SAA radiation. A comparison of measured values averaged over the orbits of different spacecraft is given in table III.4. [15, 16, 18, 29]

Mission	**Orbit parameters**	**Dose equivalent H** [mSv/day]	**Absorbed dose D** [mGy/day]	**Neutron dose** [mSv/ day]	**Quality factor Q**	**Max. absorbed dose** [mGy/h]
SKYLAB 2 (1976)	430 km 50°		0.57			0.828
Discovery (1985)	297 x 454 km 28.5°		0.544			
Atlantis (1985)	380 km 28.5°		0.208			
Salyut 6 (1980)	350 km 51.5°	0.216	0.146			
Cosmos 936 (1983)	419 x 224 km 62.8°		0.256	0.071		
MIR (12-1988)	350 km 51.5°	0.617	0.322	0.023	1.9	0.755
MIR (3/4-1989)	350 km 51.5°	0.799	0.451			

Table III.4: **Radiation Monitoring During Different Missions [18]**

III.1.5 Radiation Effects

When high energy particles encounter atoms or molecules within the human body, an atomic interaction, i.e., an ionization, may occur. A direct interaction occurs when a particle is suddenly stopped by collisions resulting in a release of energy which may remove electrons from nearby atoms or molecules, and ions result. Indirect encounters occur when the high energy particle, usually an electron, is deflected by another charged particle. The deflection causes a release of energy, i.e., radiation, which also may produce ionization. The close encounter process is commonly referred to as Bremsstrahlung. In either interaction, the effects of the ionizing radiation are proportional to the amount of energy absorbed by the surrounding material. As mentioned earlier, to quantify this absorbed radiation, a unit of measurement called Gray [Gy] was defined. To express the effects of radiation on humans, the dose equivalent H is used (see section III.1.3).

On Earth, a human experiences on the average a dose of about 0.4 mSv each year from radioactive elements in soil, rock, and wood around him. This figure varies from place to place. Cosmic rays passing through the body provides another 0.4 mSv annual dose (1.6 mSv if one lives high up in the mountains). Inescapable sources in food and water provide an additional 0.2-0.5 mSv which brings the yearly dosage for an earthbound person to about 1.7 mSv. For each transatlantic flight 0.04 mSv may be added. [27]

In contrast to this, the space traveler will receive considerably more radiation. The dose received varies with mission duration, orbital profile, and shielding. Crew members on missions to Moon or Mars will be unavoidably exposed to ionizing radiation as they pass through the Van Allen belts and the GCR flux, and there is the possibility for exposure to proton radiation from Solar Particle Events (SPE). Using absorbed doses and LET-dependent quality factors (ICRP 26), the following dose-equivalents may be estimated:

- In a spacecraft with 0.75 cm aluminum walls (2 g/cm^2) at solar minimum, the lunar round trip dose equivalent is less than 0.05 Sv.

- During a Mars mission the estimated dose equivalents are:

- Outbound (Van Allen Belts)	< 0.02	Sv
- Earth to Mars (205 days exposure to free space GCR)	0.32	Sv
- 30 days on the Martian surface (GCR)	0.023	Sv
- Mars to Earth (225 days exposure to free space GCR)	0.35	Sv
- Inbound (Van Allen belts)	< 0.02	Sv
	≅ 0.73	Sv

Conventionally, the total of 0.73 Sv over 460 days could be expected to increase the risk of cancer mortality in a 35-year old male astronaut by about 1 %. However, three-fourth of the dose equivalent in free space is contributed by high LET heavy ions (Z > 3) and target fragments with average quality factors of 10.3 and 20, respectively. As mentioned earlier, the RBE of these radiations is poorly understood, and so the quality factors are set at conservatively very high values. The entire concept of absorbed dose - quality factor - dose equivalent as applied to GCR must be reconsidered. [13]

To further complicate the difficulty in assessing biological damage for a given dose, studies show that individuals have varying degrees of tolerance based on sex and stamina. To circumvent this problem, most estimates of radiation effects are based on a population sample with the number of affected individuals expressed as a percentage. Thus, in table III.5 an overview of the probable effects of certain radiation doses on human health is given. In this table the long term effects, such as cancer are not included. [27]

Dose [Sv]	Probable effects
0 - 0.5	No obvious effects; possibly minor blood changes
0.5 - 1	Radiation sickness in 5 - 10 % of exposed personnel; no serious disability
1 - 1.5	Radiation sickness in about 25 % of exposed personnel
1.5 - 2	Radiation sickness in about 50 % of exposed personnel; no deaths anticipated
2 - 3.5	Radiation sickness in nearly all the personnel; about 20 % deaths
3.5 - 5	Radiation sickness; about 50 % deaths
10	Probably no survivors

Table III.5: **Probable Radiation Dose Prompt Effects [27]**

Thus, effects of radiation on humans can generally be placed in two categories:

- *Acute, early effects of radiation exposure*
  These occur within a few days or less. They are usually associated with exposure to a high dose of radiation over a short period of time. Acute radiation exposure is indicated by symptoms of radiation sickness, e.g., nausea, vomiting, accompanied by discomfort, loss of appetite, and fatigue. At higher dosages (> 2 Sv) also diarrhea, hemorrhaging, and hair loss may occur after a latent period of up to two weeks.

- *Delayed, late effects of radiation exposure*
  These occur many years after prolonged exposure to radiation at a low dose rate. Delayed effects include cancers of the lung, breast, digestive system, and leukemia. As a rough rule of thumb, an astronauts chance of fatal cancer is increased approximately 2 % to 5 % for each 0.5 Sv exposure during his/her career. [10]

Effects of Radiation on Plants

Studies have shown that general lethality and the summarized damage D of seedlings grown from irradiated seed embryos are enhanced significantly. Enlargement of deletions in the DNA and of chromosome breakages are suggested to be the cause. Thus, increasing of gene mutations and chromosomal aberrations can be supposed to make the stay in space risky. The effects of radiation on plant growth are discussed in more detail in section VI 4.1.2.2. [8]

Effects of Radiation on Materials

Radiation can also have many effects on materials, e.g., gas evolution, change in mechanical, electrical, and optical properties, and even complete mechanical breakdown. Materials damage depends not only on radiation dosage and material, but also on the kind of radiation, rate of application, and load and temperature of the material. One has to differentiate between dose effects, which are resulting from total absorbed radiation, rate effects, resulting from absorption per unit of time, and transient effects, resulting from rapid changes in absorbed radiation. [22]

III.1.6 Radiation Protection

The two main radiation sources of concern for space missions are solar flares (SCR) and GCR. Crews have to be protected from the potentially life-threatening (SCR) and career-limiting (GCR) effects of these heavy, charged-particle radiations. To provide necessary radiation protection at the minimum cost in terms of mass which has to be put into orbit several active and passive shielding methods for spacecraft have been considered:

- *Passive Bulk Shielding*
  Shielding stops or alters the trajectory of high energy particles before they encounter the more sensitive human tissue. Aluminum is used extensively as shielding material, since it combines both high density and lightness. Spacecraft exteriors typically have several grams per cm^2 of aluminum shielding. Low orbits, especially those confined to the equatorial plane, are substantially less hazardous than polar orbits. The former make use of Earth's natural shielding, while the latter expose an individual to ambient energetic particles in a region where natural shielding is of limited value. Shielding is particularly important at geosynchronous orbits. The magnitude of the shielding effect varies with shielding composition and the energy spectrum of the radiation. Light elements such as hydrogen, carbon, and oxygen, and their compounds, such as

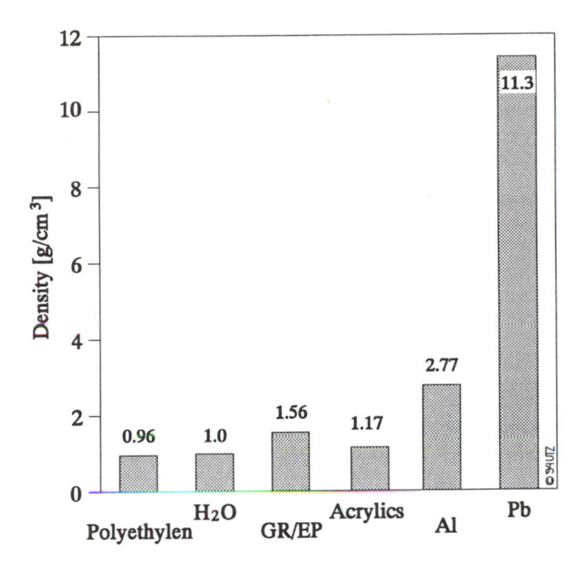

Figure III.10: **Material Densities**

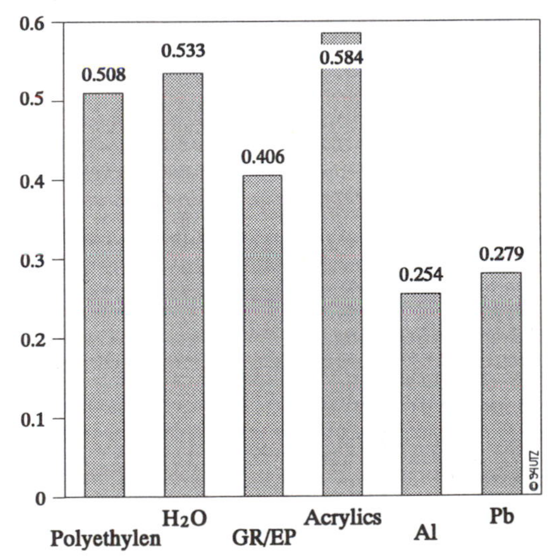

Figure III.11: **Shielding Thickness Values**

water and plastic, are the most effective shields per unit mass. In charged particle shielding applications lead (Pb) and other heavy elements are poor shields. This may also be seen from the values in figures III.10 and III.11. For the given materials and shielding thicknesses the best mass per area unit value can be obtained for polyethylene (0.192 g/cm²), the worst for lead (1.246 g/cm²).

The radiation contribution of solar flares (SCR) is of greatest importance for shorter missions and thinner shields and becomes less significant in comparison to the GCR component as mission times and layer thicknesses increase. A study of Nealy clearly indicates that the typically shielded spacecraft of 1 to 2 g/cm² aluminium would not provide sufficient protection for crew members on a long-term space mission. Besides special passive shielding, the design of a spacecraft shall make optimal use of onboard mass for radiation shielding. On the surface of Moon or Mars, it is possible to protect habitats from GCR with layers of regolith. The protection equivalent to Earth's atmosphere would be five meters of soil (see sections IX.3 and IX.4).
[7, 10, 14, 27, 28]

- *Electromagnetic Shielding*
  During the early 1960's, the possibility of electromagnetic fields to shield spacecraft from SCR was recognized. Hence, a cutoff energy, depending on particle charge, mass and kinetic energy, and the magnetic field intensity, was defined for a torus magnetic shield configuration. Particles with kinetic energies greater than the cutoff energy are transmitted through the magnetic shield, while particles with kinetic energies lower than the cutoff energy are deflected by the magnetic field and do not intercept the shielded interior volume. For a proposed shield (200 MeV cutoff for protons) it was found that multiply charged, heavy ions passed through the magnetic field with little deflection. Hence, GCR exposures will be largely unaffected by the presence of a magnetic shield. Anyway, for the same shield, most protons of a solar flare were deflected. This lead to the conclusion that magnetic shielding appears to be very effective for the protection from solar flares exposure but has virtually no effect on GCR exposure. Also, it was estimated that through the proper selection of materials and proper design of a bulk shielding configuration, lower exposures could be obtained from passive shielding with the same total shield mass as needed for the magnetic shield configuration. [28]

- *Electrostatic Shielding*
  Obviously, extensive research on the electrostatic shielding from cosmic radiation has been conducted. Unfortunately, no detailed information on these studies could be obtained. [7]

- *Chemical Radioprotection*
  For man, chemical radioprotectants (CRP), like APAETF (Aminopropyl-aminoethyl thiophosphoric acid), have been developed both in the U.S. and the former Soviet Union. The highly effective APAETF demonstrates a dose reduction factor of 3. Small doses of this compound protect against relatively high levels of radiation. The major criteria for the design of CRP are that they have to be active taken internally, they have to be rapidly absorbed and distributed to the tissues from the gastrointestinal tract, that they are free of negative side effects, and effective during fractioned and prolonged radiation. [7]

III.2 GRAVITY

Gravity fields play a very important role in manned spaceflight. From the beginning of their existence humans, animals and plants are used to the effects of gravity on Earth. Thus, the most dramatic environmental characteristic of spaceflight is the state of microgravity ("weightlessness") when flying in free space or when being exposed to the reduced gravity on the other planets of the solar system. This results in extensive physical, physiological and psychological effects. [30]

III.2.1 Gravity on the Planets and in Free Space

Gravity constants for the planets and some other bodies in the solar system relative to the gravity level on Earth are given in table III.6.

Planet	Gravity Constant [m/s²]	Gravity Relative to Earth's Gravity
Mercury	3.53	0.36
Venus	8.83	0.9
Earth	9.81	1
Mars	3.73	0.38
Jupiter	26	2.65
Saturn	11.18	1.14
Uranus	10.5	1.07
Neptune	13.24	1.35
Pluto	2.16	0.22
Earth's Moon	1.67	0.17
Mars' Moon (Phobos)	0.02	0.002
Large Asteroids	0.02	0.002

Table III.6: **Gravity on Other Planets of the Solar System [17]**

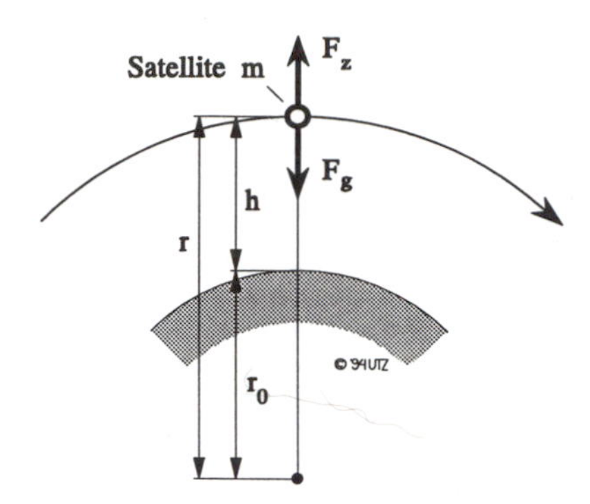

Figure III.12: **Forces on a Satellite in Orbit**

For flights in interplanetary orbits, basically constant microgravity conditions may be assumed. For satellites orbiting near the surface of a planet, however, it may be noted that there is a gravity gradient in the radial direction. On a satellite or space station in a planetary orbit, and with a given velocity v, basically two forces are acting: the centripetal force and the gravity force (see figure III.12).

In the center of mass these two forces are in a balance, principally yielding zero gravity in this point. In all other points of the satellite, though, some small accelerations do occur. This may be demonstrated for a point 1 which is at a distance h radially above the center of mass in point 0. For any point, the equilibrium may be written as follows:

Centripetal force: $F_z = \frac{mv^2}{r}$ Gravitational force: $F_g = -\gamma \frac{mM}{r^2}$

Mass of the satellite:	m	Mass of the planet:	M	Radius of the orbit:	r
Velocity of the satellite:	v	Gravity constant:	γ	Acceleration:	a_*

As mentioned before, in the center of mass $a_0^* = 0$ and for any other point may be written:

$$a_1^* = \frac{v^2}{r_0 + h} - \frac{\gamma M}{(r_0 + h)^2}$$

These accelerations are yielding gravity stabilization, but may also have to be taken into account when designing a satellite or a space station.

III.2.2 Effects of Lower Gravity

Here, only the effects of microgravity on the human physiological processes are examined. The effects of microgravity on plant growth are discussed in section VI 4.1.2.2. A significant shift of intravascular and extravascular fluids in the human body takes place in microgravity. This shift is evidenced by a calf girth decrease of nearly 30 %, head congestion and associated facial puffiness. In adapting to the shift, the body's homeostatic systems responds by an increasing urine output. This leads to a severe orthostatic hypotension and an extensive diuresis upon return to Earth. Inflight, the heart is overloaded with blood and triggers an excretion of fluids and salts by the kidneys. This diuresis of fluids in combination with a decrease in thirst leads to significant reduction of the intravascular space. After compensation has occurred, a decrease in heart chamber size is associated with a 10 % atrophy of the heart muscle. This causes an increased heart rate throughout the mission.

Past manned missions have also shown that bone and muscle atrophy occurs in space crews, which is proportional to the length of time spent in microgravity. In general, muscle atrophy precedes skeletal atrophy with the greatest reduction in muscle mass occurring during the first month. Calcium loss (osteoporosis) begins slowly in the first week and increases gradually over the next several months (average rate: 0.5 % per month; peak rate in some bones: 3-5 % per month). A loss of this magnitude is conceivable only for a one- to two-year period. The overwhelming majority of the calcium loss is from the weight bearing bones. Unlike other physiologic adaptations, this calcium loss does not seem to reach a plateau. Thus, measures have to be taken for long-term manned missions, like the provision of artificial gravity (see section III 2.3).

It is currently not known, though, what amount of stress, i.e., gravitational force, must be maintained to prevent osteoclastic demineralization of the weight bearing bones. Finally, microgravity is known to cause an atrophy of the bone marrow and the immune system leading to the so-called "space-anemia" and a defect in the T-lymphocytes. The significance of the possible space-immunosuppression goes beyond the danger of a simple infection. In particular, T-lymphocytes also help fight off growing neoplasms, and a suppression of this system may hamper the astronaut's ability to fight off a cancer while in orbit. This is a serious hazard in the high radiation environment of space. It was discovered, though that T-cells can maintain their normal functioning level in space by being provided to artificial gravity. [5]

III.2.3 Artificial Gravity

As discussed in the preceding chapter it might be necessary to introduce artificial gravity to space stations or even bases on Moon or Mars. Artificial gravity can basically be obtained by rotating a habitat or space station. There are four major features that cause artificial gravity to be different from the known terrestrial gravity. These are:

- Artificial Gravity Level
- Gravity Gradients
- Coriolis Forces
- Cross-coupled Angular Accelerations

What is called artificial gravity is in fact a centripetal acceleration α [m/s²] produced by rotation:

$$\alpha = \omega^2 \cdot r$$

Angular velocity: [1/s] Radius: r [m]

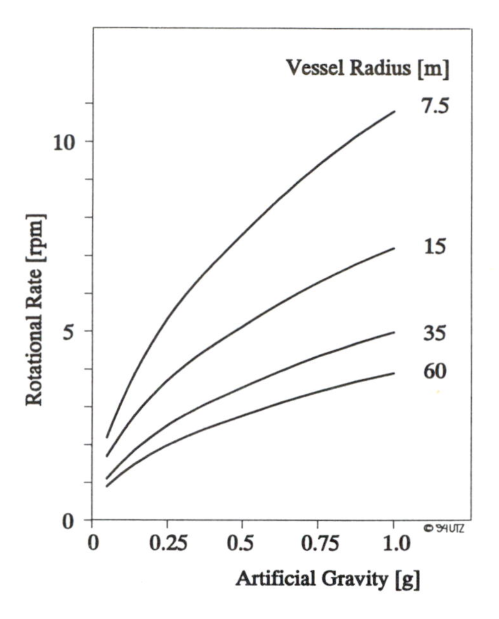

Figure III.13: **Induced Gravity**

As indicated in this equation, a specific increase in artificial gravity level can be achieved either by increasing the radius, or the angular velocity. This translates to a trade-off between cost and complexity, which depend on r, vs. physiological/psychological concerns, which depend on ω. Induced gravity levels as a function of rotational rate and diameter are given in figure III.13. [5]

Since the centripetal acceleration is a linear function of the radius, a linear gravity gradient runs from the center of the habitat to the outer rim. Thus, any object would be basically weightless in the center of the habitat, and would have maximum weight at the outer rim. The percentage weight change $\Delta G/G$ that an object undergoes during translation from any outer radius R_a to any inner radius R_b can be described by the equation:

$$\frac{\Delta G}{G} = \frac{R_a - R_b}{R_b}$$

By far the most remarkable effects in a rotating habitat are caused by the Coriolis force. The Coriolis force is applied to any object moving linearly with the velocity *v* within a rotating system and its magnitude is:

$$\bar{F}_c = 2\,(\bar{w} \times \bar{v})$$

Any object moving in a direction not parallel to the axis of rotation will, thus, experience Coriolis force in a way indicated in table III.7. The Coriolis acceleration may be defined as:

$$A_c = \frac{F_c}{m}$$

In figure III.14, this Coriolis acceleration is given as a function of the radius of rotation, level of artificial gravity, and rotational rate.

Object moving	Direction of Coriolis force
From R_a radially towards R_b	Direction of rotation
From R_b radially towards R_a	Counter to direction of rotation
Tangentially in direction of rotation	Radially outwards ("object gets heavier")
Tangentially counter to direction of rotation	Radially inwards ("object gets lighter")

Table III.7: **Directions of Coriolis Force for an Object Moving in a Rotating Habitat**

Movement of a subject within a rotating environment gives rise to peculiar stimulations of the bodies sensory systems. Coriolis cross-coupled angular accelerations occur within a rotating environment when an angular motion is made about an axis not parallel to the system axis of rotation ($\varpi_1 \times \varpi_2$). These cause gyroscopic forces which produce symptoms of vertigo, disorientation, or nausea. These symptoms are referred to as motion sickness. The most complete simulation of a rotating habitat, yet, was conducted by North American Rockwell Space Division which rotated four men for seven days at a radius of 22 m. Concerning maximum angular velocity sustainable by humans three points of reference could be established:

- A speed of 1 rpm is not disturbing even to those subjects who are highly susceptible to vestibular effects.
- At a speed of 4 rpm some individuals will be naturally immune to motion sickness, while others will have motion sickness, but will adapt after a few days.
- A speed of 10 rpm will cause Coriolis sickness in most individuals if the transition from a stationary environment to the 10 rpm environment is abrupt. Subjects can adapt to 10 rpm if a schedule of stepwise increases in angular velocities is followed.

It is obvious that there are several questions to be examined if artificial gravity is seriously considered for future space habitats:

- Which gravity levels and gravity gradients will be required or acceptable for any biological system?
- What are the physiological limits to radius and angular velocity, i.e., Coriolis effects?
- What are the engineering/cost limits to radius and angular velocity?
- Will all engineered systems still function properly under these conditions?
- What countermeasures will be necessary? [17]

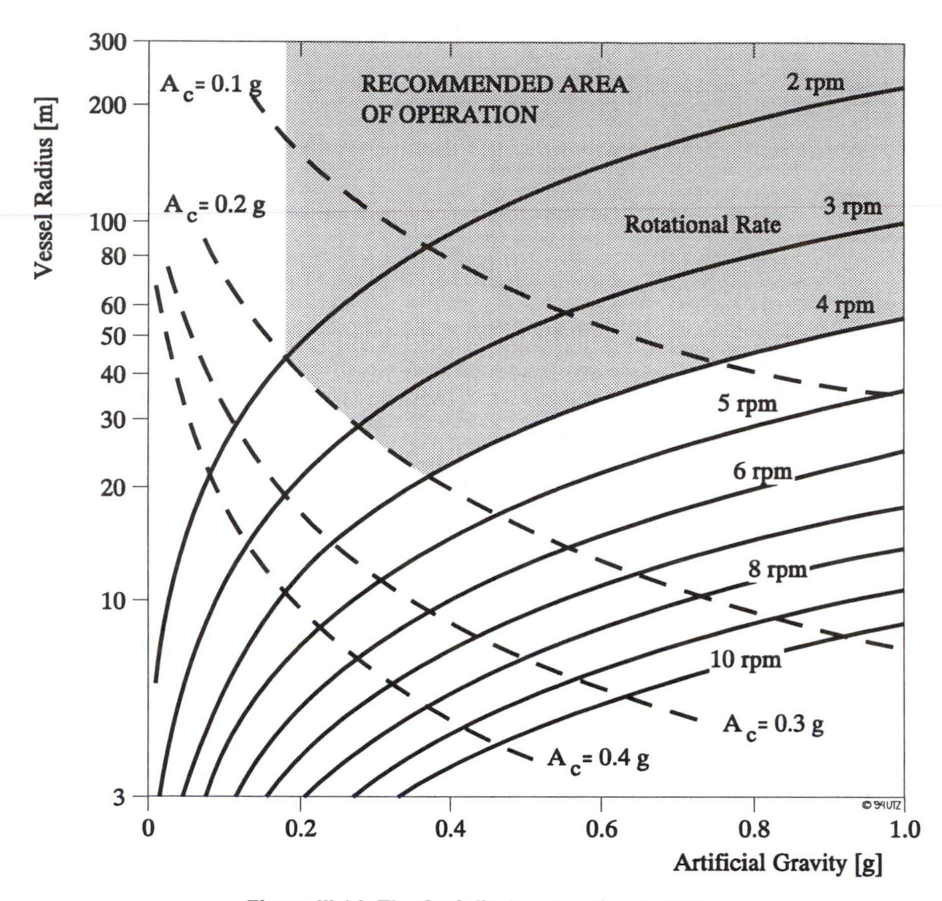

Figure III.14: **The Coriolis Acceleration A_c [30]**

III.3 VACUUM

Due to the vacuum in free space it has to be considered that:

- Heat transfer is only possible via radiation
- Pressurized shells are necessary to house equipment, materials and man
- Materials are affected in several ways

III.3.1 Temperature Effects

One of the main effects of vacuum in free space is the fact that heat transfer is only possible via radiation. A way how to determine the temperature of a spacecraft or space station is given here:

Planet	Albedo
Mercury	0.096
Venus	0.76
Earth	0.39
Moon	0.07
Mars	0.25
Jupiter	0.51
Saturn	0.42
Uranus	0.66
Neptune	0.62
Pluto	0.63

Table III.8: **Planetary Albedo**

First, the equilibrium temperature is influenced by the energy created within the vehicle, energy coming from the outside and energy given to the outside. Basically, in the free space of the solar system the Sun is the only source of electromagnetic radiation. The solar radiation input may be written as a function of the distance from the Sun:

$$I = \frac{I_0}{r^2}$$

Radiation energy flux at 1 AU: $I_0 = 1390\ W/m^2$
Distance from the Sun: r [AU]

Energy can be radiated away from the spacecraft. It can be assumed that the background temperature of free space $T_{space} = 4$ K. However, in lower planetary orbits the planetary-reflected solar input also has to be taken into account. Thus, the energy balance for any spacecraft flying in a low orbit around any planet may be written as: [6,22]

$$Q_{sun} + Q_p + Q_i = Q_{s,space} + Q_{s,p}$$

Solar input to spacecraft: .... $Q_{sun} = \alpha_s A_\perp I_{sun}$
Planetary-reflected solar input: .... $Q_p = a\,\alpha_s f_{s,p} A_s I_{sun}$
Internally generated energy: .... Q_i
Energy radiated from spacecraft to space: ..... $Q_{s,space} = \sigma \varepsilon_s f_{s,space} A_s (T_s^4 - T_{space}^4)$
Energy radiated from spacecraft to planet: ...... $Q_{s,p} = \sigma \varepsilon_s f_{s,p} A_s (T_s^4 - T_p^4)$
Planetary albedo: .... a
Spacecraft surface absorptivity: .... α_s
Spacecraft surface emissivity: .... ε_s
View factor spacecraft-to-space: .... $f_{s,space}$
Temperature of the spacecraft: .... T_s
Space background temperature: .... T_{space}
Solar energy flux: .... I_{sun}

Area of the spacecraft: .... A_s
Projected area of the spacecraft: .... $A_\perp$
View factor spacecraft-to-planet: .... $f_{s,p}$
Temperature of the planet: .... T_p
STEPHAN BOLTZMANN constant: σ
($\sigma = 5.67 \cdot 10^{-8}\ W/(m^2 \cdot K^4)$)

Assuming that $T_{space} \cong 0$ and $F_{s,space} + F_{s,p} = 1$, in the equilibrium condition the energy balance equation becomes:

$$\sigma \varepsilon_s A_s F_{s,p} T_p^4 + Q_{sun} + Q_p + Q_i = \sigma \varepsilon_s A_s T_s^4$$

In a free space environment, this equation may be simplified to

$$Q_{sun} + Q_i = \sigma \varepsilon_s A_s T_s^4$$

yielding the equilibrium temperature of the spacecraft:

$$T_s = \sqrt[4]{\frac{\alpha_s A_\perp I_{sun} + Q_i}{\sigma \varepsilon_s A_s}}$$

III.3.2 The Necessity of a Pressurized Environment

Because of the vacuum conditions in free space it is necessary to provide pressurized shells in order to house equipment, materials, and, of course, man. The design of such a pressurized shell should be fail-safe. In particular, a hole should not lead to complete disintegration of the cabin, since man can survive for only about 15 seconds after an explosive decompression in space. Another important aspect is the leakage of any pressurized spacecraft or space station in the free-space vacuum. Like any pressure vessel it loses gas by several processes, e.g., through diffusion through the walls, leaks through seals, operation of air locks, and holes. It is important to note that, in general, the loss rate is proportional to internal pressure, which should be, hence, as low as possible from that point of view. The fact of any gas leakage leads to the necessity of gas resupply, even in future regenerative life support systems. Of course, this resupply has to be taken into account in mass calculations. [22]

III.3.3 Material Effects

Due to the vacuum in free space materials are affected in several ways:

- *Outgassing*
  The gas layer adhering to the surface of materials is lost.

- *Sublimation evaporation*
  The higher the vapor pressure of a material, the more mass sublimates from its surface per second per unit area.

- *Diffusion*
  With no gas layer between them, solid materials can come into close contact. This can lead to a kind of cold welding process by diffusion of the materials into each other.

In order to avoid failures of materials, the selection, surface condition, and surface treatment of the materials that are exposed to the vacuum of free space are critical. [22]

III.4 MAGNETIC FIELDS

Earth has a magnetic field that can be approximately described as the field of a dipole, sitting about 500 km from Earth's center and being 11.5° inclined towards the rotational axis. The field is fairly constant over a period of some months, but disturbances cause changes for days, hours, or minutes. The main effect of Earth's magnetic field on spaceflight is the trapping of charged particles, leading to the Van Allen belts around Earth. A secondary effect of the magnetic field of Earth is the electromagnetic and magnetic induction in moving objects. [22]

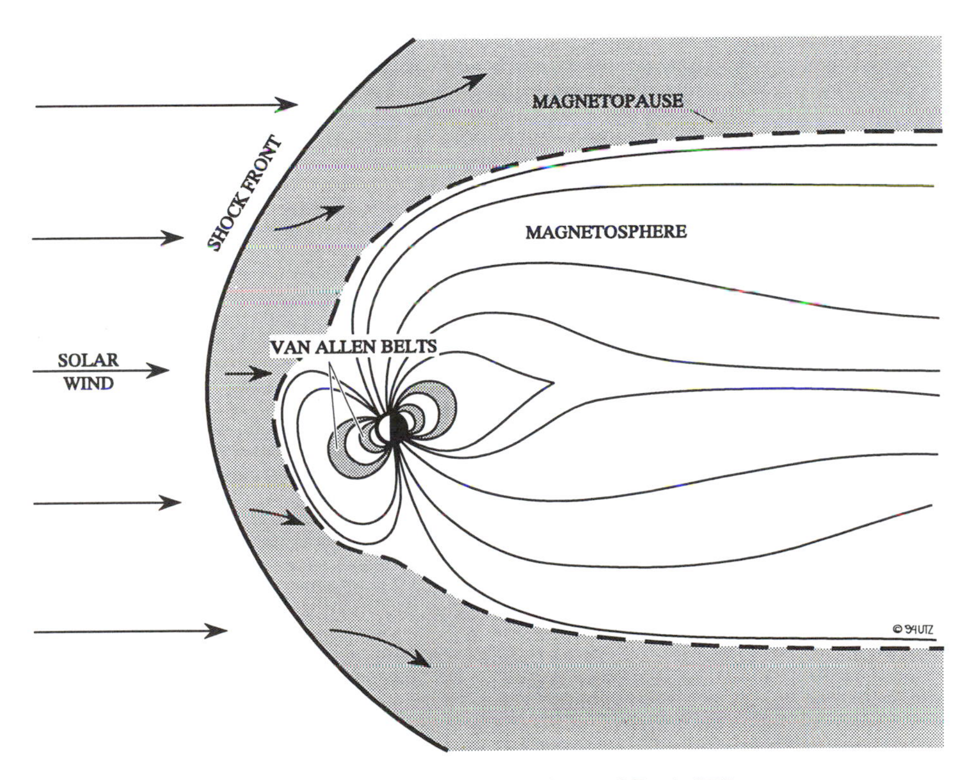

Figure III.15: **The Magnetosphere of Earth [26]**

In the absence of an interplanetary plasma, Earth's dipole magnetic field would extend indefinitely in all directions. However, the geomagnetic field produces a semipermeable obstacle to the solar wind and the resulting interaction produces a cavity around which most of the plasma flows: the magnetosphere. The magnetopause is the "surface" where the outward force of the compressed geomagnetic field is balanced by the force of the plasma wind. As the solar wind encounters the magnetosphere, a shock wave, called the bow shock, forms in front of the magnetopause. The bulk of the solar wind plasma is directed around the magnetosphere and does not approach Earth any closer than the magnetosphere. The resulting magnetosphere is shaped like a bullet, fairly blunt on the sunward side, and nearly cylindrical for a long distance in the antisolar direction. The magnetopause occurs approximately 10 Earth radii (10 RE) on the sunward side, and the geomagnetic tail extends well beyond the orbit of Moon (60 RE). The nose of the bow shock forms at about 15 RE (see figure III.15). [26]

Intermixed with the solar wind plasma is the background coronal magnetic field of the Sun which is trapped-in or "frozen-in" by the ever expanding solar wind. The field lines of this so-called interplanetary magnetic field (IMF) follow a spiral pattern. This is due to the fact that the ends of the field, being carried outward with the solar wind, remain attracted to the rotating Sun. At Earth the IMF strength is about $5 \cdot 10^{-5}$ Gauss, i.e., about 10^{-4} times Earth's surface magnetic field.

III.5 LOCAL PLANETARY ENVIRONMENTS

In this chapter an overview is given about the environmental parameters and physical properties of Moon and Mars. These two celestial bodies are subject to major efforts for future manned spaceflight. The other planets and moons of the solar system are not examined here, because manned missions to these are either impossible or not considered for the foreseeable future.

III.5.1 Moon

III.5.1.1 Lunar Parameters

Moon is orbiting Earth once in about four weeks. Since Moon's orbital and rotational period coincide, always the same side of Moon is facing Earth. The exact lunar and lunar orbital characteristics can be found in tables III.9 and III.10. [23]

Mass	Mean Density	Radius	Gravity	Escape Velocity	Albedo
$7.359 \cdot 10^{22}$ kg	3340 kg/m³	1738 km	1.62 m/s²	2.38 km/s	0.07

Table III.9: **Lunar Characteristics [23]**

Mean value of semi-major axis	384400 km
Perigee	364400 km
Apogee	406730 km
Elipticity	0.002
Inclination of axis to ecliptic	1° 32′
Inclination of lunar equator to lunar orbital plane	6° 41′
Inclination of orbital plane to ecliptic	5° 9′
Sidereal month	27.32 days
Synodic month	29.53 days

Table III.10: **Lunar Orbital Parameters [23]**

III.5.1.2 Environment of the Lunar Surface

The typical characteristics of the environment on the lunar surface are summarized below:

- *Electromagnetic Radiation and Lighting Conditions*
  The total input of solar radiation on Moon's surface is equal to the radiation above Earth's atmosphere, i.e., 1390 W/m^2. On the equator, and for all latitudes except very close to the poles there is a 14-day-night, 14-day-light cycle. On the poles, due to orbital parameters, the Sun elevation is only ± 1° 32′ with a 1/2-year-day, 1/2-year-night cycle.

- *Atmosphere*
  A lunar atmosphere is basically non-existent (pressure of less than 10^{-13} bar). This is due to the fact that the low gravity of Moon cannot retain light atoms such as H or O. The light atoms that can be found on Moon are basically resupplied by the solar wind.

- *Temperature*
  At the equator the temperature ($T_{Equator}$) changes between 80 K and 390 K during one lunar day. The temperature change depends on a thermal inertia parameter γ. This parameter determines the rate of heating or cooling

of material. Below one meter depth, temperature can be assumed constant over time at about 230 K. At high latitudes, NASA recommends the following approximation for a given latitude β:

$$T = T_{Equator} \cos^{1/4} \beta$$

The temperature at the poles is basically unknown, but guessed to be as low as 40 K in some permanently shaded areas.

- *UV Radiation*
  The total UV radiation input on the lunar surface is about twice that on Earth's surface and, thus, the same as for LEO. Hence, some materials, like plastic, can be destroyed by the UV radiation and plants may require shielding.

- *Ionizing Radiation*
  Since Moon has no atmosphere, ionizing radiation is not naturally shielded that way from the lunar surface. Nevertheless, Moon itself serves as a shield because on its surface the radiation does not arrive from all directions like in free space. Thus, the approximate dose rates on the lunar surface during solar minimum are as follows:

  Solar wind: 0.5 Sv/year
  GCR: 0.2 Sv/year
  Solar flares: 1 - 50 Sv/event

  Still, shielding will be required for men in a lunar base. It is likely that plants require less shielding. For shielding purposes lunar regolith may be used. It is not an ideal shielding material, but it is abundant and freely available.

- *Meteorite Environment*
  For the lunar environment, NASA gives the following annual cumulative meteoroid model:

  $\log N_t = - 14.597 - 1.213 \log m$ for $10^{-6} < m < 1$
  $\log N_t = - 14.566 - 1.584 \log m - 0.063 \log m^2$ for $10^{-12} < m < 10^{-6}$

  Number of particles of mass m or greater per square meter per second: N_t
  Mass: m [g]

  A lunar base would only receive half of this flux, because of the shielding by Moon.

- *Magnetic Field*
  Moon's actual magnetic field is negligible. [23]

III.5.1.3 Physical Properties of the Lunar Surface

Lunar soil is basically composed of oxygen (42 %), silicone (21 %), iron (13 %), calcium (8 %), aluminum (7 %), and magnesium (6 %). The thermal inertia parameter γ determines the rate of cooling of the soil, e.g. rocks cool down faster and heat up faster than regolith. The thermal conductivity is very low (comparable to styrofoam). The specific heat of soil is comparable to that of bricks, and about one fifth of that of water. A few numbers are indicated in table III.11.

Thermal inertia parameter γ [$cm^2 s^{1/2}$ K/J]	Density ρ [kg/m^3]	Specific heat c [J/(kg·K)]	Conductivity k [W/(m·K)]
5.97 - 334	500 - 3000	755 - 1007	$2.14 \cdot 10^{-3}$ - 1.13

Table III.11: **Lunar Surface Properties [23]**

	Marial Rocks	Highland Rocks
SiO_2	37.6 - 48.8 %	44.3 - 48.0 %
TiO_2	0.29 - 12.1 %	0.06 - 2.1 %
Al_2O_3	7.64 - 13.9 %	17.6 - 35.1 %
FeO	17.8 - 22.5 %	0.67 - 10.9 %
MnO	0.21 - 0.29 %	0 - 0.07 %
MgO	5.95 - 16.6 %	0.8 - 14.7 %
CaO	8.72 - 12.0 %	10.7 - 18.7 %
Na_2O	0.12 - 0.66 %	0.12 - 0.8 %
K_2O	0.02 - 0.096 %	0 - 0.54 %
P_2O_5	0 - 0.15 %	-
S	0 - 0.15 %	-
Cr_2O_3	0 - 0.7 %	0.02 - 0.26 %

Table III.12: **Chemical Composition of Lunar Soil (Mare and Terrae) [23]**

Geologically, one can distinguish the marae and the terrae (highlands). The marae are dark, level plains (floor of the basins). They can be generally found only on the near side of the Moon. The terrae are lighter and older than the marae and densely cratered. They cover all of the far side and parts of the near side. The soil of both the marae and the terrae consists of well graded sandy silts with average particle sizes of 0.04 - 0.13 mm. A summary of the results of the analyses of sample compositions taken during the APOLLO missions is given in table III.12. [23, 26]

III.5.1.4 Lunar Resources

As mentioned earlier, many materials are, at least theoretically, attainable on Moon, e.g., water, cements ($CaO : SiO_2 : Al_2O_3$), glass, and metals (Al, Cr, Fe, Mg, Ni, Ti). Regolith may be used for radiation shielding and thermal insulation.

	Methods	Remarks
Hydrogen extraction	Microwave techniques	Large power requirements
	Microbial extraction	Dependent on whether the molecular H is accessible to the hydrogenase; also the lunar soil must not be toxic to the bacteria
	Benefaction / Thermal release of gases	Large power requirements
Oxygen extraction	Carbothermal processing (using carbon instead of oxygen in the ilmenite process)	Large power requirements, high pressures
	Electrolysis of silicate	Large power requirements
	Destructive distillation	Extremely high temperatures
Water extraction	Hydrogen reduction of Ilmenite	Large power requirements

Table III.13: **Techniques for the Extraction of Hydrogen, Oxygen and Water from Lunar Soil**

A very promising potential processing method is the reduction of ilmenite. The two proposed reactions with hydrogen and methane, respectively, yield iron, water, carbon dioxide:

Hydrogen Reduction: $FeTiO_3 + H_2 \rightarrow Fe + TiO_2 + H_2O$

Carbomethyl Reduction: $4FeTiO_3 + CH_4 \rightarrow 4Fe + 4TiO_2 + 2H_2O + CO_2$

The water can be split by electrolysis to yield hydrogen and oxygen. A few principal techniques for the extraction of hydrogen, oxygen and water are summarized in table III.13. [23]

III.5.2 Mars

III.5.2.1 Martian Parameters

Mars, the Red Planet, is the fourth planet from the Sun. It can be described as a small, cold, dry planet with a thin atmosphere. Nonetheless, it has the most clement non-terrestrial environment in the solar system and is the most likely candidate for long-term manned exploration. The exact Martian and Martian orbital characteristics can be found in tables III.14 and III.15. [9,12]

Mass	Mean Density	Radius	Gravity	Escape Velocity	Albedo
$6.418 \cdot 10^{23}$ kg	3933 kg/m³	3397 km	3.72 m/s²	5.04 km/s	0.25

Table III.14: **Martian Characteristics [12]**

Mean value of semi-major axis	$227.94 \cdot 10^6$ km
Perihel	$206.65 \cdot 10^6$ km
Aphel	$249.18 \cdot 10^6$ km
Elipticity	0.093387
Orbit inclination	1° 50' 59.28'
Obliquity of rotation axis	25°
Inclination of orbital plane to ecliptic	5° 9'
Sidereal year	686.98 days
Synodic period (time between Earth-Mars oppositions)	779.95 days
Sidereal day	24h 37m 22s
Solar day	24h 39m 35s
Earth-Mars opposition distance: max **min**	$10.1 \cdot 10^7$ km $5.6 \cdot 10^7$ km

Table III.15: **Martian Orbital Parameters [12]**

III.5.2.2 Environment of the Martian Surface

The typical characteristics of the environment on the Martian surface are summarized below:

- *Electromagnetic Radiation*
  The total input of solar radiation on the Martian surface is at an average of about 615 W/m^2. The maximum radiation input is 718 W/m^2 at perihelion, the minimum radiation input is 493 W/m^2 at aphelion.

- *Atmosphere*
  The nominal pressure of the Martian atmosphere is 7 mb. The pressure varies seasonally by about 25 % as a result of condensation of the polar caps. The Martian atmosphere is basically composed of CO_2. The detailed composition of the Martian lower atmosphere is given in table III.16. The mean horizontal wind velocity at the surface is 2-9 m/s. The maximum wind speed in the atmosphere is at about 60-80 m/s. The scale height is 11 km.

Gas	CO_2	N_2	Ar	O_2	CO	H_2O	Ne	Kr	Xe
Volume	95.3%	2.7%	0.16%	0.13%	0.07%	0.03%	2.5 ppm	0.3 ppm	0.08 ppm

Table III.16: **Composition of Martian Lower Atmosphere [12]**

- *Temperature*
  On the Martian surface there are large ambient-temperature contrasts from one region to another. The nominal temperature value is at 215 K, with a temperature range from 130 K to 300 K.

- *UV Radiation*
  The nominal total solar UV flux is 10 $J/(m^2 \cdot s)$

- *Ionizing Radiation*
  The free-space environment surrounding Mars is comprised of a continuous flux of solar wind particles and GCR, augmented on occasion by random solar flare events (SCR). Since Mars is devoid of a magnetic field strong enough to deflect the charged particles, many of the high energetic particles are able to reach the outer atmosphere. The Martian atmosphere provides protection from GCR and SCR with the amount of protection depending upon the atmospheric composition and structure. For a low-density model a surface pressure of 590 Pa is assumed, providing a shielding of 16 g/cm^2 in the vertical direction. For a high-density model the values are 780 Pa and 22 g/cm^2.

- *Magnetic Field*
  Mars' magnetic dipole moment is less than 10^{22} G/cm³. [4, 9, 12, 21, 24]

III.5.2.3 Physical Properties of the Martian Surface

Present knowledge of the mineralogy and composition of the Martian surface rocks and sediments is poor by terrestrial or lunar standards. At the VIKING 1 landing site about 50 % of all elements of the Martian soil were not directly determined. Of the detected elements the main components are silicone (21 %), iron (13 %), magnesium (5 %), calcium (4 %), aluminum (3 %), and sulfur (3 %). A few parameters of the Martian soil are indicated in table III.17.

Thermal inertia parameter γ [$cm^2 s^{1/2} K/J$]	Density ρ [kg/m^3]	Specific heat c [J/(kg·K)]	Conductivity k [W/(m·K)]
24 - 144	3933	625 - 800	$8.5 \cdot 10^{-3}$ - $85 \cdot 10^{-3}$

Table III.17: **Martian Surface Properties [12]**

Geologically, one can distinguish two hemispheres north and south of a great circle on the Martian surface inclined by about 35° to the equator. South of this boundary, the ancient cratered highlands are dominating, north of this boundary, mainly the young cratered and smooth plains are occurring. A summary of the results of the analyses of samples carried out by VIKING 1 is given in table III.18. [4, 12, 25]

Compound	SiO_2	Al_2O_3	Fe_2O_3	MgO	CaO	K_2O	TiO_2	SO_3	Cl
Mass	44.7%	5.7%	18.2%	8.3%	5.6%	< 0.3%	0.9%	0.77%	0.7%

Table III.18: **Chemical Composition of Martian Soil at VIKING 1 Landing Site [12]**

III.5.2.4 Martian Resources

Basically, the principles and techniques for the use of lunar resources, as described in section III.5.1.4, are also applicable for Martian resources. The major difference to Moon is that the carbon dioxide that makes up 95 % of the Martian atmosphere can be a valuable starting material for the manufacturing of critical products. CO_2 can be had by merely compressing the atmosphere and might be used in the following ways: [26]

- Direct support of plant growth
- Oxygen production by passing CO_2 through a zirconia electrolysis cell at 800 to 1000° C (dissociation of 20 - 30 % of the CO_2), yielding:

$$2\,CO_2 \rightarrow 2\,CO + O_2$$

- The oxygen can then be converted into water if hydrogen is supplied from Earth (combustion):

$$O_2 + 2\,H_2 \rightarrow 2\,H_2O$$

Maybe it will also be possible to obtain water from the environment of Mars, which could then also be used as a source of hydrogen (electrolysis):

$$2\,H_2O \rightarrow 2\,H_2 + O_2$$

III References

[1] Bork U. et al
Defective Embryogenesis of Arabidopsis induced by Cosmic HZE-Particles
Life Science Research in Space, ESA SP-307, p. 571-572, 1990

[2] Bourdeaud´hui J. et al
Radiation Protection Strategies in HERMES Missions
Acta Astronautica, Vol. 23, p. 233-244, 1991

[3] Bücker H.; Facius R.
Radiation Problems in Manned Spaceflight with a View Towards the Space Station
Acta Astronautica, Vol. 17, No. 2, p. 243-248, 1988

[4] Chicarro A. et. al.
Mission to Mars
ESA SP-1117, 1990

[5] Diamandis P.
Contermeasures and Artificial Gravity
International Space University, Space Life Sciences Textbook, p. 291-322, 1992

[6] Griffin M.; French J.
Space Vehicle Design
AIAA Education Series, Washington, 1991

[7] Helmke C.
Synopsis of Soviet Manned Spaceflight Radiation Protection Program
USAF Foreign Technology Bulletin, FTD-2660P-127/105-90, 1990

[8] Kranz A.
Genetic Risc and Physiological Stress Induced by Heavy Ions
Life Science Research in Space, ESA SP-307, p. 559-563, 1990

[9] Kuznetz L.
Space Suits and Life Support Systems for the Exploration of Mars
NASA Ames Research Center, 1991

[10] Letaw J.
Ionizing Radiation Hazards in Space
International Space University, Space Life Sciences Textbook, p. 17-31, 1992

[11] McCormack P.
Radiation and Shielding for the Space Station
Acta Astronautica, Vol. 17, No. 2, p. 231-241, 1988

[12] McKay C.
A Short Guide to Mars
The Case for Mars, Vol. 57, Science and Technology Series, American Astronautical Society, p. 303-310, 1981

[13] Nachtwey D.; Yang T.
Radiological Health Riscs for Exploratory Class Missions in Space
Acta Astronautica, Vol. 23, p. 227-231, 1991

[14] Nealy J. et al
Deep-Space Radiation Exposure Analysis for Solar Cycle XII (1975-1986)
20th Intersociety Conference on Environmental Systems, SAE Technical Paper 901347, 1990

[15] Nguyen V. et al
New Experimental Approach in Quality Factor and Dose Equivalent Determination During a Long Term Manned Space Mission
Life Science Research in Space, ESA SP-307, p. 555-558, 1990

[16] Nguyen V. et al
Real Time Quality Factor and Dose Equivalent Meter CIRCE and its Use On-Board the Soviet Orbital Station MIR
Acta Astronautica, Vol. 23, p. 217-226, 1991

[17] Olson R. et al
CELSS for Advanced Manned Mission
HortScience, Vol. 23(2), p. 275-286, 1988

[18] Petrov V.
Principle and Realization of the Instrument used for the CIRCE Experiment on Board the Space Station MIR
Life Science Research in Space, ESA SP-307, p. 577-580, 1990

[19] Pissarenko N.
Radiation Environment Due to Galactic and Solar Cosmic Rays During Manned Missions to Mars in the Periods between Maximum and Minimum Solar Activity Cycles
Advances in Space Research, Vol. 14, No. 10, p. 771-778, 1994

[20] Reitz G. et al
Radiation Biology
Life Sciences Research in Space, ESA SP-1105, p. 65-79, 1989

[21] Ruppe H.O.
Die grenzenlose Dimension Raumfahrt, Band 1 + 2
Econ Verlag, Düsseldorf, 1980

[22] Ruppe H. O.
Introductions to Astronautics, Vol. I + II
Academic Press, New York, 1966

[23] Schwartzkopf S.
Lunar Base Controlled Ecological Life Support System (LCELSS)
Lockheed Missile & Space Company, NASA Contract NAS 9-18069, 1990

[24] Simonsen L. et al
Ionizing Radiation Environment at the Mars Surface
Proceedings of SPACE 90, American Institute of Civil Engineers, 1990

[25] Simonsen L.; Nealy J.
Radiation Protection for Human Missions to the Moon and Mars
NASA Technical Paper 3079, 1991

[26] Sullivan T.; McKay D.
Using Space Resources
NASA Johnson Space Center, 1991

[27] Tascione T.
Introduction to the Space Environment
Orbit Book Company, Malabar, Florida, 1988

[28] Townsend L. et al
Radiation Protection Effectiveness of a Proposed Magnetic Shielding Concept for Manned Mars Missions
20th Intersociety Conference on Environmental Systems, SAE Technical Paper 901343, 1990

[29] Vana N. et al
Dosimir - Radiation Measurements inside the Soviet Space Station - First Results
International Space Year Conference, ESA ISY-4, p. 193-197, 1992

[30] von Puttkamer J.
Der Mensch im Weltraum
Umschau Verlag, Frankfurt, 1987

[31] Wertz J., Larson W.
Space Mission Analysis and Design
Kluwer Academic Publishers, 1991

[32] Columbus Human Factors Engineering Requirements
ESA, COL-RQ-ESA-013, 1989

IV FUNDAMENTALS OF LIFE SUPPORT SYSTEMS

IV.1 DEFINITIONS

Thermodynamically speaking, man as a living creature is an open system, i.e., he exchanges matter and energy with his environment that maintains its own structure. He is living in his closed terrestrial life support system known as the "biosphere". The biosphere is a basically closed system in terms of matter, but an open one in terms of energy. For spaceflight purposes the goal is therefore to develop those techniques necessary to ensure the biological autonomy of man when isolated from his original biosphere, i.e., to provide a controlled and physiologically acceptable environment for a crew of any kind of spacecraft, space station, or planetary base - a life support system (LSS). [41]

The traditional components of life support are air, water, and food. Beyond these obvious requirements habitability aspects also have to be considered. These refer to those human factors which make a living pleasant and/or desirable. In addition to biological factors, certain physical factors are important. These include vibration, noise, thermal and pressure requirements, ionizing and non-ionizing radiation, electromagnetic exposure, and gravitational effects. Principally, life support systems can be divided into the five main areas as shown in table IV.1.

Life Support Area	Purpose
Atmosphere Management	Atmosphere composition control, temperature and humidity control, pressure control, atmosphere regeneration, contamination control, ventilation
Water Management	Provision of potable and hygiene water, recovery and processing of waste water
Food Production and Storage	Provision and, potentially, production of food
Waste Management	Collection, storage, and processing of human waste and trash
Crew Safety	Fire detection and suppression, radiation shielding

Table IV.1: **Main Areas of Life Support**

The only practically available ways for the provision of all these life support requirements of man in space are:

- Launching of all required consumables at the start of the mission
- Resupplying of consumables during the mission
- Recycling of life support materials in-flight
- Utilization of in-situ resources, e.g., for planetary bases

Historically, air, food and water have been carried onboard and waste was stored and returned to Earth. These open-loop life support systems have been used very successfully for short-duration space missions. As space missions get longer, however, the supply load gets heavier and resupply becomes prohibitive. It is therefore essential to recycle consumables and, thus, introduce regenerative life support systems for future long-duration space missions.

IV.2 CLASSIFICATION OF LIFE SUPPORT SYSTEMS

Life support functions can be divided into two categories: *non-regenerative* and *regenerative*. Non-regenerative functions refer to processes that are not subject to recycling, such as system monitoring or makeup for system leakage losses. Regenerative functions involve life support resources such as water, oxygen, and food that potentially may be reused.

Systems which provide regenerative functions with no recovery of life support resources are said to be *open loop*, while systems which achieve recycling of these resources are *closed loop*. In open loop systems matter continuously flows in and out of the system. In this scenario, all food water, water, and oxygen are from stored sources. The quantities of resources that are resupplied must equal the quantities of resources used during the mission. Open loop technologies tend to be simple and highly reliable and have been extensively used in manned spaceflight to date. The big disadvantage of open loop systems in general is that resource requirements continue to increase linearly as mission duration and crew size increase. In a closed loop scenario an initial supply of resources is brought from Earth and the non-useful waste products are processed to recover useful resources. As loop closure increases, quantities of resources to be resupplied are reduced. The big advantage of closed loop systems is the one time mass transport of mass to orbit, of processing hardware and initial resource supply, with minor subsequent resupply of non-recoverable losses and processor expendables. The disadvantage are lower technology maturity to date and increased power and thermal requirements.

Closed loop technologies which provide regenerative functions can use *physico-chemical* and/or *biological* processes. Systems including both physico-chemical and biological processes are called *hybrid* life support systems. Physico-chemical processes include use of fans, filters, physical or chemical separation, concentration processes, etc. Biological or *bioregenerative* processes employ living organisms such as plants or microbes to produce or break down organic molecules. Traditionally, physico-chemical processes have been used to provide life support. They are well understood, relatively compact, low maintenance and have quick response times. On the other hand, these processes consume a lot of energy that is expensive to produce and cannot replenish food stocks, which must still be resupplied. Consequently, solid wastes have to be collected, pre-treated, and stored. Biological processes are less well understood, tend to be large volume, power and maintenance intensive, with slow response time, but have the potential to provide food. [13, 15, 41]

IV.3 DESIGN AND DEVELOPMENT CONSIDERATIONS

The design of life support systems for spacecraft is governed by one common objective, namely, to maintain in an isolated volume an environment, suitable for the well-being of men and systems during the mission. Although being derived from this one objective, the design of life support systems will strongly depend on several specific, constraining requirements. Some constraints are related to the technologies, while others come from the human, mission specific, system, safety, and test requirements or cost considerations. In general, simple designs are most reliable and easiest to operate. For safety reasons it is important to design in tolerance to failures, such that the failure of one component will not result in hazardous situations. The importance of designing an Environmentally Controlled Life Support System (ECLSS) to be fail-safe or fail-operational cannot be overemphasized. *Fail-safe* is defined as "the ability to sustain a failure and retain the capability of safe crew and mission operation". *Fail-operational* is "the ability to sustain a failure and retain full operational capability". Anyway, the primary consideration for the designer is to meet the specified requirements in the most appropriate manner and the goal to keep in mind is minimizing the total system mass to mission mass ratio, power consumption, volume, and resupply and storage requirements in balance with safety and maintainability requirements. [17, 48]

Developing an ECLSS is an iterative process involving evaluation of technologies and system configurations, manual and computerized analyses, and hardware and software testing. Initially, simplified scenarios are evaluated for their requirements and constraints, such as mass limits due to launch vehicle limitations, and are later refined as more detailed information becomes available on the missions and the technologies. Early evaluation involves top-level trade studies where simple models of the candidate technologies are used to identify the most suitable technologies based on mission requirements. As hardware matures, the suitability for a particular mission can be more readily determined and further development and testing is done to verify performance of the subsystems prior to integrated system-level testing. As the design and development process advances, each aspect of the mission becomes defined in greater detail and ultimately results in a detailed mission scenario where all the aspects interplay and mesh into an optimum balance. This requires a thorough understanding of the mission and system requirements and constraints, as well as those of each component. [48]

Step	Method	Relative Supply Mass
0	Open loop	100 %
1	Waste water recycling	45 %
2	Regenerative carbondioxide-absorption	30 %
3	Oxygen recycling from carbondioxide	20 %
4	Food production from recycled wastes	10 %
5	Elimination of leakage	5 %

Table IV.2: **Reduction of Relative Supply Mass by Successive Loop Closure [46]**

As mentioned before, the role of man in space has emerged from the periodic, relatively short missions, to the quite frequent but still short Space Shuttle missions of the U.S., and the Russian space stations, requiring continuous mission support. In order to avoid the large resupply penalties, especially for oxygen and water, on future space stations and possible planetary bases, the mostly open loop, non-regenerative systems of today will have to be successively replaced by closed loop, regenerative systems. In this stepwise process, the loops for water, CO_2, O_2, N_2 and, finally, food will be closed. Loop closure becomes especially important, the more crew members participate in a mission and the longer the mission lasts. Successive loop closure will lead to a reduction of the required relative supply mass as indicated in table IV.2. [38, 45]

Nevertheless, reduced resupply of closed loop systems has to be traded against system costs, power requirements, volumes and other limited resources, e.g., with an increasing closure of life support systems loops, the energy need increases drastically. Thus, the optimum degree of system loop closure is mission dependent. Tailoring a life support system to a particular mission is a multiple step process. It includes determination of human requirements and critical life support functions, identification of technologies to provide these functions, and selection of the technologies most appropriate for the specific mission. Criteria which must be evaluated for a particular mission include relative cost of power, weight, and volume, availability of in-situ resources, resupply capability, as well as crew size and mission duration. A summary of the effects of several characteristics on the design of the life support system is given in table IV.3. [13]

Mission Characteristics	Impact on Life Support System Design
Crew Size	Increased size increases consumables required
Mission Duration	Increased duration increases consumables and reliability required
Cabin Leakage	Increased cabin leakage increases the required resupply
Resupply Capability	More difficult resupply increases requirement for storage of consumables and reliability
Power Availability	Limited power availability drives to passive or low energy use systems
Volume Availability	Restrictions of space drive to more volume efficient systems
Transportation Costs	Increased transportation costs drive decreases in system and resupply weight
Gravity	Selection of processes that work in anticipated gravity
Contamination Source	Contaminations require countermeasures and increased robustness of the system
In-Situ Resource Utilization	Decreases resupply needs

Table IV.3: **Mission Characteristics and their Impact on Life Support System Design [13]**

Of basic importance in the design of life support systems is the definition of the loops and interrelationship of the several subsystems. Humans and their needs are the central point of each life support system. In figure IV.1, a diagram of the four main subsystems for atmosphere, water, waste, and food management, and their interrelationship, involving man, are presented. Since not all the loops can be completely closed due to leakage and process conditions, it is indicated that a continuous input of external food, O_2, and N_2, and a permanent output of trace contaminants, heat, and non-recoverable waste can be expected. Through appropriate waste processing procedures it may be possible to basically close the loops for nutrients, CO_2, and water, and, thus, avoid external resupply.

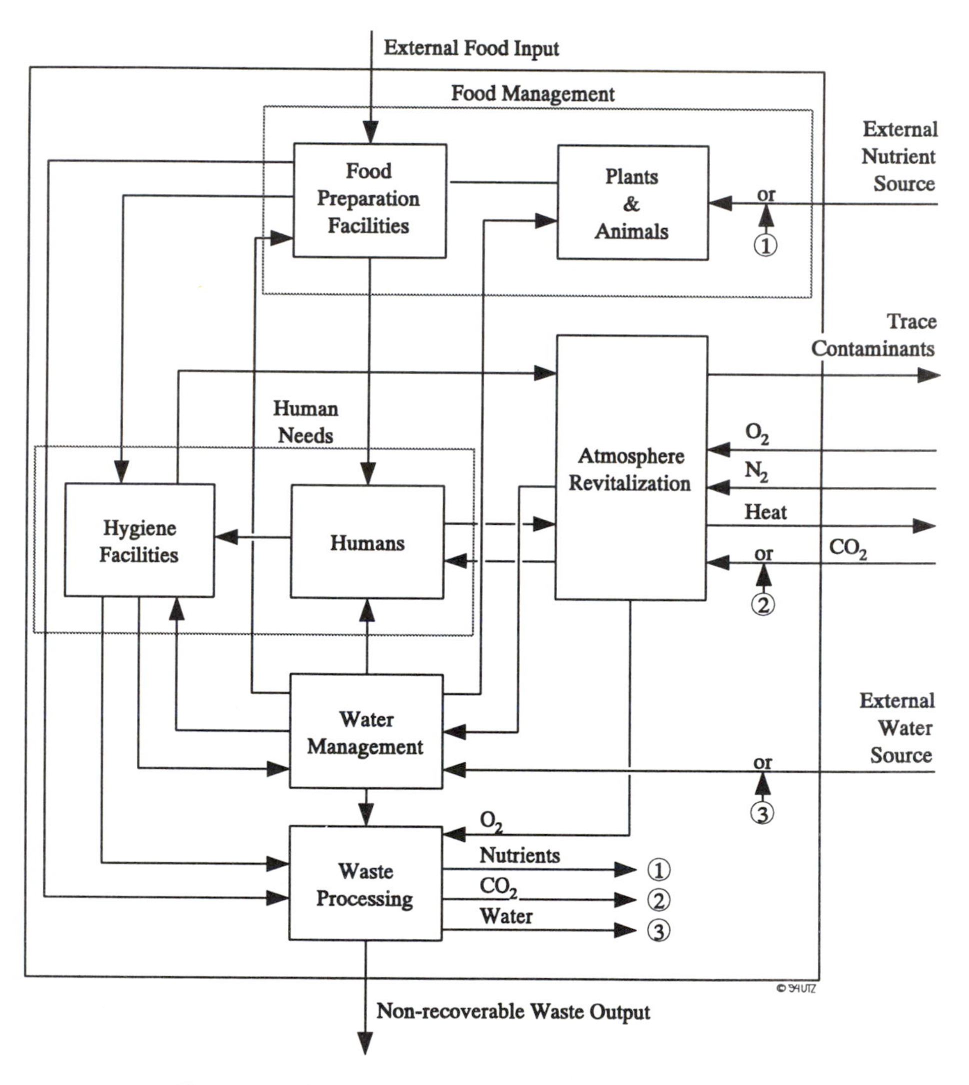

Figure IV.1: **Life Support Functions and Interrelationships [13]**

A survey of the most important non-regenerative and regenerative life support subsystems and their functions is given in tables IV.4 and IV.6. In the tables is also indicated for which mission durations the respective subsystem has to be applied.

Subsystem	Function	Mission Duration for which Function is required
Provision of Oxygen	Meeting metabolic needs of the crew (Storage or regeneration of oxygen)	Always
Carbon Dioxide Reduction	Reduction of removed carbon dioxide to close oxygen loop	Months - Years
Provision of Potable Water	Meeting the metabolic needs of the crew (Storage or recovery of water)	Always
Provision of Hygiene Water	Water for personal hygiene, flush water, cleaning of dishes, washing of clothes	Always Months - Years
Urine Processing	Recovering water from urine to reduce water resupply	Months - Years
Waste Processing	Recovering water, nutrients etc. from solid waste	Years
Provision of Food	Food stored, resupplied, or produced onboard	Always

Table IV.4: **Regenerative Life Support Functions [13]**

Urine Processing	Water processing	Waste processing	Oxygen Generation	CO_2 Removal	CO_2 Reduction

- TIMES
- Flash Evaporation
- Vacuum Distillation/ Pyrolysis
- Biocatalytic Reactor
- Air(wick) Evaporation
- Urine Electrolysis
- VCD
	- Reverse osmosis
	- Multifiltration
	  - carbon
	  - ion exchange
	- Ultrafiltration
	- RITE process
	- SCWO
	- Plants
	- Oxidation
	  - heat
	  - catalytic
	  - biological
		- Compacting and storage
		- Incineration
		- Wet oxidation
		- RITE process
		- SCWO
		- Electrochemical incineration
		- Biological decomposition
		- Photocatalytic oxidation
			- Static feed electrolysis
			- Water vapor electrolysis
			- CO_2 electrolysis
			- In-situ resources
			- Plants
			- Cryogenic storage
			- High pressure storage
				- Molecular sieve
				- EDC
				- SAWD
				- Photochemical
				- LiOH
				- Electroreactive carries
				- Plants
					- Bosch
					- Sabatier
					- Carbon formation reactor
					- Photocatalysis
					- Direct electrolysis
					- Catalytic decomposition
					- Ultraviolet photolysis
					- Plants

Table IV.5: **Life Support Technologies for Regenerative Functions [15]**

Subsystem	Function	Mission Duration for Which Function is Required
Atmosphere Composition Monitoring	Monitoring gas partial pressures, humidity, particulates etc.	Always
Pressure Control	Maintenance of correct partial and total pressures	Always
Temperature Control	Maintenance of temperature within allowable limits	Always
Humidity Control	Maintenance of humidity within allowable limits	Always
Particulate Control	Filtering of the air to remove particulates	Weeks - Years
Trace Contaminant Control	Monitoring and removal of trace contaminants	Weeks - Years
Refrigeration	Food preservation and possibly medical resupply	Months - Years
Fire Detection and Suppression	Detection and suppression of fires	Always
Water Quality Monitoring	Monitoring the quality of processed water	Months - Years
CO_2 Removal	Storing or venting of CO_2	Always
Urine and Feces Removal	Storing or venting of urine and feces	Days - Years
Waste Removal	Storing of wastes	Days - Years

Table IV.6: **Non-Regenerative Life Support Functions [13]**

The decision concerning the technology to select for a particular application also depends largely on the mission characteristics, like crew size, mission duration, mission location, etc. A list of technologies available to provide non-regenerative and regenerative life support functions is given in tables IV.5 and IV.7. A further description of these technologies is given in chapter V. [15]

Trace Contaminant Control	Temperature Control	Humidity Control	Nitrogen Makeup	Particulate Removal
- Sorbents - Catalytic oxidation - Vacuum exposure carbon molecular sieve - Low temperature plasma process	- Heat exchanger - Electric / Resistive heating - Phase change processes - Recupera-tion of waste heat - Direct solar heating	- CHX - Desiccant - Molecular sieve - Membrane separation	- Cryogenic storage - High pressure storage - Hydrazine decom-position - Electrolytic decomposi-tion of urine - In-situ resources	- Filters - Electrostatic precipitation

Table IV.7: **Life Support Technologies for Non-Regenerative Functions [15]**

In general, it is important to point out that there is not one life support system that is "best" for a given mission, much less for all missions. Also, realistic design and evaluation of life support systems cannot be done without considering their interaction with other mission systems.

IV.3.1 Human Requirements

Per definition, sustaining human life is the basic goal of any life support system. Human requirements address the physical needs of the crew and, to a lesser extent, the psychological requirements. They are the basic driver in the design and development process. In this context it may be interesting to know that humans can only survive about four minutes without oxygen, three days without water, and up to thirty days without food. Habitability is a general term which denotes a level of environmental acceptability. The term includes quality standards to support the crew's health and well-being during the stay time, i.e., duty and off-duty periods, aboard any space habitat. The requirements which represent the basic level of habitability can be subdivided into three categories: Humans require an adequate environment, certain consumables, and produce several waste products. These three categories and the interfaces of a human with other life support subsystems are indicated in figure IV.2, and are further discussed in sections III.3.1.1 to III.3.1.3. [48]

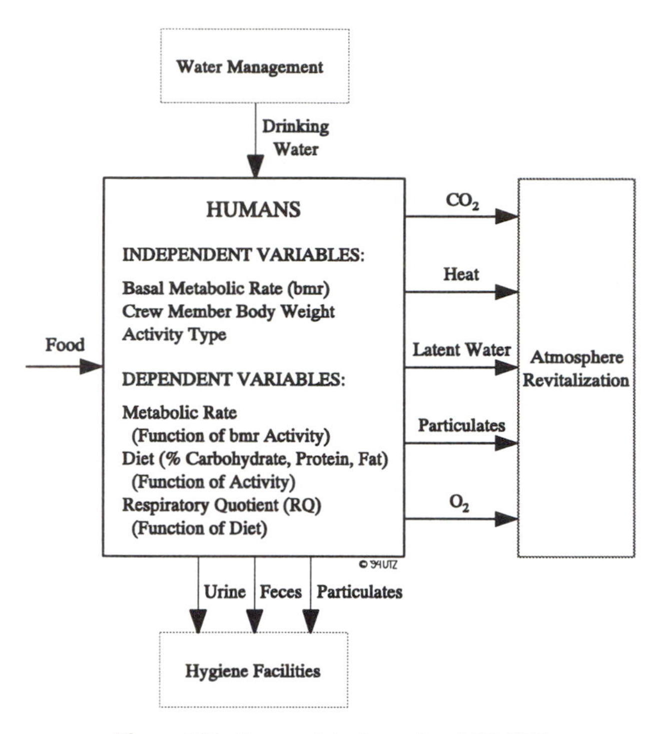

Figure IV.2: **Human Interfaces in a LSS [13]**

For permanently manned systems, like space stations or planetary bases, further habitability and human factors will also have to be considered. The extended level of habitability is introduced to take care of the long-term condition of the stay time. It shall support not only the individuals' physical health, because experience has shown that with the passage of time deleterious effects of isolation and confinement gain prominence. A survey of basic and long-term habitability needs is listed in table IV.8. [16]

Basic Habitability Aspects	Long-Term Habitability Aspects
- Climate - Illumination - Colors and Surfaces - Decor - Radiation - Contamination control - Odor - Noise - Vibration - Acceleration - Interior Space / Layout - Hygiene - Food	- Crew composition - Interpersonal dynamics - Crisis management - Motivation - Communication - Meal periods - Privacy - Mental care - Off-duty functions

Table IV.8: **Habitability Aspects**

IV.3.1.1 Environmental Aspects

The term "environmental aspects" summarizes the human needs for a respirable and comfortable atmosphere, for protection from all kinds of radiation, and possibly the provision of artificial gravity. In this chapter, radiation, metabolic and environmental requirements and boundary values are presented. Since not much is known about the long-term effects of microgravity on the human organism, no specific requirements can be given here. As described in section III.2., it seems that a stay of more than a year under microgravity conditions leads to biomedical indications that make a frequent exchange of the crew of, e.g., a space station necessary. In this case, a continuous life under microgravity conditions might be impossible or it will be necessary to invent artificial gravity by the rotation of a space station or habitat.

Radiation Requirements

As ionizing radiation exposure limits, the numbers of the U.S. National Council on Radiation Protection and Measurement may be applied for space activities involving astronauts. These values are given in figure IV.9.

Duration of exposure	BFO [Sv] 5.0 cm depth	Eye [Sv] 0.3 cm depth	Skin [Sv] 0.01 cm depth	Testes [Sv] 3 cm depth
Daily	0.002	0.003	6	0.001
30 days	0.25	1	1.5	0.13
90 days	0.35	0.52	1.05	0.18
Annual	0.5	2	3	0.38
Career	1 - 4 *	4	6	1.5

BFO = Blood Forming Organs

* The career depth dose equivalent limit is based upon a maximum 3% lifetime excess risk of cancer mortality. The career dose-equivalent limit is approximately equal to: 2 + 0.075 Sv (Age -30) for males, up to 4 Sv maximum, 2 + 0.075 (Age -38) Sv for females, up to 4 Sv maximum

Table IV.9: **Ionizing Radiation Exposure Limits [52]**

Radiation protection design aspects, i.e., shielding, radiation monitoring, and dosimetry, and radiation sources are further described in section III.1. Radiation protection has to ensure that the crew dose rates are kept "as low as reasonably achievable" (ALARA) and that the maximum allowable doses are not exceeded. The radiation dose equivalent accumulated by any astronaut shall be monitored throughout the active career of the astronauts. Thus, career, as well as mission dose equivalent levels shall be kept ALARA, thereby ensuring that the maximum career dose equivalent limit shall not be exceeded. A radiation dose management system shall be provided for keeping track of an astronaut's cumulative radiation exposure records and alerting personnel that are approaching their radiation dose limits. For crew protection during radiation emergencies contingency plans shall be provided.

For non-ionizing radiation the radiofrequency (RF) electromagnetic field exposure limits of the American National Standards Institute (ANSI) Radio Frequency Protection Guides (RFPG) for exposure, as shown in table IV.10, may be applied.

Frequency f [MHz]	E^2 [V^2/m^2]	H^2 [A^2/m^2]	Power Density [mW/cm^2]
0.3 - 3	400000	2.5	100
3 - 30	$4000 \cdot (900/f^2)$	$0.025 \cdot (900/f^2)$	$900/f^2$
30 - 300	4000	0.025	1.0
300 - 1500	$4000 \cdot (f/300)$	$0.025 \cdot (f/300)$	f/300
1500 - 100000	20000	0.125	5.0

Table IV.10: **Radiofrequency Electromagnetic Field Exposure Limits [49]**

Wavelength [nm]	200	210	220	230	240	250	254	260
TLV [J/cm^2]	100	40	25	16	10	7	6	4.6
Rel. Spectr. Eff. S_λ	0.03	0.075	0.12	0.19	0.3	0.43	0.5	0.65
Wavelength [nm]	270	280	290	300	305	310	315	
TLV [J/cm^2]	3	3.4	4.7	10	50	200	1000	
Rel. Spectr. Eff. S_λ	1	0.88	0.64	0.3	0.06	0.015	0.003	

Table IV.11: **Threshold Limit Values for Radiant Exposure of Actinic UV upon Unprotected Skin or Eye [52]**

For radiation in the UV-A spectrum (315-400 nm), the total irradiance incident upon unprotected skin shall be less than 10 W/m^2 for periods of exposure longer than 1000 seconds. For exposure times less than 1000 second, radiant exposure shall be less than 1 J/cm^2. The threshold limit values (TLV) for the actinic UV spectrum (200-315 nm) indicate the radiation exposure limits for radiant exposure incident upon unprotected skin or eye within an 8-hour period are given in table IV.11. Permissible UV exposures, depending on the duration exposure per day and the effective irradiance relative ($E_{eff} = \sum E_\lambda S_\lambda \Delta\lambda$ with E_λ = spectral irradiance, S_λ = relative spectral effectiveness) to a monochromatic source at 270 nm in [W/cm^2] are indicated in table IV.12.

Exposure per Day	**8h**	**4h**	**2h**	**1h**	**30m**	**15m**	**10m**
Effective Irradiance [W/cm²]	0.1	0.2	0.4	0.8	1.7	3.3	5
Exposure per Day	**5m**	**1m**	**30s**	**10s**	**1s**	**0.5s**	**0.1s**
Effective Irradiance [W/cm²]	10	50	100	300	3000	6000	30000

Table IV.12: **Permissible UV Exposures [52]**

In order to protect astronauts against non-ionizing radiation, monitoring and warning systems, protective measures, and safety plans for the safe operation of RF and optical radiation have to be provided. [52]

Crew Metabolic Rates and Environmental Requirements

Crew metabolic rates and environmental requirements are also important determinants in the definition of a life support system. The following paragraphs present requirements for respirable atmospheres of spacecraft cabins. Relevant parameters are atmosphere composition and pressures, humidity, and temperature and the amount of contaminants in the cabin air. Oxygen has to be provided at a partial pressure sufficient for metabolic needs. The temperature and relative humidity must be appropriate and trace contaminants must be sufficiently low to avoid adverse effects. Representative space cabin environmental requirements are given in table IV.13, a typical range of metabolic values for normal spacecraft operations in table IV.14, and nominal metabolic loads in table IV.15. Figure IV.3 shows the temperature and humidity levels allowed for the U.S. space station.

Atmosphere Parameter	Nominal Value
Total Pressure	99.9 - 102.7 kPa
pOxygen	19.5 - 23.1 kPa
pNitrogen	79 kPa
pCarbondioxide	0.4 kPa
Temperature	18.3 - 26.7° C
Relative Humidity	25 - 70 %
Ventilation	0.076 - 0.203 m/s

Table IV.13: **Space Cabin Atmosphere Requirements [36]**

The nominal oxygen partial pressure in the lungs and the arterial blood is 13.4-13.9 kPa, for CO_2 and water the partial pressure is 9.3-12 kPa. This yields a total required oxygen gas pressure of at least 22.7-25.3 kPa in the cabin air. If the oxygen partial pressure is too high this may lead to toxic effects (hyperoxy). Aboard a spacecraft the oxygen partial pressure will, thus, have to be at 22 kPa and the total pressure of 101 kPa will be reached by adding nitrogen. A pO_2 of about 19 kPa is the minimum which allows proper respiration to occur. The pO_2 is constrained by a 30 % upper limit to minimize flammability of materials and a minimum of 13.4 kPa (two hour emergency specification) for life support.

Parameter	Input / Output [kg/man-day]
Metabolic Oxygen Consumption	0.636 - 1
Carbon Dioxide Production	0.726 - 1.226
Potable Water	2.27 - 3.63
Urine Production	1.27 - 2.27
Food (dry ashes based)	0.5 - 0.863
Hygiene water	1.36 - 9

Table IV.14: **Metabolic Values for Normal Spacecraft Operation of one Astronaut [36]**

Earth's atmosphere contains 0.03% CO_2. Onboard a space station CO_2 is a product of the human metabolism. Therefore a CO_2 removal system is required. The physiological effects of high levels of CO_2 include increased respiration and heart rate, increased blood flow to the brain, hearing losses, mental depression, headache, dizziness, nausea, decreased visual discrimination, and unconsciousness. The allowable level on spacecraft ranges from a high of 1.01 kPa, acceptable for short-duration missions, to 0.4 kPa for long duration missions. During emergency conditions up to 1.59 kPa is allowable for short periods. [48]

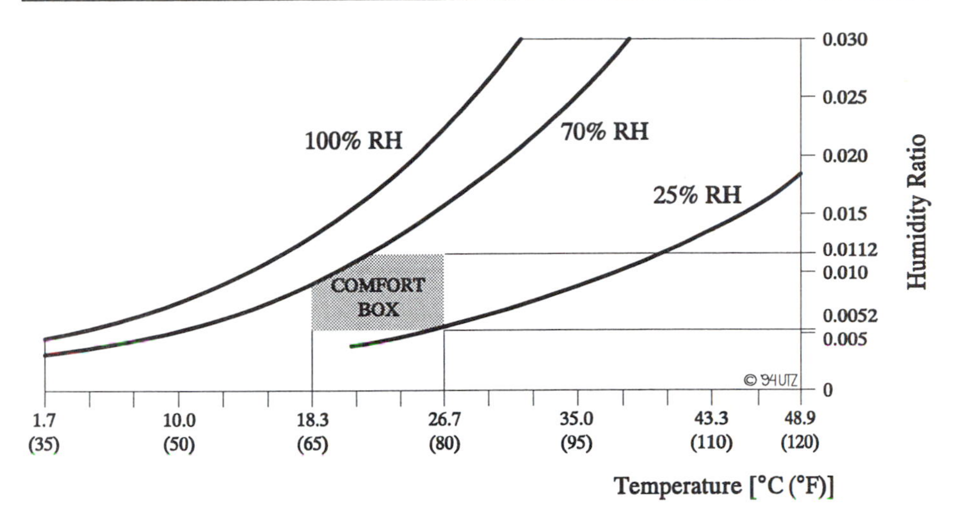

Figure IV.3: **Temperature and Humidity Ranges for the U.S. Space Station [48]**

	[kJ/h]	Duration [h]	Total [kJ]
Sensible Heat Load			
Sleep	243	8	1944
Light work	333	5	1665
Medium work	344	9	3096
Heavy work	354	2	708
Subtotal			7413
Latent Heat Load			
Sleep	74	8	592
Light work	143	5	715
Medium work	238	9	2142
Heavy work	492	2	984
Subtotal			4433
Total metabolic rate [kJ/man-day]			**11846**

Table IV.15: **Nominal Metabolic Load at a Cabin Temperature of 21° C per Man-day [52]**

Most of the data concerning trace contaminants is obtained from submarine research. There may be several origins of contaminants in the atmosphere like tank leakage, metabolic waste products, dust and fluid particles originating from food, leakage of environmental and flight control systems, products of thermic reactions, and outgassing of cabin materials. As a reference, the maximum allowable contaminant concentrations for the Space Shuttle orbiter are given in table IV.16. [46]

Trace Contaminant	Molar weight	Maximum Allowable Concentration [mg/m³]
Alcohols	32	10
Aldehyds	56	0.1
Aromatic Carbohydrates	78	3
Esters	102	30
Ethers	68	3
Chloridcarbons	93	0.2
Chloridfluorcarbons	68	24
Fluorocarbons	70	12
Carbohydrates	72	3
Inorganic Acids	20	0.08
Ketons	142	29
Merkaptans	48	2
Nitrogenoxides	46	0.9
Organic Acids	60	5
Organic Nitrogens	46	0.03
Organic Sulfides	90	0.37
Ammoniak	17	17
Carbonmonoxide	28	17
Cyanhydrogen	27	1

Table IV.16: **Maximum Allowable Trace Gas Concentration aboard the Space Shuttle [52]**

The allowable levels of trace contaminants also depend upon the duration of the mission. Spacecraft Maximum Allowable Concentrations (SMAC) for durations of 1 hour, 24 hours, 7 days, 30 days, and 180 days are being revised by the medical group at the NASA Johnson Space Center.

IV.3.1.2 Food and Nutritional Aspects

In order to maintain the functions and the structure of the human body a continuous supply with energy as nutrients is necessary. Nutrients are energy-rich substances that are reduced to lower-energy substances by the organism. The major nutrients are carbohydrates, lipids and proteins. The energy that is obtained by the reduction of the nutrients can only be partially transferred into work and, thus, according to the second law of thermodynamics is released as heat. The energy of the human metabolism depends on the kind of nutrients. The equation for the complete oxidation of glucose is:

$$C_6H_{12}O_6 + 6\,O_2 \rightarrow 6\,CO_2 + 6\,H_2O + \Delta G$$

Energy: ΔG

Under physiological conditions ΔG = - 2.86 MJ/mol. That means when oxidizing 1 mol glucose, 2.86 MJ are gained and 6 mol = 6 x 22.4 l = 134 l O_2 are burned. The Respiratory Quotient (RQ) indicates the participation of the several nutrients in the metabolic processes:

$$RQ = \frac{CO_2\text{ - Release}}{O_2\text{ - Intake}}$$

The RQ is different for different sources of energy, like carbohydrates, proteins, and lipids. The average values are given in table IV.17. Also, in this table the required amount of major nutrients and composition of food, which are depending individually on activity, sex, and age, are indicated.

Energy Source	RQ	Energy Contents [kJ/g]	Average Daily Need [g/man-day]	Percentage
Carbohydrates	1.0	17.22	300-600	10-18 %
Proteins	0.8	17.22	50-300	20-40 %
Lipids	0.7	39.06	50-150	50-60 %
European Food	0.8	-	-	-

Table IV.17: **Respiratory Quotients, Energy Contents, and Daily Needs for Different Sources of Energy [11, 43]**

	Energy consumption [kJ/d]
General Budget	7140
Resting	8400
Leisure	9660
Working	10080 - 20160

Table IV.18: **Daily Energy Budget for a Male Adult [11]**

The general budget of the human organism depends on the individual height, mass, age and sex. The general budget for a male adult is 4.2 kJ/kgh. The budget of a female adult is 10 % lower. This general budget is measured in the morning. This yields the overall energy budgets for a male adult as indicated in table IV.18. For comparison, the recommended nutrient intakes of astronauts and cosmonauts are given in table IV.19.

Nutrients	Cosmonauts	Astronauts
Energy [kJ]	13400	9600-12950
Protein [g/kg bodyweight]	1.5	0.8
Fat [g/kg bodyweight]	1.4	1.3
Carbohydrates [g/kg bodyweight]	4.5	4.8
Phosphorus [g]	1.7	0.8
Sodium [g]	4.5	3.5
Iron [mg]	50	18
Calcium [g]	-	0.8
Magnesium [g]	-	0.35
Potassium [g]	-	2.7

Table IV.19: **Recommended Daily Nutrient Intake of Astronauts and Cosmonauts [11]**

The diet is the basis for the energy supply of the human organism and the biosynthesis of several body substances (metabolism). As a rule of thumb it can be estimated that during one year a human needs an amount of approximately three times his body weight as food, four times his weight as oxygen, and eight times his body weight as drinking water. Over the life span of one human being, this means more than a thousand times his body weight. The mechanical and chemical work that is constantly done by the organism requires a constant supply with adequate amounts of energy in form of nutrients. Also, hormones, enzymes and antibodies have to be synthesized over and over again. Concerning its energy budget and metabolism the living organism is in a dynamic equilibrium.

Human nutrition consists of organic substances taken in as vegetables, fruits and meat. The organism resorbs the nutrients, vitamins, minerals and water.

Carbohydrates, lipids and proteins can partly be replaced by each other concerning their energy contents. The, so-called, isodynamic amounts of nutrients are:

1 g Carbohydrates = 1 g Proteins = 0.44 g Lipids

But the statement of isodynamy is very restricted because the same amount of energy yields a difference of up to 20 % concerning the ATP-production, depending on the combination of carbohydrates, proteins and lipids. ATP is the most important form of cellular energy storage. Also, proteins can not completely be replaced by carbohydrates or lipids because without them the essential aminoacids can not be synthesized. The main properties and required quantities of the major nutrients, and the also required vitamins and trace elements, are listed below:

- *Carbohydrates*
  are mostly Polyhydroxycarbonyl-compounds ($C_x(H_20)_n$). The daily demand is 5 - 6 g per kg bodyweight. About 50 - 55 % of the energy need of the human organism should be covered by carbohydrates. Basic food for the carbohydrate supply are crop, potatoes, rice, but also sugar, meat, etc.

- *Lipids*
  are supplying the organism with energy and essential fatty acids, but they are also the basis for many biosynthetic processes. Essential fatty acids are parts of the nutrients that are required for certain biochemical processes, but cannot be synthesized within the organism. About 25-30 % of the energy need of the human organism should be covered by lipids. The daily demand is 1 g per kg bodyweight. The minimum amount of essential fatty acids required is 10 g/day. Basic food for the supply with lipids are butter, peanut oil, corn oil, fish oil, etc.

- *Proteins*
  are required for the biosynthesis of proteins for the human metabolism. Depending on the combination of the proteins the daily demand is 0.8 - 1 g per kg bodyweight. Only surplus proteins are used for energy production within the organism. Essential amino acids are amino acids that cannot be synthesized within the organism. Very important essential amino acids are Isoleucin, Leucin, Lysin, Methionin, Phenylalanin, Threonin, Tryptophane and Valin. Eggs, milk and meat contain up to 50% essential amino acids, crops up to 30%. An overview about the required amounts of the most important essential amino acids, separately listed for male and female, is given in table IV.20.

Amino Acid	Male [mg/day]	Female [mg/day]
Histidin	700	450
Isoleucin	1100	620
Leucin	800	500
Methionin - without Cystin - including 810 mg Cystin	 1100 200	 550 180
Phenylalanin - without Tyrosin - including 1100 mg Tyrosin	 1100 300	 200
Threonin	500	300
Tryptophan	250	160
Valin	800	800
Lysin	800	800

Table IV.20: **Daily Need of Essential Amino Acids for Male and Female [46]**

Trace Element	Stock Inside the Organism [g]	Daily Need [mg]
Fe	4 - 5	10 - 12
Zn	2 - 3	15
Cu	0.10 - 0.15	1 - 5
Mn	0.01 - 0.03	2 - 5
Mb	0.001	0.2 - 0.5
J	0.01 - 0.02	0.1 - 0.2
Co	0.01	<1
Cr	0.006	0.02
F	3	0.5 - 1.0

Table IV.21: **Daily Need and Stocks of Trace Elements [43]**

- *Trace elements*
  are elements that appear only in very small amounts in food and the organism. One can distinguish essential, dispensable (e.g., Al, Au, Ag) and toxic (e.g., As, Pb, Hg) trace elements. The most important essential trace elements, their stocks in the organisms, and the daily need are given in table IV.21.

- *Vitamins*
  are vital, physiologically important, organic compounds that can not, or only under very special conditions, be synthesised

within the organism. Thus, the diet has to contain vitamins and provitamins. In table IV.22 the most important vitamins, their sources, and daily needs are given. [43]

Vitamin	Food Containing the Vitamin	Daily need [mg]
A	Green vegetables, carrots, fruits, milk, liver	0.8 - 1.2
D	Liver, animal oil	0.005 - 0.01
E	Plant oil, crops	10 - 15
K	Green plants, liver	0.07 - 0.15
B1	Liver, crops, yeast	1.0 - 1.5
B2	Liver, milk, yeast	1.5 - 2.0
B6	Green vegetables, yeast, liver, crops	1.8 - 2.0
B12	Liver, eggs, milk	0.005
Niacin	Liver, yeast, milk	15 - 20
Folacid	Green leaf vegetables	0.4
Pantothenacid	Liver, eggs, yeast	8
H (Biotin)	Liver, eggs, yeast	0.1 - 0.3
C	Citrus fruits, potatoes, green leaf vegetables, paprica	75

Table IV.22: **Daily Need and Sources of Vitamins [43]**

On the present, relatively short-term spaceflights a variety of fresh and prepackaged foods are available. Menu items include fresh, dehydrated foods (soups, vegetables), intermediate moisture foods, e.g., fruits, thermo-stabilized foods, e.g., meat or yogurt, and natural foods, e.g., nuts. Dehydrated foods require the addition of hot or cold water and a period of time for rehydration to occur. In the past, thermostabilized foods were primarily canned or in squeezable tubes. Today, they are prepackaged in a laminated pouch. The utilization of fresh foods is limited by the availability for refrigeration. A prepackaged meal can be rapidly set up by one person, although reconstitutioning and heating may require another 20-30 minutes. On the space station missions of the former Soviet Union the diet is also supplemented with fresh foods during periodic resupply by the PROGRESS supply vehicle, or when visiting crews arrive.

On long-term missions, i.e., at a lunar outpost or on a Mars mission, where some level of bioregeneration may occur, food variety will be very important. Even though the major nutrients, carbohydrates, proteins, and lipids may be produced in bioregenerative systems, it is likely that condiments, special flavors, and spices will have to be carried in sufficient quantity for the entire mission. It will be necessary from a nutritional and dietetic point of view to convert a relatively limited variety of biomass into a nutritionally balanced, aesthetically pleasing variety of dishes. Prior to that time, more information must be gathered regarding many facets of space nutrition, including nutritional requirements, diet acceptability, and food processing and storage with limited mechanical and power resources.

There are no hard data indicating that space travel will modify either macro- or micronutrient requirements. The little information which is available has been obtained from pre- vs. post-flight blood or tissue sample analysis. Given that many of the blood parameters in question can change rapidly, post-flight data is only of marginal value as a predictor of status during flight. One obvious feature of gaining experience in providing nutrition to space travelers has been the recognition that an increase in total energy intake is necessary to maintain body mass. But where data are available, it appears that during most spaceflights crew members consumed less calories than were provided and recommended. Another important problem present in spaceflight is the prevention or amelioration of the negative nitrogen balance and amino acid loss. Based on current knowledge, it may seem that both diet and exercise may play a role. Also, a hypoglycemia and decreased insulin have been seen in space. Flight data relating to lipid levels and metabolism show no changes from pre- and post-flight ground controls. In pre- and post-flight blood vitamin concentrations and urinary vitamin excretions no obvious changes were noted. The negative calcium balance and osteoporosis-like changes in the skeleton during spaceflight have been mentioned earlier. They are a result of the fact that calcium is mobilized from mammalian bones and excreted in the urine. It does not seem likely that nutritional approaches alone will be able to correct the effect of weightlessness on calcium loss. The body water and electrolyte metabolism is particularly modified in early spaceflight. A new balance is achieved after a few days in flight. An additional aspect of overall nutritional success involves gastrointestinal function. If ingested nutrient digestion and absorption is altered, nutrition may be inadequate. There appear to be significant chronological changes early in flight, some of which seem to be related to space motion sickness (SMS).

Besides food, the other most important consumables are, of course, water and oxygen. As a guideline, the values for the U.S. Space Station ECLSS design are given in table IV.23. [30]

Consumable	Design Load [kg/man-day]
Metabolic oxygen	0.85
Drinking water	1.6
Food preparation water	0.75
Clothes wash water	12.5
Hand wash water	4.1
Shower water	2.7
Food water	1.15
Food solids	0.62
Urinel Flush	0.5
Dish wash water	5.45

Table IV.23: **Average U.S. Space Station Design Loads for Water and Oxygen [17]**

Both the potable and the hygiene water have to match certain quality requirements. In addition to such factors as flavor and clarity, trace levels of organic and inorganic contaminants are of great concern. Certain contaminants are of greater concern for longer duration missions than for shorter ones due to the capacity of some compounds to accumulate in body tissues. Also of concern are micro-organisms which can clog water lines and filters, and which may become pathogenic under some conditions. The potable water quality requirements, limits for physical parameters, and chemical and biological constituents, are given in tables IV.24 and IV.25.

Physical Parameters	Limits (Potable Water)	Limits (Hygiene Water)
Total solids [mg/l]	100	500
Color true [Pt/Co units]	15	15
Taste [TTN]	3	3
Odor [TON]	3	3
Particulates (max. size) [microns]	40	40
pH	6.0 - 8.4	5.0 - 8.4
Turbidity [NTU]	1	1

Table IV.24: **Limits for the Physical Parameters of Potable Water and Hygiene Water [48]**

Personal hygiene water is the water which is used for external body cleaning. Personal hygiene water requirements, limits for physical parameters, and chemical and biological constituents, are also given in tables IV.24 and IV. 25. [48]

Inorganic Constituent	Limit [mg/l]	Bactericide	Limit [mg/l]
Ammonia	0.5	Residual Iodine	0.5 - 6.0
Arsenic	0.01	**Aesthetics**	**Limit [mg/l]**
Barium	1.0	Cations	30
Cadmium	0.005	Anions	30
Calcium	30	Carbon dioxide	15
Chloride	200	**Bacteria**	**Limit [CFU/100 ml]**
Chromium	0.05	Total count	1
Copper	1.0	Anaerobes	1
Iodine	15	Aerobes	1
Iron	0.3	Gram negative	1
Lead	0.05	Gram positive	1
Magnesium	50	Coliform	1
Manganese	0.05	Enteric	1
Mercury	0.002	**Organic Parameters**	**Limit [g/l]**
Nickel	0.05	Total acids	500
Nitrate (NO_3-N)	10	Cyanide	200
Potassium	340	Halogenated Hydrocarbons	10
Selenium	0.01	Phenols	1
Silver	0.05	Total Alcohols	500
Sulfate	250	Total Organic Carbon	500 (Potab. Water)
Sulfide	0.05	(TOC)	10000 (Hyg.Water)
Zinc	5.0	Uncharacterized TOC (UTOC)	100 (Potab. Water) 1000 (Hyg. Water)
Virus [PFU/100ml]	1	**Yeast & Mold [CFU/100ml]**	1

Table IV.25: **Limits for Biological and Chemical Constituents of Potable Water and Hygiene Water [48]**

IV.3.1.3 Waste Production and Waste Categories

When designing a waste management subsystem, the types of waste, e.g., liquid, gas, solid, or metabolic, to be dealt with and the amount of each must be determined. The availability of power, resupply, and storage volume are key factors which must be considered. As a reference the U.S. Space Station ECLSS design average waste loads are given in tables IV.26 and IV.27. [17]

Waste Products	Amount [kg/man-day]
Metabolic carbon dioxide	1.00
Perspiration and respiration water	2.28
Urine solids, dry	0.06
Urine water	1.50
Sweat solids	0.02
Fecal solids, dry	0.03
Fecal water	0.09
Metabolic water production	0.354
Urinal flush water	0.50
Hygiene latent water	0.43
Hygiene water	12.58
Hygiene solids	0.01
Food preparation latent water	0.04
Experiments latent water	0.454
Laundry latent water	0.06
Laundry water	11.90
Laundry solids	0.08
Charcoal required	0.059
Food packaging	0.454
Trash	0.817

Table IV.26: U.S. **Space Station Daily Design Average Waste Loads [17]**

Waste Load Parameter	Amount
Trash volume [m^3/man-day]	0.0028
Metabolic sensible heat [kJ/man-day]	7400
Leak air per module [kg/day]	0.45

Table IV.27: **Further U.S. Space Station Waste Load Parameters [17]**

IV.3.2 Subsystem Interfaces and Integration

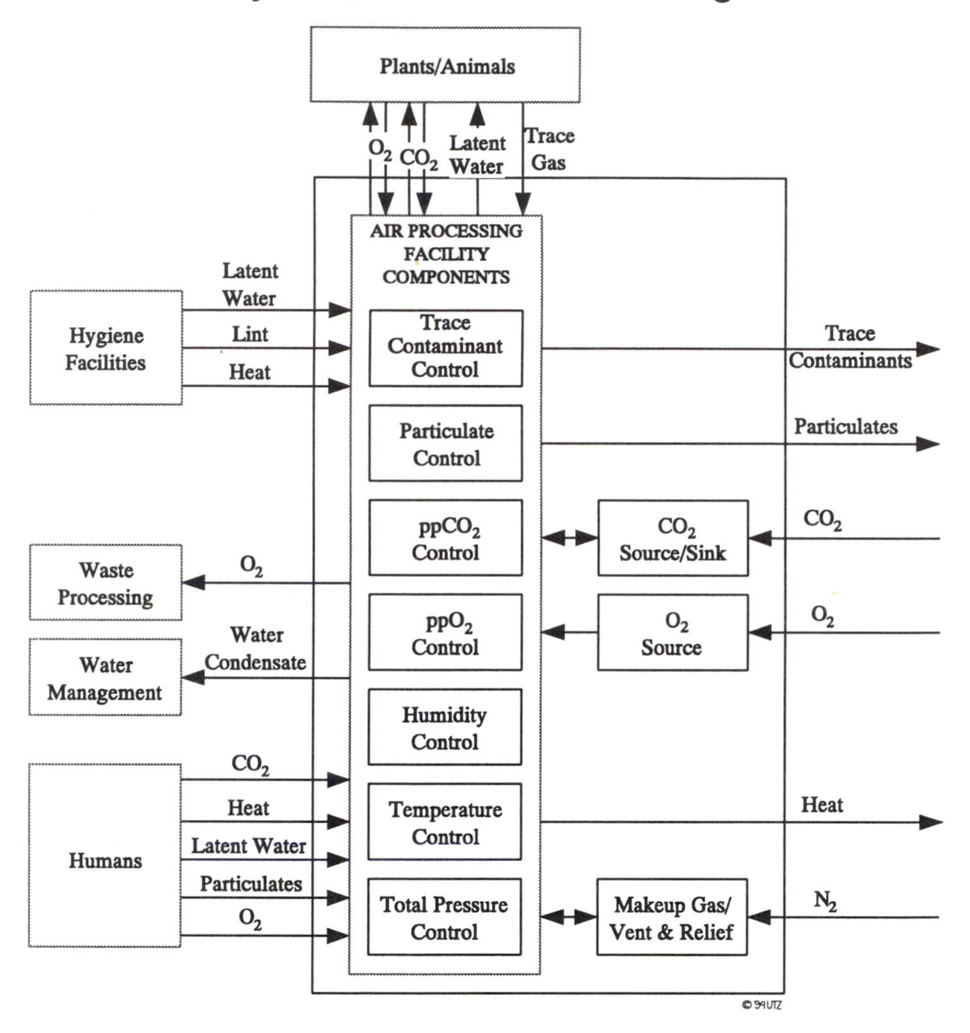

Figure IV.4: **Atmosphere Revitalization Interfaces [13]**

One of the most critical items in the design of a life support system is the integration of the several subsystems. First of all, it is most important to define the several interfaces of the life support subsystems. These interfaces are shown below in figures IV.4 to IV.9. Once the subsystem interfaces are defined, the basis for mass flow calculations or simulations is given. Of course, a life support system also has to be very well integrated into the other spacecraft systems and the impacts of a particular technology on the overall system have to be considered. A realistic evaluation of life support systems cannot be done without considering these interactions with other mission systems. Fuel cells, for example, produce electrical energy, but also water and heat as waste products. Biological waste that is produced onboard may be used as fertilizer, radiation shielding or fuel in future space habitats. [30]

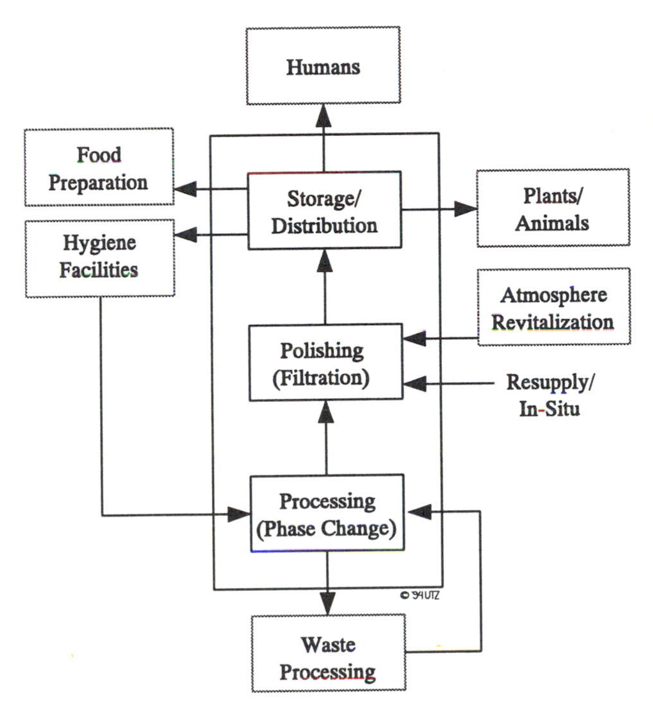

Figure IV.5: **Water Management Interfaces [7]**

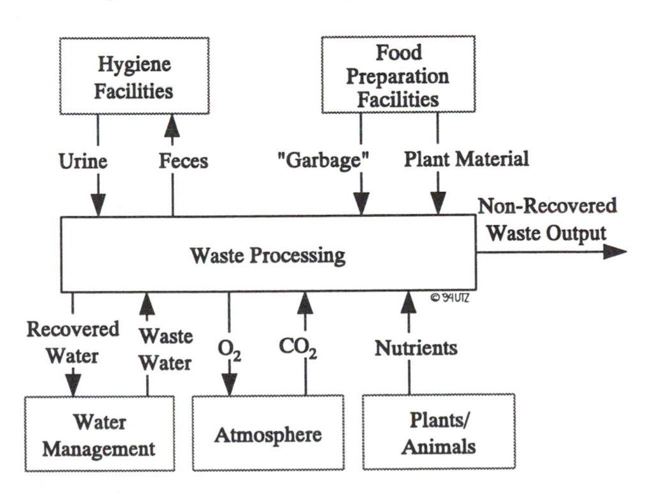

Figure IV.6: **Waste Processing Interfaces [13]**

Integration concerns relating to atmosphere revitalization include the following aspects:

- *Gas Stream Purity*
- *Appropriate Connections* - For example, some of the technologies available for CO_2 removal prefer that the inlet atmosphere be cool, while others prefer that it be warm. Pressure regulation of the streams is also important.
- *O_2 Consumption* - The consumption and production of O_2 have to be balanced.
- *Synergistic Effects* - An increase in the level of integration can reduce the resupply requirements significantly. The potential for a failure to adversely impact a connected system must also be considered to ensure safe and reliable operation of critical functions.

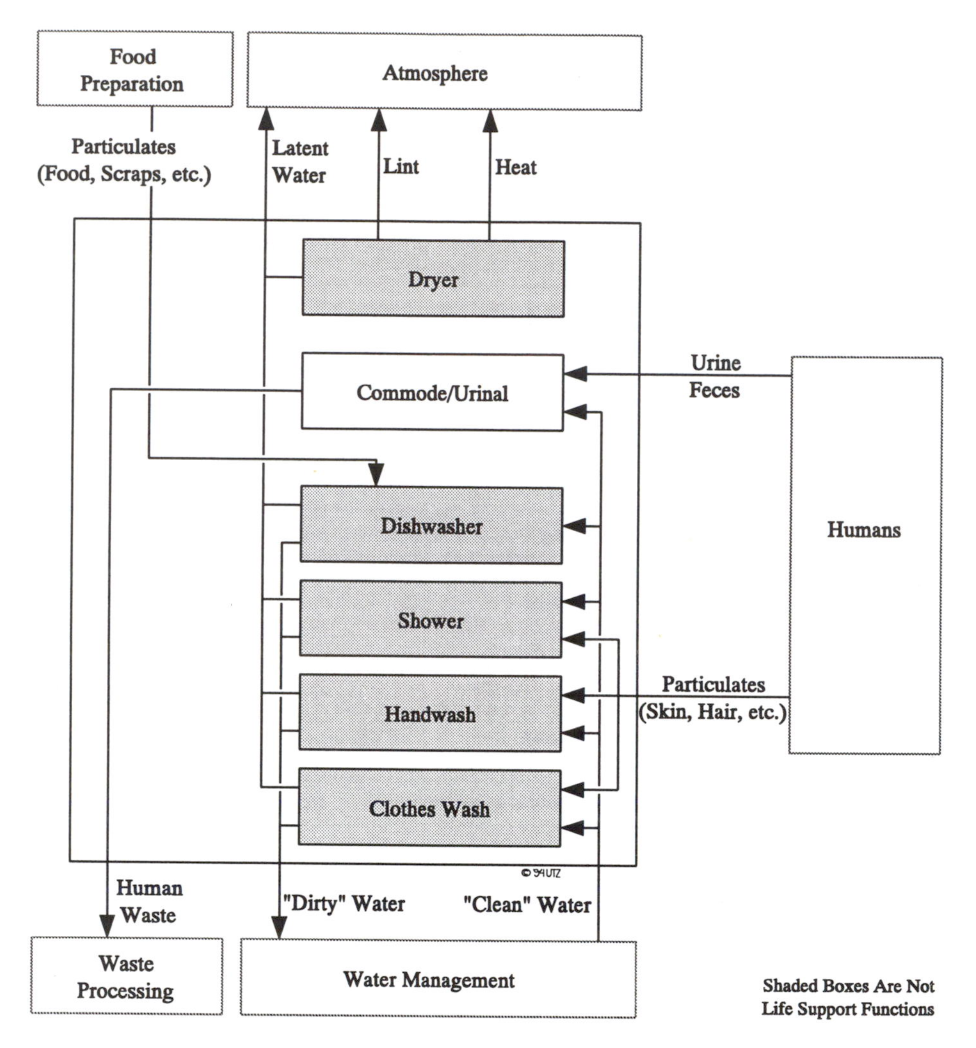

Figure IV.7: **Personal Hygiene Facilities Interfaces [13]**

Integration concerns relating to water management include the following aspects:

- *Water Balance* - Water balance management is especially complicated in a system incorporating separate potable and hygiene water loops. In such a two-loop system, net transfer of water between the loops must be considered. A single-loop system avoids this problem, but requires that all water be purified to potable specifications, which may require more power or consumables than a two-loop system.
- *Waste Water Quality* - One of the most significant factors in determining the design of a water recovery system is the contaminant load of the incoming waste water. Although a full accounting of all waste water

contaminants by species and quantity is impossible because of analytical limitations and the inherent variability in waste water contaminant loads, important types of contaminants typically drive the design of water recovery systems and therefore must be understood early in the design process. Also, the ability of a water recovery to effectively remove potentially hazardous chemicals must be considered when evaluating the acceptability of bringing such chemicals onboard spacecraft with payloads.

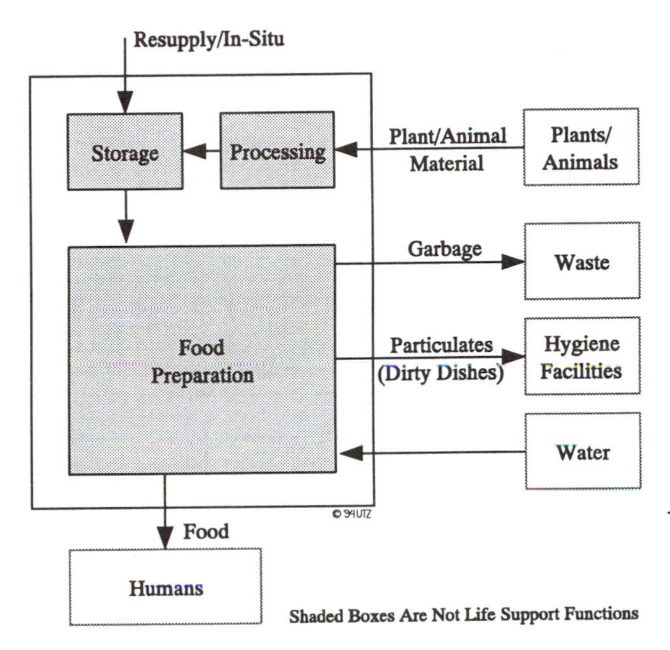

Figure IV.8: **Food Preparation Facilities Interfaces [13]**

- *Contaminant Recovery* - As mission durations increase, the likelihood that a water subsystem will become contaminated in some fashion increases. Depending on the safety and mission criticality of the water supply for a given mission, provisions may be required to restore the cleanliness of contaminated water subsystems inflight.

- *Microorganisms Control* - Microorganisms are of increasing concern as mission durations increase due to greater potential of growth of pathogenic microorganisms and the formation of biofilms which can lead to clogging of filters and tubing. [13,48]

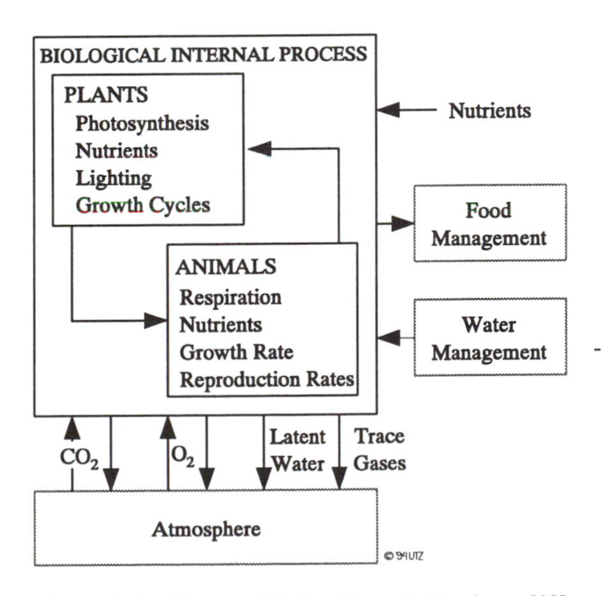

Figure IV.9: **Plant and Animal Growth Interfaces [13]**

IV.3.3 Mission Related Aspects

As mentioned earlier, the design of life support systems for spacecraft is governed by the objective to maintain in an isolated volume an environment, suitable for the well-being of men and systems for the whole duration of the mission. The decision concerning the technologies to select for a particular application depends largely on the mission characteristics, like crew size, mission duration, mission location, available resources, etc. Crew size and mission duration affect system sizing, although closed loop systems are relatively insensitive to mission duration. Mission location establishes physical distance from Earth, which determines cost of transportation, duration of travel, and availability of resupply opportunities, as well as the external operating environment. A survey of several mission characteristics that have to be taken in account when designing a life support system for a specific mission, has been given before in section IV.3. Whatsoever, in the next phases of manned spaceflight, including space stations, and possibly lunar bases and Mars flights, a reduced dependence on expendable supplies and, thus, an increased loop closure, will have to be achieved. The required improvements in life support technology might be implemented in conjunction with a scenario for mission growth. This evolutionary process is outlined in figure IV.10.

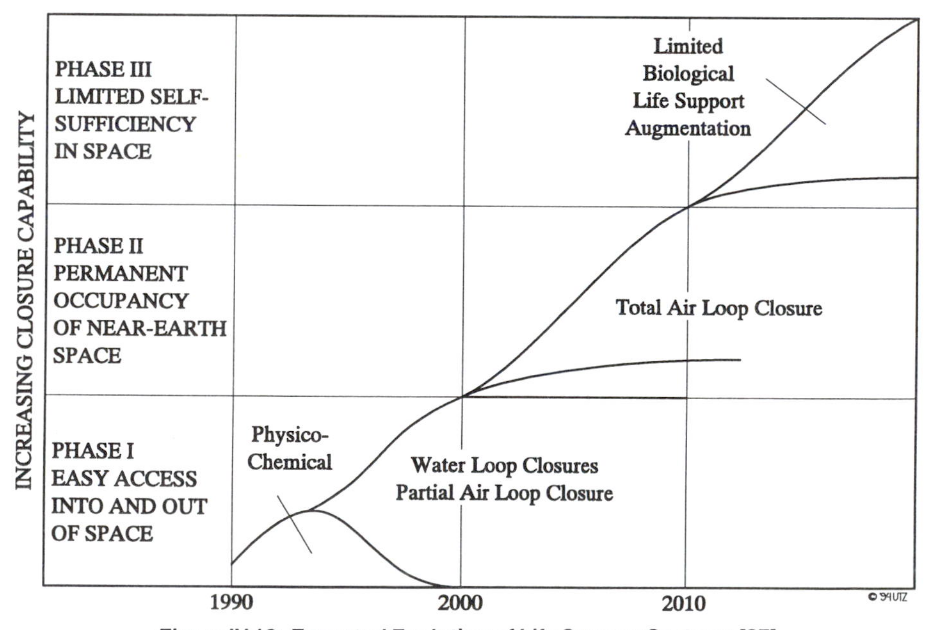

Figure IV.10: **Expected Evolution of Life Support Systems [37]**

Traditionally, life support system analysis has used optimization of system mass as the principal criteria for technology selection. For regenerative systems, though, the parameters that have to be considered are mass, power, and volume.

This analysis method called "equivalent mass" has been developed for comparing these unlike parameters. A simple formula for the determination of the equivalent mass is given by:

Equivalent Mass = System Mass + (Power x Conversion Factor)

Mass and mission aspects related with the development of life support systems are further outlined in sections IV.3.3.1 and 3.3.2. Again, it is important to point out that there is not one life support system that is "best" for a given mission, much less for all missions. [15,17,37]

IV.3.3.1 Mission Duration and Location

The specific life support needs depend strongly on the destination and duration of any mission. This is true for both the overall design of the ECLSS and the selection of the technologies used for a respective mission. As mission durations increase, factors such as reliability and maintainability become increasingly important. Hardware must be readily maintainable for those times when components must be repaired or replaced. For long duration missions, technologies which require fewer expendables become much more attractive. Also a very critical point is the degree of mass loop closure for a specific mission. Maximum degree of mass loop closure which is economical can be defined as soon as a mission scenario has been sufficiently defined. The "breakeven" points for the available technologies can be determined on plots of mission duration vs. cumulative mass. The sample plot shown in figure IV.11 compares four approaches of performing the ECLSS functions: non-regenerable physico-chemical (P/C), regenerable P/C, hybrid P/C - biological, and completely biological (CELSS). These four approaches bound the range from a totally open mass loop to closed water, oxygen, and food loops. The relative initial mass of each system is indicated by the height of each line at the left of the graph. As mission duration increases, the cumulative mass increases due to resupplied expendables. The breakeven points, in which the lines intersect, indicate that as mission duration increases an increased level of closure will require less mass, and, therefore, have lower total cost. In order to determine these breakeven points with confidence, the candidate technologies must

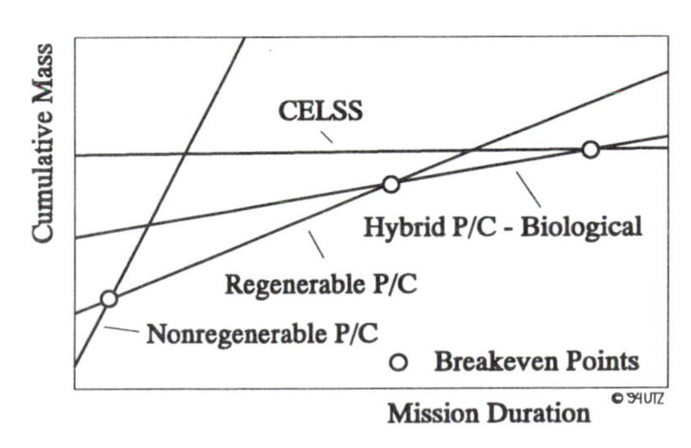

Figure IV.11: **Mission Duration vs. Mass Breakeven [48]**

be well understood and must have a high level of maturity, and the mission scenarios and assumptions must be sufficiently well defined. The trade study methods described in section IV.3.5 are used in preparing such a plot.

A basical overview about which technologies will have to be applied for future mission scenarios is given below:

- For short-term missions in an Earth orbit, open-loop ECLSS, similar to those used in the past, may be applied. These will include CO_2 removal via expendable LiOH canisters or SAWD, contaminant removal via activated charcoal and catalytic oxidation, humidity control via CHX and water separator, and will use water from fuel cells, while waste will be stabilized and returned to Earth.

- On medium-term, long-duration missions in an Earth orbit, i.e., space stations using advanced technologies, closed water and air loops, based on physico-chemical concepts, i.e., CO_2 concentration and reduction, oxygen reclamation, substantial water recovery from condensate, hygiene water and urine, water treatment and storage, may be introduced. Also, solid waste may be treated on a limited level, and habitability issues associated with extended periods in space, e.g., exercise, food preparation and presentation, entertainment, and social contacts, will have to be considered more seriously.

- In the long run, closed air and water loops, waste processing, extensively using biological systems, limited food production, and testing of prototype biological air revitalization and recovery systems will be introduced on space stations. For lunar and planetary bases, operational CELSS, possibly with physico-chemical ECLSS for emergency back-up, and, finally, medical facilities have to be developed. [31, 48]

IV.3.3.2 Mass and Cost Aspects

The reclamation of water and recovery of oxygen, i.e., mass loop closure, desirable for a given mission depends on the scope of the mission, the services to be provided, and the technologies available to perform the functions. From a hardware point of view the system has to be designed as light as possible and the number of parts which must be periodically replaced has to be minimized. In this context, a system can be designed for a given mission with little or no repair capability, or for indefinite repair and maintenance. Thus, for short duration missions redundant hardware may be used in the event of failure of primary hardware, and for long duration missions spares are provided for all critical functions. As an example, a mass balance for a space station with closed water and atmosphere loops is given in figure IV.12. [48]

A mass analysis for life support systems is used to analyze the requirements for ECLSS carbon dioxide, latent water, crew potable water, crew hygiene water, crew urinal water, crew waste water, hydrogen, oxygen, nitrogen, and various solid wastes. The goal is to reduce the complexity of the ECLSS analysis by concisely defining the sources, sinks, and net changes in mass for each fluid.

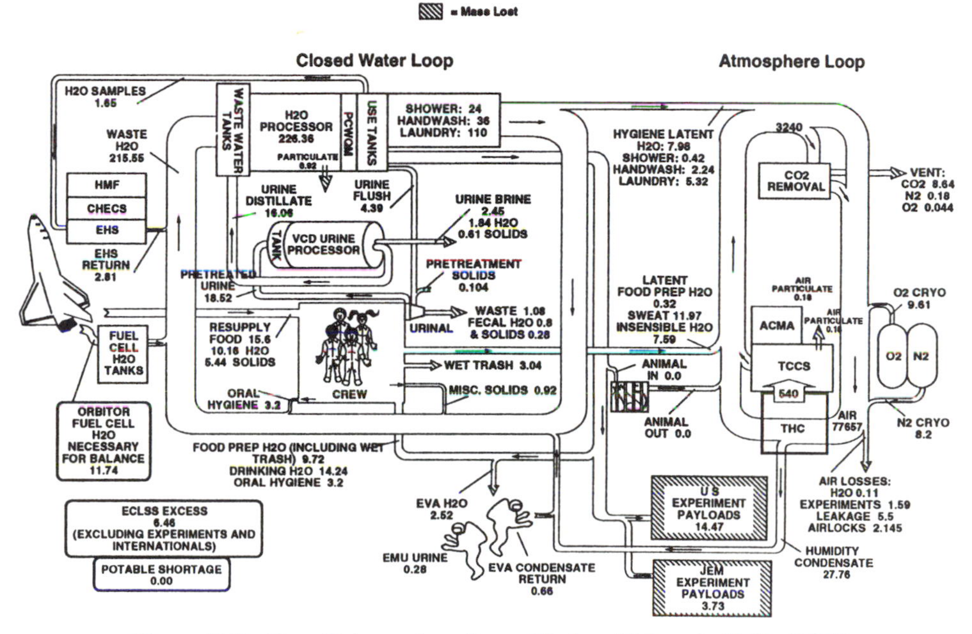

Figure IV.12: **Mass Balance for a Space Station (Values in [lb/day]) [25]**

First of all, the ECLSS has to be divided into several subsystems and functions. According to the NASA Architectural Control Document, the following seven categories can be distinguished [7]:

1. Temperature and Humidity Control (THC) Subsystem, including:

- Air Temperature Control
- Humidity Control
- Ventilation
- Equipment Air Cooling
- Thermally Conditioned Storage

2. Atmosphere Control and Supply (ACS) Subsystem, including:

- O_2 / N_2 Pressure Control
- Vent and Relief
- O_2 / N_2 Distribution

3. Atmosphere Revitalization (AR) Subsystem, including:

- CO_2 Removal
- CO_2 Reduction
- O_2 Generation
- Contamination Control and Monitoring

4. Fire Detection and Suppression (FDS) Subsystem

5. Water Recovery and Management (WRM) Subsystem, including:

- Urine Processing
- Hygiene Water Processing
- Potable Water Processing
- Water Storage and Distribution
- Water Thermal Conditioning

6. Waste Management (WM) Subsystem, including:

- Return Waste Storage
- Fecal Waste Processing
- Trash Processing

7. EVA Support

The mass balance for each subsystem has to be ensured by using the equation:

Sources = Sinks + Net Change

In this equation, the net change reflects the corresponding change of tank level. It is the quantity that the ECLSS interfaces with other systems. In the following tables IV.28 to IV.31, an example for the definition of the sources and sinks for the different subsystems is given.

Nitrogen Sources	Nitrogen Sinks
Integrated nitrogen system	Module leakage
	EVA airlock loss
	Ventilation loss
	Emergency repressurization

Table IV.28: **N_2 Sources and Sinks**

CO_2 Removal Sources	CO_2 Removal Sinks
Crew metabolic	Electrochemical Depolarized Concentrator
Animal metabolic	Four-bed molecular sieve
Lab waste	
CO_2 Reduction Sources	**CO_2 Reduction Sinks**
Carbon dioxide removal	Lab usage
	Carbon dioxide reduction
Latent Water Sources	**Latent Water Sinks**
Crew sweat and respiration	Recovered water condensate
Food preparation latent	Unrecovered water condensate
Handwash latent	
Shower latent	
Laundry latent	
Dishwash latent	
Oxygen generation latent	
Carbon dioxide removal	
Hydrogen Sources	**Hydrogen Sinks**
Oxygen generation	Carbon dioxide removal
	Carbon dioxide reduction
Crew Potable Water Sources	**Crew Potable Water Sinks**
Recovered water condensate	Crew drinking
CO_2 Removal (EDC liquid)	Crew food preparation
Carbon dioxide reduction	EVA drinking and sublimation
Fuel cells	Crew potable water (waste transferred)
Logistics	Integrated water system (internal water)
Dishwash latent	
Oxygen generation latent	

Table IV.29: **Sources and Sinks of CO_2 Removal, CO_2 Reduction, Latent Water, H_2, and Crew Potable Water**

Crew Hygiene Water Sources	Crew Hygiene Water Sinks
Recovered hygiene water Crew potable waste Crew potable transferred	Handwash latent and liquid Shower latent and liquid Laundry latent and liquid Dishwash latent and liquid O_2 Generation latent and liquid Crew urinal flush Carbon dioxide removal
H_2O Crew Hygiene Waste Sources	**H_2O Crew Hygiene Waste Sinks**
Handwash liquid Shower liquid Laundry liquid Dishwash liquid Recovered urine and flush	Recovered hygiene waste Unrecovered hygiene waste
Crew Urinal Water Sources	**Crew Urinal Water Sinks**
Crew urine Urinal flush	Recovered urine and flush Unrecovered urine and flush
Animal Potable Water Sources	**Animal Potable Water Sinks**
Fuel cell Recovered animal respiration, urine, and fecal Recovered cage wash	Animal drinking Animal potable transferred
Animal Hygiene Water Sources	**Animal Hygiene Water Sinks**
Fuel cells Animal potable transferred Recovered animal respiration, urine, and fecal Recovered cage wash	Cage wash

Table IV.30: **Sources and Sinks of Crew Hygiene Water, Crew Hygiene Waste Water, Crew Urinal Water, Animal Potable Water, and Animal Hygiene Water**

Waste Water Sources	Waste Water Sinks
Unrecovered condensate water Unrecovered crew urine and flush Unrecovered hygiene waste Unrecovered cage wash Unrecovered animal respiration, urine, and fecal Crew fecal water	Return to Earth
Oxygen Nominal Sources	**Oxygen Nominal Sinks**
Oxygen generation	Module leakage EVA airlock loss Ventilation loss Crew metabolic Carbon dioxide removal (EDC) EVA suit leakage Animal metabolic
Oxygen Energy Sources	**Oxygen Energy Sinks**
Integrated oxygen system backup	Module repressurization Hyperbaric airlock repressurization
Solid Waste Sources	**Solid Waste Sinks**
Crew urine and fecal solid Crew sweat solid Handwash solid Shower and Laundry solid Dishwash solid Carbon Trash	Return to Earth

Table IV.31: **Sources and Sinks of Waste Water, Nominal O_2, Emergency O_2, and Solid Waste**

If several configurations for a life support system are proposed, energy and material balances can be used to calculate the consumable needs of each proposed configuration. Such balances can also provide a clear picture of the multi-component composition of feed and product streams which is of importance in anticipated actual system performance. Nevertheless, for overall mass calculations it is useful to determine also the equivalent weights and weight penalties of proposed life support systems configurations. Using energy and material balance data and estimates of equipment weights it is possible to calculate the equivalent weight of each proposed design and to investigate the effect that various power supply and heat rejection technologies will have on the selection of a preferred concept. The equivalent weight of a life support system may be defined as:

Equivalent Weight	=	*Weight of Equipment (BOL=Beginning of Life) +* *Weight of Initial Flight Consumables +* *Weight of Resupply Consumables*

and:

Weight of Equipment (BOL)	=	*Mass of Process Equipment +* *Mass of Redundant Process Equipment +* *Mass of Associated Power Supply/Storage System (BOL) +* *Mass of Associated Heat Rejection System (BOL) +* *Mass Impact on Other Support Systems*
Weight of Initial Flight Consumables	=	*Mass of Process Consumables +* *Mass of Operating Spares +* *Mass of Orbital Decay Fuel to Compensate for Drag*
Weight of Resupply Flight Consumables	=	*Mass of Process Consumables +* *Mass of Operating Spares +* *Mass of Orbital Decay Fuel to Compensate for Drag*

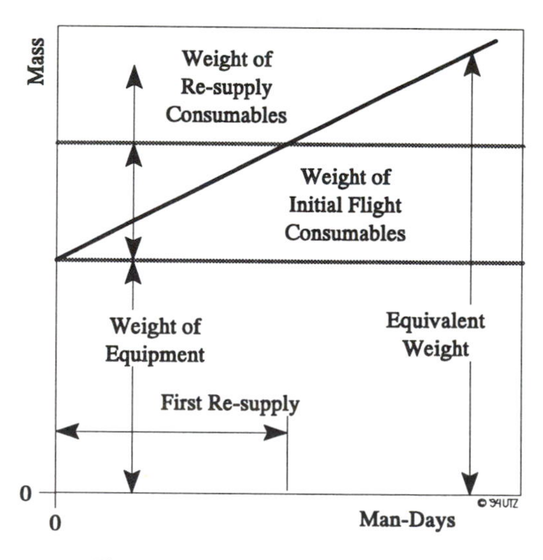

Figure IV.13: **Equivalent Weight [19]**

In case of an orbiting installation, an overall energy balance for solar powered electrical systems can be expressed as:

$$P_{sa}\,(t_0 - t_e) = \frac{1}{\eta_{pd}}\left[P_{eq}\,(t_0 - t_e) + \frac{P_{eq}}{\eta_{ps}}t_e\right]$$

Power at solar array:	P_{sa} [W]
Power supplied to equipment:	P_{eq} [W]
Efficiency of power distribution/supply system:	η_{pd}
Orbital time:	t_o [h]
Efficiency of power storage charge/discharge system:	η_{ps}
Eclipse time:	t_e [h]

and rearranged:

$$P_{sa} = \frac{P_{eq}}{\eta_{pd}}\left[1 + \frac{t_e}{\eta_{ps}}(t_0 - t_e)^{-1}\right]$$

This equation can also be modified to take into account the degradation in performance of a solar array due to ionizing radiation, typically 25 % over 10 years. This term, expressed as a ratio between beginning (BOL) and end of life (EOL), can be represented as follows:

$$P_{sa}\,(BOL) = \frac{P_{eq}}{\eta_{pd}}\left[1 + \frac{t_e}{\eta_{ps}}(t_0 - t_e)^{-1}\right]\frac{BOL}{EOL}$$

For an Earth orbit the typical orbiting and power system parameters are as follows:

$t_o = 1.5$ h $\quad t_e = 0.6$ h $\quad \eta_{pd} = 0.9 \quad \eta_{ps} = 0.7$

Thus, a continuous supply of 1 kW to process equipment requires the following power and energy needs:

Solar Array Power	=	2.170 kW/kW
Stored Energy Capacity	=	0.6 kWh/kW
Power Supply Heat Load	=	0.503 kW/kW

Depending on the type of power generation and energy storage equipment used, a range of weight penalties can then be calculated. Typically, a weight penalty ranging around 180 to 330 kg/kW, including solar cells, batteries, and heat rejection equipment, can be expected. The total capacity and, thus, weight penalties of a heat rejection system can be expressed as:

Total Heat Rejection Capacity = *Process Sensible Heat Load +*
Process Latent Heat Load +
Power Supply System Heat Load

While sensible and latent heat loads can be derived from process energy and material balances, the heat generated by an electrical power network will depend on the efficiency of the supply and energy storage system. The heat load generated by a power supply and energy storage system on a LEO installation in a 90 minute ecliptical orbit will typically vary between:

Solar panel supply and battery storage:	$\eta_{pd} = 0.9$, $\eta_{ps} = 0.78$:	394 W/kW
	$\eta_{pd} = 0.9$, $\eta_{ps} = 0.7$:	503 W/kW
Solar panel supply and fuel cells storage :	$\eta_{pd} = 0.9$, $\eta_{ps} = 0.6$:	680 W/kW
	$\eta_{pd} = 0.9$, $\eta_{ps} = 0.55$:	790 W/kW

To translate these heat rejection loads into weight penalties, details of the weight to power ratio of thermal control systems must be determined. In the absence of a more detailed analysis of the technology used to reject heat from the power supply of a life support system, an assumed value of 200 kg/kW may be used. This value which will vary from mission to mission, is a function of attitude, size of radiators, capacity of any thermal storage system, etc. Also, equipment redundancies and additional fuel needs due to the compensation of drag (only in LEO) have to be taken into account. It may be assumed that many items of equipment will be duplicated. As installations become larger, it should be possible to utilize an (n+1) approach, e.g., 3 units of 50 % or 4 units of 33 % capacity. The amount of fuel needed to compensate for the drag imposed by solar arrays depends on a number of factors that include size of array, orbiting altitude and the specific impulse of the fuel used. These factors will be highly mission dependent. [19]

The costs to be considered when selecting technologies include development and operational costs. These costs, then, include direct and indirect costs. Indirect costs are, e.g., those which support infrastructure including test facilities. Direct costs include hardware procurement and manufacturing costs. The costs incurred to develop equipment to operational status are referred to as design, development, test, and evaluation (DDTE) costs. Life cycle costs include the costs to operate the equipment as well as DDTE costs. [48]

IV.3.4 Development Phases

Every development program is divided into the following four distinct phases:

- *Phase A - Concept Study and Preliminary Analysis*
  During this phase the feasibility of a project is studied, top level objectives are identified, and the initial organizational groundwork is laid out. In order to assess the feasibility of a future program, concept studies and preliminary analyses must be performed, using manual and computer methods to evaluate options.

- *Phase B - Preliminary Definition and Design*
  During Phase B the preliminary concepts, developed during Phase A, are iteratively reviewed and analyzed, and the capabilities of each concept are compared to the system requirements using trade study techniques (see section IV.3.5). As the name implies, these studies are performed to trade two or more alternatives for accomplishing the same function. The ultimate objectives of a Phase B study are to establish vehicle and individual system requirements and designs to be carried forward into Phase C/D.

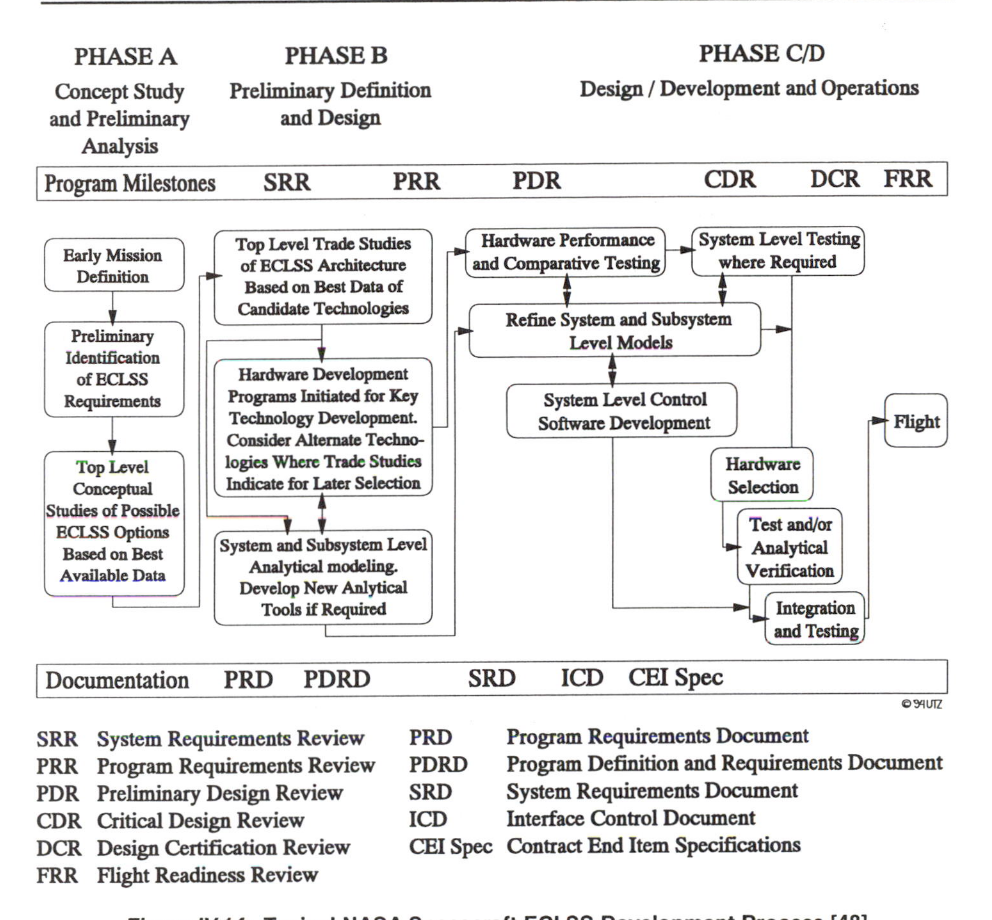

Figure IV.14: **Typical NASA Spacecraft ECLSS Development Process [48]**

- *Phase C/D - Design, Development and Operation*
  This is the final and most time consuming phase of spacecraft development, in which most of the development and integration activities are performed. The end result of Phase C/D is the final design, fabrication, test, and verification of the spacecraft.

As an example, the typical NASA spacecraft ECLSS development process and Phase C/D baseline design review milestones and objectives are outlined in figure IV.14 and table IV.32. [48]

Milestone	Objectives to be Established
Program Requirements Review (PRR)	- Configuration concepts and requirements Qualification approach - System requirements baseline - Safety assessment plan - Determination of required support
Preliminary Design Review (PDR)	- Basic design approach - Compatible design / requirements - Test planning - Safety assessment - Commonalty - Producibility - Baseline part I Contract End Item (CEI) specification
Critical Design Review (CDR)	- Design configuration - System compatibility - Reliability assessment - Maintainability assessment - Safety assessment - Approved design baseline - Producibility - Authorize release of baseline design
Design Certification Review (DCR)	- Assure that design and performance meet requirement specifications
Configuration Inspection (CI)	- Configuration complies with design documentation - Baseline part II CEI specification
FinalAcceptance Review (FAR)	- Functional performance complies with CEI specification - Baseline part II specification

Table IV.32: **Typical NASA Phase C/D Baseline Design Review Milestones and Objectives**

IV.3.5 Trade Studies and Simulation Models

Choices are inherently part of the engineering design process between including or not including particular functions, between technologies to perform particular functions, regarding the requirements to be met, and numerous other aspects. Analysis techniques, both manual and computerized, are important tools in making these choices.

Trade studies are performed to make selections between two or more alternatives. There are two general approaches which may be used, depending on the amount of information available. The "advantage / disadvantage" method can be used when the options are not well described or when it is difficult to quantify how well each option satisfies the selection criteria. The "weighed factor" method can be used when much information is available and the alternatives are well characterized. Trade studies involve qualitative as well as quantitative information to help choose between alternate approaches. Trade studies can be performed by "manual" or analytical techniques using computers. The steps in a trade study can be defined as indicated in table IV.33.

Step	Objective
1	Derive the relevant life support system functional requirements from a given mission scenario.
2	Develop a tradeoff decision methodology consisting of a set of evaluation criteria and, for the weighed factor approach, corresponding weighting functions.
3	Synthesize a set of options to be evaluated.
4	Model and analyze each option to generate data to quantitatively score each option, for weighed factors approach.
5	Evaluate each option, and repeat as necessary to optimize the selection.

Table IV.33: **The Steps in a Trade Study**

Available software for trade studies includes the ECLS System Assessment Program (ESAP), a spreadsheet developed by MSFC for evaluation of potential space station technologies, the Life Support Systems Analysis (LISSA) tool, developed by JPL and used to trade different options for SEI, the Boeing Engineering Trade Study (BETS) for the evaluation of long duration planetary missions, and the Life Support Operations Program, developed by Lockheed for JSC.

For the MSFC method, technologies to be considered must be at the breadboard stage of development or beyond. For comparing these different technologies, first of all, some parameters have to be chosen. Sample parameters and their descriptions are given in table IV.34. One can distinguish quantitative (No. 1-4) and qualitative parameters (No. 5-8). Of course, such parameters differ in relative importance depending on the mission scenario. To reflect relative importance, a weighting factor must be assigned to every parameter for each mission scenario. Quantitative parameters are assigned numerical weight factors, e.g., 1 to 10, but since analogous numerical data for qualitative parameters are often not available, assigning numerical scores to these parameters was considered too subjective. Thus, for qualitative parameters only relative weight factors, e.g., high - medium - low, are assigned. Since weighting factors are rather subjective, they are often determined by averaging the opinions of several engineers. The LISSA computer analysis tool mentioned above basically works the same way.

No.	Parameter	Parameter Description
1	Weight	Equipment weight in [kg]. Includes also the impact of technology on the weights of other system hardware.
2	Power	Equipment power consumption in [W]
3	Volume	Equipment volume in [m³]
4	Resupply	Weight of expendables to be resupplied, e.g. filters, water, chemicals etc.
5	Development Potential	Indicates a technology's potential to improve with time. Low maturity technologies have the highest development potential.
6	Emergency Operation	Evaluation of how well a technology is expected to operate in a degraded / contingency environment.
7	Reliability	Measure of the amount of unscheduled maintenance and downtime expected during operational lifetime.
8	Safety	Chance that a technology will lead to or cause a hazardous situation.

***Table IV.34:* Sample Trade Study Parameters**

For trade studies specific subsystems are generally not much more than black boxes with controlled inputs and outputs. In order to adequately model the processes within a subsystem, e.g., chemical processes, more specialized and detailed models, like Aspen Plus®, which is used to model any type of process involving a continuous flow of materials and energy from one processing unit to another, are required. In order to predict integrated ECLSS performance, system level models are useful to predict system ability to meet the design requirements, predict functional interfaces between individual components, size fluid accumulation devices, study alternate component technologies, and to uncover potential interface problems or test plan deficiencies. For simulating true transient characters of a given system, five software programs are widely used for system level ECLSS modeling:

- *G-189A,* developed by McDonnell Douglas, is a generalized environmental thermal control program for modeling steady-state and transient performance of an ECLSS. The heat and mass flows between the subsystems and assemblies can be evaluated, and the effects on an ECLSS of specific conditions or equipment configurations can be determined. Subroutines of the basic components or of combined components which are used to construct an ECLSS model include habitat, human, general thermal control and piping equipment, and assembly simulations. These component subroutines use heat and mass transfer and chemical reactions to compute mass and energy balances for the steady-state and transient conditions.

- *SINDA´85 / FLUINT* (Systems Improved Numerical Differencing Analyzer and Fluid Integrator), developed by Martin Marietta, is designed to solve lumped parameter representations of physical problems governed by diffusion-type equations. Physical systems are described and modeled using a resistor-capacitor representation. The SINDA´85 portion of the program is applied to solving thermal network problems, while the FLUINT portion is designed to analyze fluid flow networks.

- *CASE/A* (Computer Aided System Engineering and Analysis), developed by McDonnell Douglas, is evolved from G-189A and has many similarities with regard to how a model is conceived and constructed. It also incorporates features of SINDA´85/FLUINT. A major difference is the graphical user interface for model construction and an improved data management system. CASE/A provides transient tracking of the flow stream constituents and determination o their thermodynamic state throughout an ECLSS simulation, performing heat transfer, chemical reactions, mass and energy balances, and system pressure drop analyses, based on user-specified operating conditions. Basically, CASE/A consists of:

  1. The schematic management system, allowing the user to graphically construct a system model by arranging icons representing system components and connecting the physical fluid streams.
  2. The database management system, supporting the storage and manipulation of component data, output data, and solution control through interactive edit screens.
  3. The simulation control and execution system, initiating and controlling the iterative solution process. Simulation time and diagnostic messages are displayed.

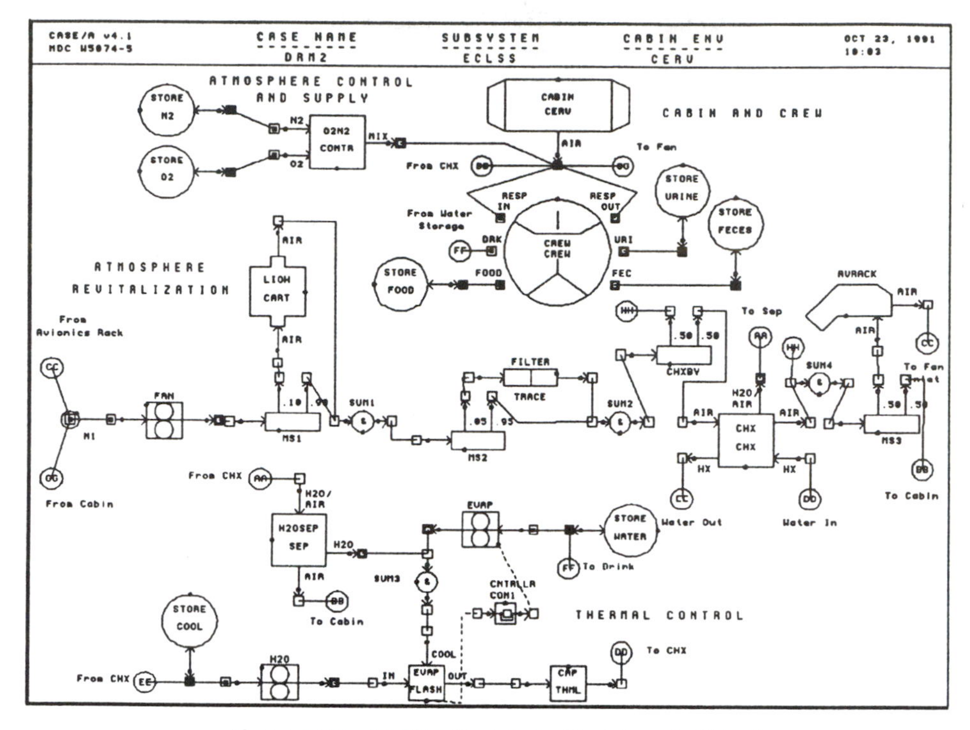

Figure IV.15: **CASE/A Model of a Vehicle ECLSS [48]**

- *TRASYS* (Thermal Radiation Analyzer System), developed by Lockheed, is a modularized computer program for computing the total thermal radiation environment of a spacecraft. Exterior and interior radiation is modeled and conductor values are calculated to represent the radiation transfer. The results are typically used as input data to SINDA for calculating the overall energy balance. [48]

- *ECOSIM* (Environmental Control and Life Support Simulation), developed by ESA, is a software package that enables experiments to be performed on mathematical models of ECLSS. It depends on external customization libraries to provide the information required for simulating areas such as air, waste, and liquid management systems. ECOSIM models are built from components, connected together by ports in any arrangement in so far as ports are compatible. The models are described as sets of differential and algebraic equations. A flow sheet editor automatically creates a skeleton source code, to which the final details then can be added. [10]

IV.4 PHYSICO - CHEMICAL VERSUS BIOREGENERATIVE LIFE SUPPORT

Traditionally, mainly nonregenerable physico-chemical processes have been used to provide life support in space. If the four basic requirements of a life support system, atmosphere management, water management, food production, and waste management, are partly or completely fulfilled by regenerative processes, a (partly) closed life support system is obtained. Regenerative processes can be either physico-chemical or biological. If exclusively biological processes are used in a regenerative system, a biological, or controlled ecological life support system (CELSS), is obtained. A combination of both physico-chemical and biological processes for life support yields a hybrid life support system.

In physico-chemical systems human is the only biological component of the system. Concerning the potential reduction of supply mass, the water and oxygen loops of the life support system can be closed by means of regenerable physico-chemical subsystems. Future space habitats, though, will require that the carbon loop, the third and final part-loop in the life support system, be closed. This will only be practical if advanced life support systems can be developed in which metabolic waste products are regenerated and food is produced. Thus, bioregenerative processes that can also synthesize food, will gain importance. In a bioregenerative system based on photosynthesis, e.g., oxygen is produced from carbon dioxide by algae or plants. A part of the edible biomass is later transformed to CO_2 and H_2O again by the human metabolism. The rest of the edible biomass will partly be oxidized and appears in the secretions, i.e., urine, feces and transpiration. If these products are later completely oxidized, together with the inedible biomass, a completely closed ecosystem, an "artificial biosphere", can be achieved. The introduction of biological techniques for food production into life support systems produces a certain number of problems to be solved, but also opens up a new area of solutions for other life support requirements. For example, many low-weight, volatile organic compounds that are found as air contaminants in small, closed inhabited volumes such as spacecraft cabins, arising mainly from human metabolic processes, equipment off-gassing, leakage from coolant loops, and fire control equipment, can also be substrates, i.e., nutrients, for a variety of microorganisms. Here, the concept of a biological air filter appears to be a possible solution for contamination control.

In general, physico-chemical processes are well understood, engineers feel comfortable with them, they are relatively compact and low maintenance and have quick response times. Nonregenerable physico-chemical systems utilize comparatively simple hardware, less subject to mechanical failure and the active substances are always fresh for the expendables, such as filters, provided they are stored properly. They also typically have low power requirements, but high life time mass and volume requirements. The same functions performed by regenerable physico-chemical systems utilize slower and less efficient processes. Also, these regenerable systems often use substances and hardware subject to a given life cycle limitation, consume a lot of energy that is expensive to produce and cannot replenish food stocks, which must still be resupplied from Earth. Consequently, solid wastes have to be collected, pre-treated, and stored. The advantages of regenerable methods for long duration methods include lower total mass/mission mass requirements.

Biological processes are less well understood. They make engineers nervous, they tend to be large volume, power and maintenance intensive, with slow response time. Nevertheless, food can only be produced by biological means. Food production in the restricted area of a space station is a complex operation, involving careful control of many parameters, including light intensity and spectral distribution, light/dark cycle times, temperature, nutrient supply, air and water composition and quality, etc. It will also be necessary, if propagation from generation to generation is to be permitted, to monitor continually the quality of the end-products to ensure that any harmful mutations are detected and eliminated. There are heated debates about whether physico-chemical or biological methods

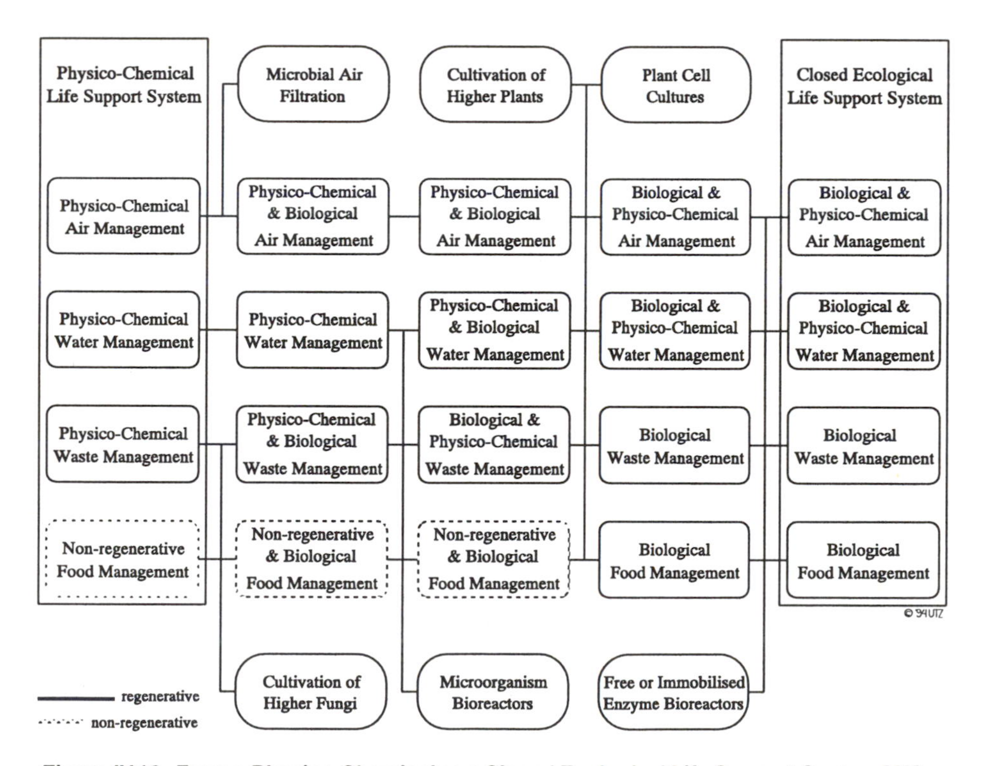

Figure IV.16: **From a Physico-Chemical to a Closed Ecological Life Support System [40]**

are most reliable with engineers claiming plants will die and biologists retorting that if they die it is usually because of the failure of some physico-chemical device. There is no definite answer to the question which technologies to select for future applications, but it is likely that hybrid life support systems will be applied in the foreseeable future. In this context, integration problems of physico-chemical and biological subsystems will be of major importance. An example for a logic for a stepwise implementation of biological components into a physico-chemical life support system is given in figure IV.16. Using this logic, finally, a CELSS is built up. [11, 15, 21, 40, 41, 42, 46, 48]

IV.5 PAST AND PRESENT LIFE SUPPORT SYSTEMS

The design of life support systems for future manned missions, i.e., for and beyond the space station must begin with the knowledge of past and present designs. This knowledge should come both from the manned space programs and the analogous terrestrial systems, like submarines, Biosphere 2, or the BIOS projects, of both the United States and the former Soviet Union. The U.S. and Soviet ECLSS design decisions from the past and present should take on equal importance when considering the design choices for future ECLSS. Thus, in this chapter an overview on the several life support systems, terrestrial and spacebound, is given.

IV.5.1 Spacebound Systems

Animals were the first pioneers of space. Keeping these animals alive during their short jaunts into space involved the true beginnings of the spacecraft ECLSS. The ECLSS for these early flights was a simple open-loop system. The first high-order living creature placed in Earth orbit was a female Eskimo dog named Laika, launched on SPUTNIK II by the Soviet Union on November 3, 1957. Laika's open-loop life support system was a hermetically sealed, air conditioned compartment complete with food and water. The first "passenger" of the U.S. was a monkey named Gordo, launched in December 1958. In this open-loop design, carbon dioxide was absorbed by pallets of baralyme, and the breathing gas was compressed oxygen from a cylinder. Temperature control was partially achieved by insulating layers of metal foil and fiber glass, and water vapor was absorbed by a porous material. The waste management system consisted of clothing the monkey in a diaper. Gordo was provided neither food nor water. The first human being in space was Yuri Gagarin, launched into Earth orbit aboard a VOSTOK capsule by an A-1 rocket on April 12, 1961.

Gagarin made one orbit around Earth completing the mission after a total of 108 minutes. Below, a survey of all manned programs and a short description of the respective life support systems of the U.S. and the former Soviet Union to date is given:

VOSTOK (1960-63)
The one-man spherical Vostok, with a habitable volume of about 2-3 m^3, was the first spacecraft to carry a man into space. The Vostok ECLSS was a simple semi-closed system with a 101 kPa air atmosphere. Cabin ECLSS equipment was responsible for CO_2 removal, and odor and humidity control, in addition to the control of cabin ventilation, temperature, and air supply. The cosmonaut wore a space suit ventilated by cabin air which did not have the capability for air purification or humidity control. In an emergency the suits could be supplied air and oxygen from tanks mounted on the Vostok exterior.

MERCURY (1960-63)
Mercury was a pressurized one-man capsule in the shape of a bell, with 1.56 m^3 of habitable space for the astronaut. The Mercury ECLSS is described by separating the system into the pressure suit and cabin subsystems. The pressure suit subsystem was primarily responsible for revitalizing the astronaut's atmosphere supply and for controlling the astronaut's temperature and humidity level. The cabin subsystem controlled cabin ventilation, cabin temperature (the cabin heat exchanger did not remove water vapor), and atmospheric pressure.

VOSKHOD (1964-65)
The spherical Voskhod capsule was basically an improved version of the Vostok, with a rearranged interior to accommodate three crewmen. To help make room for the cosmonauts, Voskhod became the first spacecraft in which the crew did not wear spacesuits. Voskhod 2 was equipped with one space suit and an inflatable decompression chamber from which the first space walk was performed.

GEMINI (1964-66)
Gemini was a pressurized two-man capsule with 2.26 m^3 of habitable space for the astronauts. Similar to Mercury, the Gemini ECLSS was divided into the pressure suit and cabin subsystems (see figure IV.17). Gemini improvements over the Mercury ECLSS included supercritical oxygen storage instead of high pressure storage, which reduced storage tank weight and volume. Also, an integrated heat exchange/water separator was applied instead of the separate heat exchanger and mechanically activated sponge-type water separator, which increased reliability and reduced power and weight.

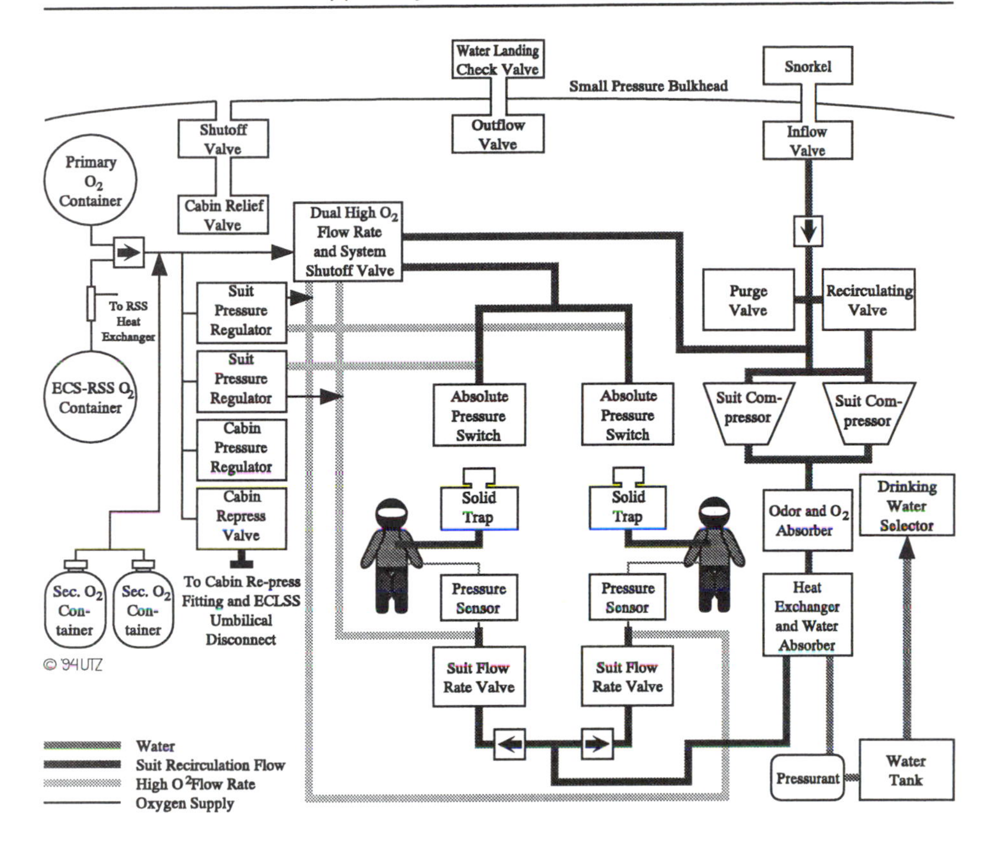

Figure IV.17: **GEMINI ECLSS [48]**

SOYUZ / SOYUZ T (since 1967)

Designed for seven day missions, the Soyuz was a three-man vehicle with two pressurized compartments, one for living and working, and the other for descent. Soyuz cosmonauts did not wear pressurized suits until after an accident on Soyuz 11, in which all three crewmen died after a pressure vent valve failed and the spacecraft depressurized. After Soyuz 11, the crew was reduced to two to make room for pressurized suits. The hermetically sealed Soyuz cabin was designed for zero leakage. Stores of gas for cabin repressurization and leakage makeup were considered unnecessary, and were not included in Soyuz. In contrary, U.S. spacecraft were designed to allow a small amount of leakage that was made up by stores of gas. The atmosphere leakage rate of approximately 1 kg/day on Apollo was a Soviet concern during the planning of the Apollo-Soyuz mission. Soyuz T, the modified version of Soyuz that retains the same basic size and shape of the Soyuz craft, has a redesigned interior to accommodate three space-suited cosmonauts and is still used to ferry cosmonauts to and from Mir space station.

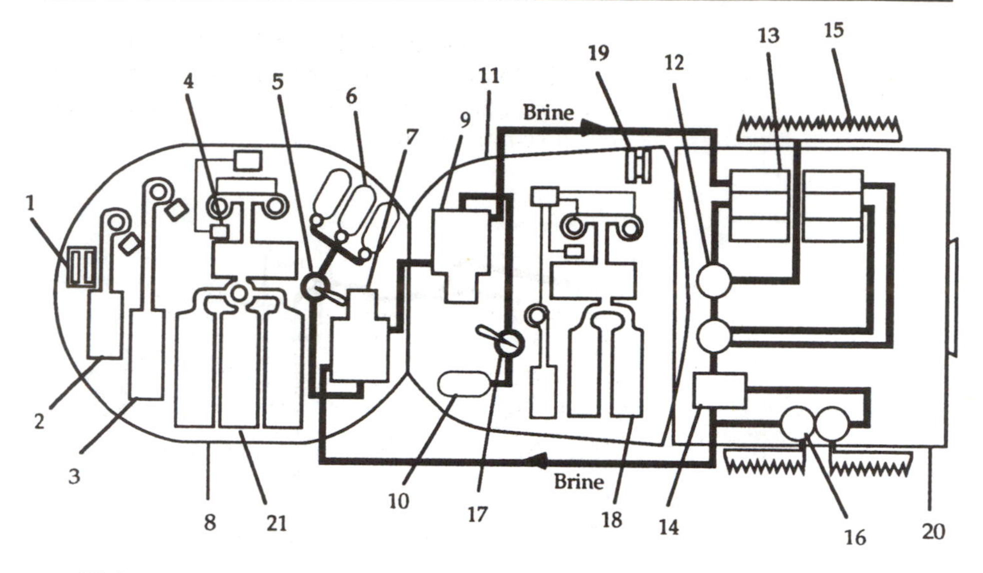

(1) Compressed Air for Leakage Makeup
(2) LiOH for Topping CO_2 Removal
(3) KO_2 Oxygen Supply and Primary CO_2 Removal Beds
(4) Flowmeter and Fans
(5) Manual Pump
(6) H_2O Storage Tanks
(7) Condensing Heat Exchanger with wick-type H_2O separator
(8) Flight Module
(9) Condensing Heat Exchanger with wick-type H_2O separator
(10) H_2O Storage Tank
(11) Landing Module
(12) Temperature Control Valves
(13) Equipment Cooler (Primary and Topping)
(14) Primary Heat Exchanger
(15) Primary Space Radiator
(16) Sequencing Space Radiators
(17) Manual Pump
(18) Trace Contaminant Control Bed
(19) Pressure Relief Valve
(20) Equipment Module
(21) KO_2 Beds for Oxygen Supply and Trace Contaminant Removal with Activated Charcoal and Bacteria Filter

Figure IV.18: **SOYUZ Life Support System [48]**

APOLLO (1968-72)

The total Apollo space vehicle included two separate life support systems, one on the command module (CM) and one on the lunar module (LM). Apollo, like Mercury and Gemini, had separate pressure suit and cabin ECLSS subsystems in both the LM and CM (see figures IV.19 and IV.20).

The Apollo command module was a pressurized three-man conical capsule with 5.9 m^3 of habitable space for the astronauts. The CM ECLSS occupied 0.25 m^3 of the cabin, and was capable of operating for 14 days. On orbit use of fuel cell potable water byproduct reduced water launch weight. Launch safety was increased by using a 60 %/40 % O_2/ N_2 cabin gas mixture during prelaunch and launch periods, although the suit circuit remained 100 % oxygen. The ascent stage of the Apollo lunar module was a pressurized two-man craft with 4.5 m^3 of habitable space. Differences between the Apollo LM and CM ECLSS included potable water from storage tanks instead of fuel cells, no overboard venting of urine on the lunar surface, and iodine bactericide instead of chlorine to avoid corrosion problems anticipated between chlorine and the LM sintered nickel sublimator plates.

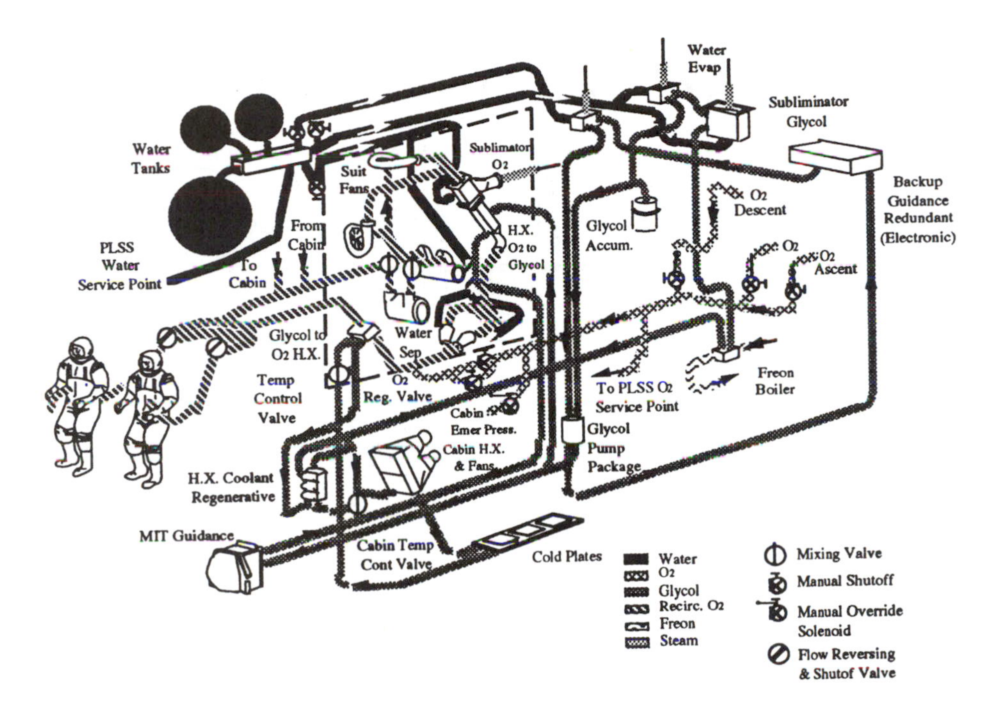

Figure IV.19: **APOLLO Lunar Module ECLSS [48]**

SALYUT SPACE STATIONS (since 1971)

Salyut was the first spacecraft designed for extended missions in space, and thus became the world's first space station. Since the beginning of the Salyut program, seven Salyut stations have been placed in orbit. Salyut, designed for a crew of 5, consists of three inseparable modules with a total usable volume of about 100 m^3. The ECLSS had remained predominantly the same on Salyut stations until Salyut 6, when a water regeneration system was added to recover condensate and wash water.

SKYLAB (1973-74)

Skylab, the first U.S. space station, was a three-man laboratory with a total habitable volume of 361 m^3. The crew lived and worked in the two-level orbital workshop (OWS), although most of the ECLSS equipment was located in the

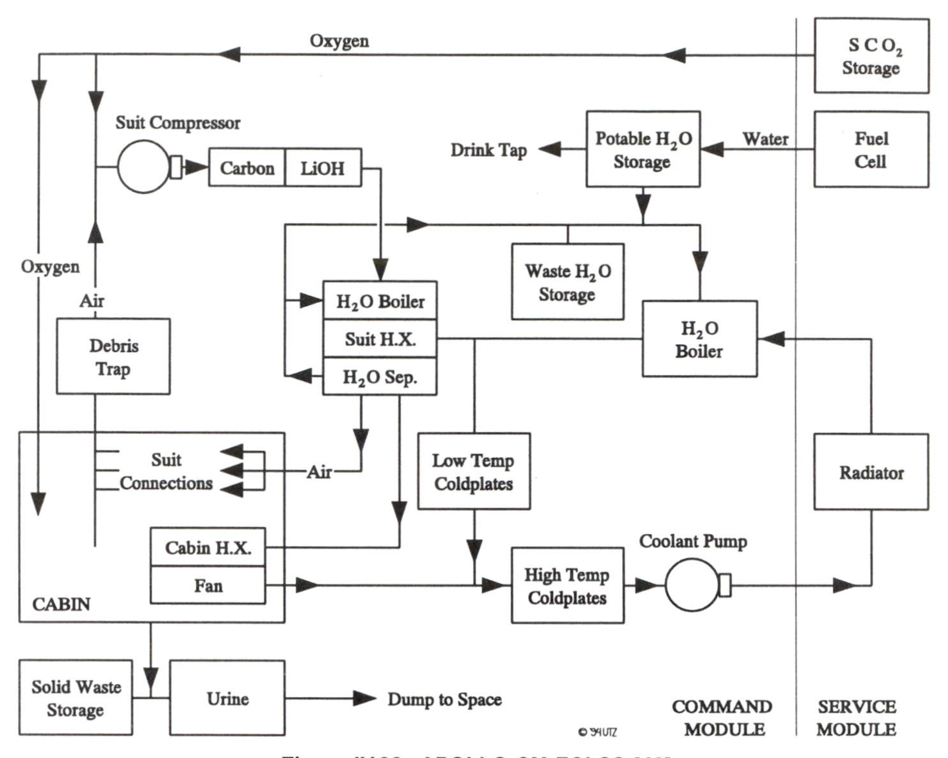

Figure IV.20: **APOLLO CM ECLSS [48]**

airlock module (AM). New ECLSS techniques on Skylab included a mixed O_2/N_2 atmosphere, a two bed molecular sieve instead of LiOH canisters to remove CO_2, a method for monitoring iodine concentration in the water supply, the storage of urine samples in a freezer for analysis on Earth, and ultraviolet fire detectors. Between missions Skylab was unoccupied, the atmosphere was depressurized to 13.8 kPa and allowed to decay down to 3.45 kPa until the next group arrived. Depressurization removed trace contaminants from the cabin and reduced the chance of fire between missions.

SPACE SHUTTLE ORBITER (since 1981)

The Space Shuttle Orbiter is designed to carry an average crew of 7 for a nominal mission of 7 days in a total habitable volume of 74 m^3. The Orbiter became the first U.S. spacecraft to use a standard sea-level atmosphere, i.e., 22 % oxygen and 78 % nitrogen at a total pressure of 101 kPa. Other Orbiter innovations included Halon 1301 fire suppressant, microbial check valves to passively adjust iodine concentration in the potable water supply on a continuous basis instead of periodic iodine injection by the crew, and a commode for fecal collection and storage instead of simple bag collection.

SPACELAB (since 1983)

The cylindrical Spacelab laboratory module, situated in the Space Shuttle cargo bay during a Spacelab module mission, provides a pressurized shirt-sleeve atmosphere for performing experiments in microgravity. Most of the Spacelab ECLSS is very similar to the Orbiter system, although Spacelab does not handle potable water or metabolic waste, and depends on the Orbiter for its metabolic oxygen supply.

MIR SPACE STATION (since 1986)

The Mir core, designed for a crew of 6 in a habitable volume of about 150 m^3, is the first space station to accommodate growth by modules. The Mir core ECLSS is similar to the Salyut 7 system, although some improvements are included in the attached modules. For example, in the attached Kvant-2 module oxygen is produced by water electrolysis (Electron device) instead by nonregenerable cartridges of potassium superoxide, and carbon dioxide is removed by a Vozdukh system and vented to space instead of being removed by chemical reaction inside oxygen regenerators and lithium hydroxide canisters.

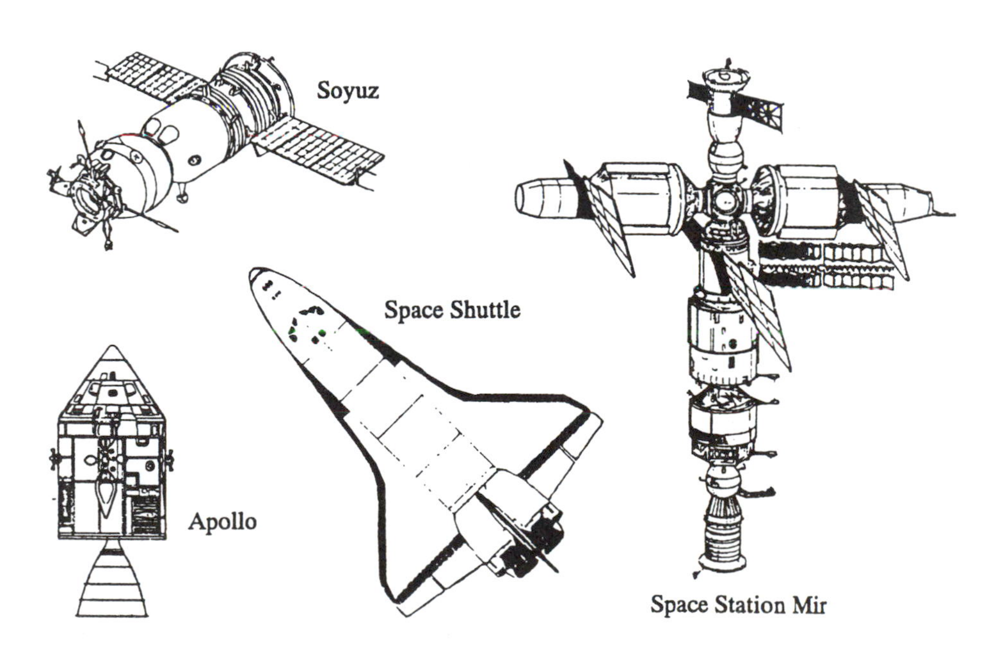

Figure IV.21: **A Survey of Past and Present Space Vehicles**

Project	Time Period	Mission Objectives	Number of Flights	Flight Durations
VOSTOK	1960-63	- Testing human behavior under micro-g and high acceleration levels (8-10g) - Testing and further developing ground-controlled automatic spacecraft guidance - Making astronomical and geophysical observations	6	1h 48 m to 4d 23h 6m
MERCURY	1960-63	- Placing manned spacecraft in Earth orbit - Testing human reactions in orbit - Testing possibility of manual spacecraft control by pilot - Recovering astronauts and capsules safely from space	6	15m to 34h 20m
VOSKHOD	1964-65	- Gathering data on group crews - Studying human behavior outside capsule	2	1d 17m to 1d 2h 2m
GEMINI	1964-66	- Testing crew and capsule behavior for a non-stop 14-day orbit - Developing the capability to rendezvous and dock with other spacecraft - Performing extravehicular activities - Developing methods for controlling spacecraft reentry flightpath - Provide a basis for scientific experiments	10	4h 53m to 13d 18h 35m
SOYUZ SOYUZ T	since 1967	- Docking and crew transfer between capsules and Salyut space station - Practicing craft orbital transfer - Performing scientific observations and experiments - Ferrying cosmonauts to and from Mir		
APOLLO	1968-72	- Landing a man on the Moon and returning him safely to Earth - Exploring the Moon from lunar surface and lunar orbit Demonstrating - that humans can move about and work in an alien environment	11	5d 22h 55m to 12d 13h 52m
SALYUT	since 1971	- Long-term space research and experimentation Establish a permanent presence in space -		
SKYLAB	1973-74	- Studying the effects of long-duration spaceflight on humans - Studying the Earth, Sun, and stars - Performing experiments in a microgravity environment	3	28 d 50m to 84 d 1h 16m
SPACE SHUTTLE	since 1981	- Replacing expendable launch vehicles with a reusable transport system		
MIR	since 1986	see Salyut		

Table IV.35: **Past and Present Manned Space Programs**

Project	Carbon Dioxide Removal	Gas Recovery and Generation	Trace Contaminant Control	Trace Contaminant Monitoring
VOSTOK	Reaction of carbon Dioxide with KOH in the oxygen rege- nerator forming K_2CO_3 and water	Nonregenerative che- mical cartridges of KO_2. KO_2 is reacted with water to produce oxygen and KOH	Activated charcoal, filters, and reaction with constituents in the oxygen regene rator	Gas analyzer for determination of oxygen and carbon dioxide percentages in cabin atmosphere
MERCURY	2 parallel LiOH canisters. Airflow through only one canister	None	Activated charcoal in LiOH canisters upstream of LiOH. Filters for removal of particulates	CO Sensor
VOSKHOD	Similar to Vostok design	Similar to Vostok design	Similar to Vostok design	Similar to Vostok design
GEMINI	Similar to Mercury design	None	Similar to Mercury design	None
SOYUZ	Similar to Vostok design. Additional LiOH beds to absorb about 20% of the CO_2	Similar to Vostok design	Similar to Vostok design	Similar to Vostok design
APOLLO CM / LM	Similar to Mercury design	None	Similar to Mercury design	None
SALYUT	Similar to Soyuz design	Similar to Vostok design	Activated charcoal, high efficient fiber- glass filters, catalytic chemical absor- bents, and reaction with constituents in the oxygen regene- rator	Similar to Vostok design. Several gas analyzers
SKYLAB	2 canister molecu- lar sieve containing Zeolite 5A for car- bon dioxide remo- val and Zeolite 13X for water removal. CO_2 vacuum desorbed to space	None	Activated charcoal, filters to remove particulates, venting of atmosphere between missions	Monitoring of CO and other contami- nants with Draeger tubes.
SPACE SHUTTLE	Similar to Mercury design	None	Activated charcoal, ATCO to convert CO in CO_2, filters to remove particulates	None
SPACELAB	Similar to Mercury design	None	Similar to Space Shuttle	None
MIR	Mir core similar to Salyut design. Carbon dioxide adsorption and venting into space in Kvant-2 module	Mir core similar to Salyut design. Oxygen production by water electrolysis in Kvant-2 module. Resulting hydrogen vented into space.	Similar to Salyut design.	Similar to Salyut design.

***Table IV.36:* Atmosphere Revitalization Methods of Past and Present Missions [12]**

Project	Water Processing	Water Monitoring	Water Storage and Distribution	Water Supply Microb.Control
VOSTOK	None	**	Water held in containers made of elastic polyethylene. Low pressure of crewman's mouth induced water flow from container.	Silver preparation added to water which was boiled before launch
MERCURY	None. Waste water vented.	None	One tank with a flexible blad- der. Squeezing an air bulb pressurized bladder to deliver water (sphygmomanometer)	Chlorine as in public water system
VOSKHOD	None	**	Similar to Vostok design	Similar to Vostok design
GEMINI	None. Waste water vented.	None	Similar to Mercury design. 7.3 liter tank was refilled from reserves in service module when empty	Chlorine added before launch
SOYUZ	None	**	Similar to Vostok design	Similar to Vostok design
APOLLO CM / LM	None. Waste water stored (CM / LM) and vented or sent to evaporators for cooling (only CM)	None	Water produced from fuel cells principle source. Palladium and silver removed dissolved H2 (CM). 3-4 potable tanks (LM). Potable tanks with oxygen (CM) and nitro- gen (LM) pressurized bladders to deliver water	Chlorine (0.5 mg/l) maintained by adding sodium hypochlorite solution every 24 hours (CM). Iodine added before launch (LM).
SALYUT	From Salyut 6 on potable water recovered from condensate and wash water by using ion exchange resins, activated charcoal, filters containing fragmented dolomite, artificial silicates, and salts, and adding minerals including Ca, Mg, chloride, and sulfate	**	Rodnik ("spring") system supplies water from tanks with a total volume of 400 liters	Water is heated and ionic silver is introduced electrolytically (0.2 mg/l)
SKYLAB	None. Waste water stored and vented.	Iodine sampler. Water samples fixed with linear starch reagent and compared to photographic standards	Ten cylindrical 272 kg stainless steel tanks fitted with pressurized steel bellows to deliver water. One 11.8 kg portable tank.	Iodine (0.5-0.6 mg/l) maintained by periodic injection of potassium iodide solution.
SPACE SHUTTLE	None. Waste water stored and vented.	None	Four 76 kg stainless steel tanks fitted with metal bellows pressurized by nitrogen. Drinking water from fuel cells	Iodine (1-2 mg/l) passively adjusted by microbial check valves (MCV)
MIR	Similar to Salyut 6 design	**	Similar to Salyut design	Similar to Salyut design

* On all of the above listed missions only potable water was used ** No information available

Table IV.37: **Water Recovery and Management Methods of Past and Present Missions [12]**

Project	Atmosphere Temperature and Humidity Control	Cabin Ventilation	Equipment Cooling
VOSTOK	Liquid-air CHX. Temperature adjusted automatically or manually (12°-25° C). Humidity controlled by dehumidifier containing a silica gel drying agent impregnated with lithium chloride and activated carbon (30-70 % relative humidity).	Cabin fan	**
MERCURY	Separate suit and cabin CHX. Mechanically activated sponge water separator removed water from CHX. (15°-22° C)	Cabin fan	Cold plates
VOSKHOD	Similar to Vostok design	Cabin fan	**
GEMINI	Separate suit and cabin CHX. Wicks removed water from CHX by capillary action. (15°-22° C)	Cabin fan	Cold plates
SOYUZ	Similar to Vostok design but humidity primarily also controlled by CHX.	Cabin fan	**
APOLLO CM/LM	Suit CHX primary method for cabin temperature and humidity control. Wicks removed water from CHX by capillary action (CM). Water circulated through pressure garment assembly to cool astronaut (LM). (25° C)	Cabin fans	Cold plates
SALYUT	Similar to Soyuz design. Condensate collected in moisture trap and periodically pumped out manually.	Cabin fans	**
SKYLAB	Four CHX, two operating at all times. (18°-24° C)	3 ventilation ducts with 4 fans each. 3 portable fans with adjustable diffusors.	Cold plates
SPACE SHUTTLE	Centralized cabin liquid/air CHX using water coolant. Air bypass ratio around CHX is used to control temperature. Condensate removed by slurper bar and centrifugal separator. (21° C)	Cabin fan with ventilation ducts	Air cooling, cold plates, liquid/liquid equipment dedicated HX.
SPACELAB	Similar to Space Shuttle design (18°-27° C)	Cabin fan	Similar to Space Shuttle design
MIR	Temperature controlled by heat pipes and manual adjustment of fans, heaters, and air conditioners. Humidity controlled by CHX.	Fans pull air through ducts to exchange gas between modules	Gas circulation in the instrument bay

CHX = Condensing Heat Exchanger ** = No Information available

Table IV.38: **Temperature and Humidity Control of Past and Present Missions [12]**

Project	Atmosphere Composition	Atmosphere Pressure	Gas Storage
VOSTOK	Sea level atmosphere (Oxygen / Nitrogen)	101 kPA	Oxygen stored chemically. Emergency tanks of high pressure oxygen and air. No nitrogen storage. Cabin hermetically sealed for zero leakage.
MERCURY	100 % Oxygen	34.5 kPa (max. ppCarbon dioxide = 1 kPa)	Oxygen stored at 51.7 MPa in two 1.8 kg carbon steel, nickel plating tanks.
VOSKHOD	Sea level atmosphere (Oxygen / Nitrogen)	101 kPa	Similar to Vostok design
GEMINI	100 % Oxygen	34.5 kPa (max. ppCarbon dioxide = 1 kPa)	Oxygen stored at 5.86 MPa as super- critical cryogenic fluid in one spherical tank. Two secondary cylindrical oxygen bottles. One small oxygen bottle under each seat.
SOYUZ	Sea level atmosphere (Oxygen / Nitrogen)	94.4 - 113 kPa (ppOxygen = 10.5-15.2 kPa)	Chemical oxygen regenerators. No additional gas storage. Complete reliance on cabin hermetic seal to prevent depressurization.
APOLLO CM / LM	100 % Oxygen (60 % Oxygen, 40 % Nitrogen during launch at CM)	34.5 kPa (max. ppCarbon dioxide =1 kPa)	Oxygen stored at 6.2 MPa and 180° C as supercritical cryogenic fluid in two 145 kg spherical Inconel Dewar tanks. One 1.7 kg oxygene tank used during reentry (CM). 21.8 kg oxygene stored as gas at 18.6 MPa in descent stage. Oxygen stored at 5.86 MPa as supercritical cryogenic fluid in two Inconel bottles in ascent stage (LM).
SALYUT	Sea level atmosphere (Oxygen / Nitrogen)	93.1 - 128 kPa (ppOxygen = 21.4 - 31.7 kPa)	Oxygen stored chemically. Cylinders of compressed air for leakage makeup. No separate nitrogen or carbon dioxide storage.
SKYLAB	74 % Oxygen 26 % Nitrogen	34.5 kPa (ppOxygen = 25 kPa; max. ppCarbon dioxide = 0.7 kPa)	Total of 2779 kg oxygen and 741 kg nitrogen stored as gases at 20.7 MPa in six bottles each.
SPACE SHUTTLE	78.3 % Nitrogen 21.7 % Oxygen	101 kPa (ppOxygen = 22 kPa; max. ppCarbon dioxide = 1 kPa)	4 spherical nitrogen tanks and 1 emergency oxygen tank for gas storage at 22.8 MPa. Metabolic oxygen supplied by power reactant and distribution system (uses supercritical cryogenic storage tanks).
SPACELAB	78.3 % Nitrogen 21.7 % Oxygen	101 kPa (max. ppCarbon dioxide = 1 kPa)	Nitrogen for leakage makeup and scien- tific airlock operation stored as gas in spherical tank at 22.8 MPa. Oxygen supplied from Space Shuttle.
MIR	Sea level atmosphere (21-40 % Oxygen, up to 78 % Nitrogen)	(max. ppOxygen = 46.9 kPa)	Similar to Salyut design

Table IV.39: **Atmosphere Control and Supply Methods of Past and Present Missions [12]**

Project	Waste Management	Fire Detection	Fire Suppressant
VOSTOK	Urine and feces entrained in an air stream and collected	**	**
MERCURY	Urine stored in in-suit collection bag. No provision for fecal handling.	Crew senses	Water from food rehydration gun. Capability to de- pressurize cabin manually
VOSKHOD	Similar to Vostok design	**	**
GEMINI	Urine collected with urine transfer system consisting of rubber cuff connected to flexible bag and directed to boiler tank to assist heat rejection. Feces collected in bags and stored.	Crew senses	Similar to Mercury design
SOYUZ	Similar to Vostok design	**	**
APOLLO CM / LM	Urine collected using transfer system. After Apollo 12 use of urine receptacle assembly. Urine vented, but no overboard dumping on lunar surface. Feces collected in bags and stored (CM).	Crew senses	Similar to Mercury design plus a portable aqueous (hydroxy-methyl cellulose) extinguisher which could expel foam.
SALYUT	Feces collected in hermetically sealed metal or plastic containers, which are ejected to space once a week	Carbon dioxide detectors doubled as fire detectors	**
SKYLAB	Urine collected using individual receivers, tubing, and disposable collection bags. Feces collected in gas permeable bags, then vacuum dried and stored.	Ultraviolet detectors	Portable aqueous (hydroxymethyl cellulose) extinguisher which could expel foam. Capability to depressurize cabin.
SPACE SHUTTLE	Commode/Urinal - Feces collected in commode storage container, vacuum dried and held. Vane compactor facilitates fecal storage and containment. Urine sent to waste water tank (vented when full).	Ionization smoke sensors	Halon 1301 - one halon tank with distribution lines in each avionics bay and three portable extinguishers. Capability to depressurize cabin.
SPACELAB	Utilizes Space Shuttle facilities	Ionization smoke sensors	Halon 1301 - one halon tank with distribution lines in each equipment rack and two portable extinguishers. Capability to depressurize cabin.
MIR	Similar to Salyut design	**	**

** = No information available

Table IV.40: **Waste Management, Fire Detection and Suppression Methods of Past and Present Missions [12]**

IV.5.2 Earthbound Systems

Research and engineering testbeds provide the laboratory for the study of life support systems. In this chapter the testbeds of ECLSS that have already been used onboard the past and present spacecraft are neglected for obvious reasons. Instead, the focus is on the research projects and testbeds for future regenerative ECLSS. These facilities are earthbound attempts to control as many variables as possible, allowing the variable of interest to shift and change as it would in the modeled system. One can basically distinguish regenerative physico-chemical and bioregenerative testbeds. For example, NASA is assembling the ECLSS testbed for the U.S. Space Station at the Marshall Space Flight Center (MSFC) in Huntsville, Alabama. This series of tests is the first NASA attempt to integrate multiple physico-chemical subsystems in series to reclaim, reuse, and recycle diverse source waters including urine, hygiene effluent, and humidity condensate in a single life support system. Ground tests of physico-chemical systems used on Mir also have been conducted in the former Soviet Union. [9]

U.S. Space Station Testbed

The ECLSS test facility at the MSFC comprises an area of about 1858 m^2. A scheme of the facility is given in figure IV.22. It has the capability of operating as a cleanroom, although this is not necessary for most of the testing, and includes the following features [48]:

- The Control Module Simulator (CMS) is the primary habitat simulator and is 4.6 m in diameter and 12.2 m long. It is able to operate at ambient and reduced pressures.
- The End-use Equipment Facility (EEF) is a sealed room used to collect waste water for realistic testing of the water purification equipment. It includes exercise equipment to generate perspiration and respiration moisture from volunteers for collection by condensing heat exchangers. Also, microwave ovens and other cooking equipment to generate cooking moisture and fumes, a shower, and clothes and dishwashers to generate hygiene waste water are included the EEF.
- The monitoring and control room contains the computer controllers and displays to operate the equipment and run the tests performed in the CMS.
- The gas and liquid sample analysis lab contains the analysis equipment necessary to perform initial evaluations of field samples to monitor the performance of the equipment being tested.
- The Predevelopment Operational System Test (POST) facility is a clean room configured for integrated testing of ECLSS equipment.

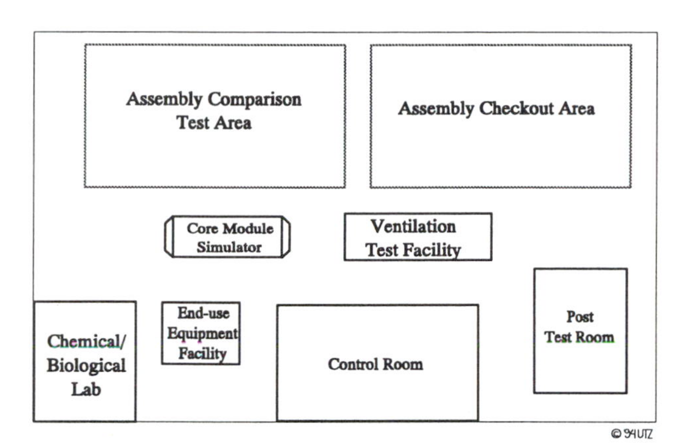

Figure IV.22: **ECLSS Test Facility at Marshall Space Flight Center [48]**

- The assembly checkout and evaluation area was configured for comparison testing of the candidate ECLSS technologies for the U.S. Space Station. For that purpose eleven assemblies were tested simultaneously.

- The intramodule ventilation test facility duplicates the interior passageway of a Space Station module and the diffusers for atmosphere circulation. It was specifically constructed for testing of intramodule concepts to ensure that all open areas will be properly ventilated.

Bioregenerative Testbeds

The Soviet closed system studies in BIOS-1, 2, and 3 near Krasnoyarsk, Siberia have provided data for almost 30 years. NASA's testbeds include the BioHome at the Stinnes Research Center in Mississippi, the CELSS Breadboard at the Kennedy Space Center in Florida, the Plant Crop Chamber at the Ames Research Center in California, and the Human-Rated Test Facility at the Johnson Space Center in Houston, Texas. Industrial test facilities include the McDonnell Douglas space station simulation that was operated between 1965 and 1970, the Lockheed plant growth chambers in Sunnyvale, California and the Boeing plant air and transpiration water recovery chambers built and operated in collaboration with the University of Alabama in Huntsville. A survey of the cornerstones of earth-bound bioregenerative life support system research is given in table IV.41. [9]

In 1967, Folsome of the University of Hawaii at Manoa initiated experiments with sealed, small, i.e., 100 ml - 5 l flasks. The laboratory flasks contained with aquatic solutions a range of microbial communities and air. Folsome exposed them to artificial light or indirect sunlight. These flasks were materially closed, i.e., there was no exchange of air or nutrients with the outside, but they were energetically open to light energy. They were also informationally open as Folsome developed non-intrusive ways of conducting measurements. These laboratory ecospheres proved to be indefinitely persistent as long as the initial sample contained a full functional representation of microbes, fulfilling the entire range of metabolic functions from biosynthesis to detritus feeding. Ecospheres initiated in 1967-1968 are still alive, exhibiting periodic changes in microbial content.

System	Investigator, Project	Characteristics
Small closed ecological systems (Microbes)	C. Folsome, University of Hawaii (1967)	- Sealed flasks (100 ml - 5l) - Multiculture aquatic solutions (biosynthetic decomposer) - Energy + information exchange
Algae-based systems	US (1961)	- Monkey / algae gas exchange - Duration up to 50 hours
(Chlorella)	USSR (1961)	- Rats and dogs, up to 7 days - First human / algae system (BIOS 1 & 2) 15 to 30 days
Higher plants	CELSS-US, Japan, ESA (since 1977)	- Controlled environment plant growth (light, carbon dioxide, temperature, photoperiod) - Focus on increased yield
	BIOS-3, USSR (1972-84)	- 2-3 people, up to six months - Food production (30-50%) - Water recovery (transpiration: filter, boil)
	Biosphere 2, Space Biosphe-res Ventures (1984-present)	- 8 people, up to 2 years - Complete ecological systems for water recovery and air purification

Table IV.41: **Survey of Terrestrial, Closed Bioregenerative Research Projects [14]**

In 1961, experiments were conducted at the U.S. Air Force School for Aviation Medicine in which monkeys were linked in gas exchange with algae tanks for up to 50 hours. During 1960-61, researchers at the Institute of Plant Physiology and the Institute of Biomedical Problems in the Soviet Union conducted experiments along the same lines with rats and dogs for periods up to seven days. In 1961, Shepelev was the first human to place himself as an experimental subject in a human/algae system. The basic oxygen/carbon dioxide exchange between Shepelev and the photosynthesizing algae *Chlorella* was successful, although a build-up of trace gas contaminants was noticed (see also section VI.3). [27]

BIOS 1 - 3

In 1965, another system was constructed by the Soviets that regenerated the atmosphere for a human with *Chlorella*. The first regeneration link comprised 8 m^2 of photosynthesizing algae which could absorb CO_2 and produce O_2 for one person. This represented 20 % of the essential substances when water and nutrition were stored in advance. In 1968, a 80 to 85 % closure was achieved by recycling the water. These early systems, located in Krasnoyarsk, Siberia, were called BIOS-1 and BIOS-2. Beginning in 1968, units with volumes of 5-20 m^3, regenerating oxygen with *Chlorella* and even higher plants, were established in Moscow. During 1969, three people lived in one of these small systems with physico-chemical plus biological life support for a full year. Food and water were usually stored ahead of time, but in the one-year experiment, oxygen and water were regenerated 100 % and a few green plants were used to provide vitamins. Whereas research in Krasnoyarsk is mostly basic, the results of studies in Moscow are applied in the former Soviet, now Russian, space program.

BIOS-3, which was built in 1972, is a sealed structure in which two or three crew members can live and produce 70 to 80 % of their food in three rooms called phytotrons, each provided with high-irradiance artificial light. A fourth room contains living facilities, control panels, and other equipment. The goal in building the system was to have it fully isolated from the environment with closed cycles that regenerated air, water, and some nutrients. Crew members inside do all of the tasks related to the operation of the facility. Although plants were grown in the phytotrons more or less continually, there were only three full-scale experiments with a crew sealed inside. These were carried out in the winters of 1972-73 (three crew members for six months), 1976-77 (three crew members for four months), and 1983-84 (two crew members for five months). BIOS-3 is completely underground and constructed of welded plates of stainless steel to provide the hermetic seal. The structure has a surface area of 14 x 9 m^2, is 2.5 m high, and has a volume of 315 m^3 divided into four main compartments, each the same size (about 7 x 4.5 m^2 and 79 m^3). Two of the compartments are for hydroponic higher plant growth (each with a growing area of 20.5 m^2), the third with three tanks for algae cultures, and the fourth for a crew of three. Each compartment has two doors, sealed tightly with a rubber gasket, so that any compartment can be sealed in combination with any other compartment. Gases and liquids required by each compartment are supplied by pipelines running between the compartments.

In the algae compartment, three *Chlorella* cultivators are located. Each has a light receiving surface area of about 10 m^2 and produces up to 800 g/day of dry algae biomass. In the two plant growth chambers about 17 m^2 of wheat and 3.5 m^2 of vegetables are grown hydroponically in metal trays. The output of each phytotron is about 1000 l/day of oxygen. Climate in the phytotrons is maintained at a temperature of 22°-24° C and a relative humidity of 70 %. Electric power and tap water coolant for heat removal from BIOS-3 are supplied from external sources to the system. The pressure is normally elevated slightly compared with ambient atmospheric pressure. Pollutants that were detected included various organic substances, CO, NH_3, H_2S, SO_2, and others. They resulted from humans, plants, technical operations, and material outgassing. Due to the good purification performance of the plants and a catalytic converter, toxins did not accumulate. Inedible biomass was burned in an incinerator. Since it was very hot, there was no smoke from the incinerator. Everything was broken down to CO_2, with no CO, and the resulting gas went directly to the phytotrons. CO_2 and oxygen were measured constantly or periodically. The dynamics of CO_2, O_2, and total oxygen concentration (TOC) in the atmosphere of BIOS-3 are shown in figure IV.23. Note the mirror image responses of O_2 and CO_2 and that, obviously, between days 0 and about 60, a periodical measurement mode was applied, while later the concentrations were measured constantly. The sharp changes during the second half of the run were caused by burning the inedible plant wastes.

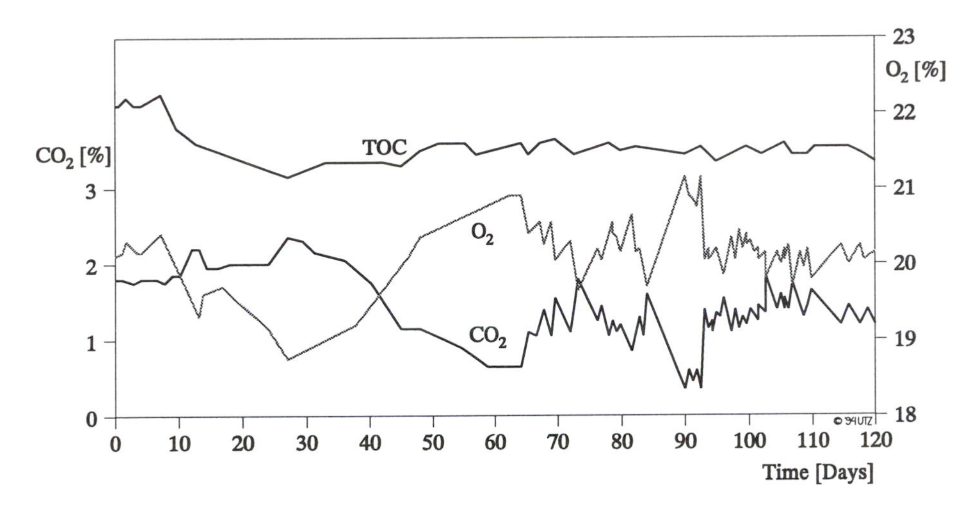

Figure IV.23: **Dynamics of CO_2, O_2, and TOC in the Atmosphere of BIOS-3 [34]**

The species grown in BIOS-3 included wheat, carrots, beets, radish, turnip, cabbage, cucumbers, onions, chufa, and sorrel. Tomatoes and potatoes were grown, but the yield was very low because of the continuous illumination and high temperatures. Urine was added in doses to the wheat nutrient solution, depending on how much nitrogen the plants needed. This led to a built up of sodium in the nutrient solutions. Plants were grown with round-the-clock lamp irradiation with 130 W/m² photosynthetically active radiation (PAR) intensity. Each phytotron has 20 vertical xenon lamps, operating at 220 V with 6 kW per lamp, and with a starting voltage of 20000 V at the moment of ignition. The lamps are surrounded by glass cylinders through which water circulates for cooling. The plants production was food for people. The 14 plant species provided 70 % of the caloric requirement. The animal part of the diet was compiled depending on the requirements of the crew. Meat, beef, pork, poultry, and fish, was introduced into BIOS-3 once in a month.

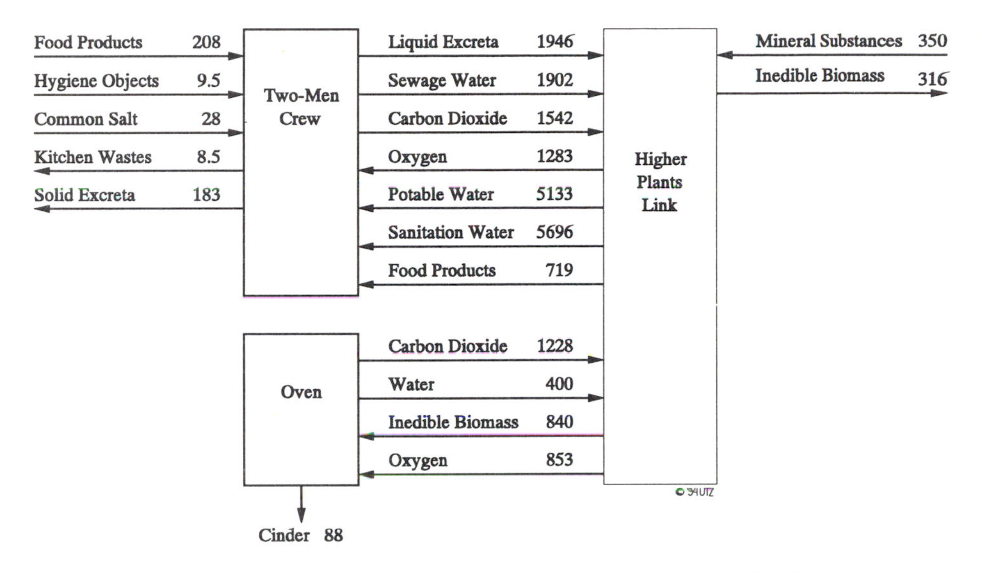

Figure IV.24: **Mass Exchange of BIOS-3 (Units in [g/day]) [34]**

The crew needed 2.2 to 2.5 l of potable water per day. Additional water was required for sanitation and everyday needs, including personal hygiene, wet cleaning, laundry and cooking. Experience over the years suggested a consumption of 6.5 l/man-day. Water exchange of the ecosystem, as well as gas exchange, were basically fully closed. Transpired water was condensed, recirculated, and additionally purified by activated charcoal and ion exchange filters for drinking, and boiled for 5 minutes for other uses. Small quantities of salts were added to the purified water for the sake of nutrition (KI) and to improve the taste (KCl). The sources of condensate water were the air conditioners, phytotron moisture condensers, a drying chamber, and the incinerator for the burning of inedible biomass. In the drying chamber, inedible biomass and kitchen wastes were dried at 100°-110° C. An overall mass and mineral exchange balance for BIOS-3 is given in figures IV.24 and IV.25. In the 1983-84 experiment, the highest closure was achieved with 91 %. Nevertheless, the mineral exchange accounted for only 1.5 % of the closure.

Health state of the crew was estimated before, during, and after the experiment. Crew members often had sensors attached to their bodies to monitor various physiological parameters, and the electrical signals were passed to the outside through sockets. The crew could communicate with the outside by phone or through the viewing ports. The first experiment (December 24, 1972 to June 22, 1973) consisted of three stages lasting two months each, with each stage involving a different arrangement for exchanging gases and water between plants and crew. Figures IV.26 to IV.28 depict the water and atmosphere exchange signs for each of the three stages.

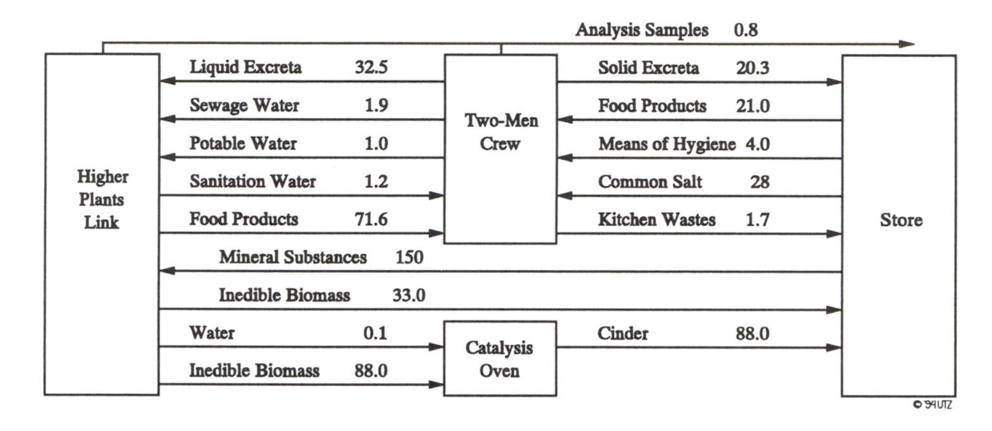

Figure IV.25: **Mineral Exchange of BIOS-3 (Units in [g/day]) [48]**

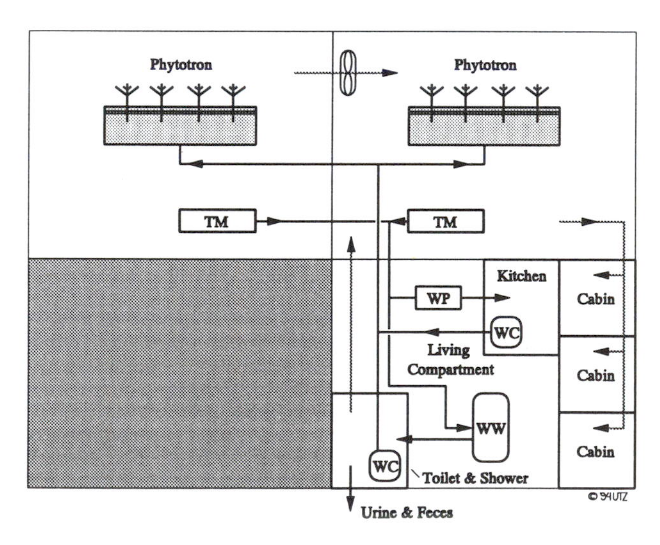

Figure IV. 26: **First Stage of BIOS-3 [48]**

The first stage consisted of a living compartment for the crew and two phytotrons with a wheat crop and a selection of vegetable plants. All crew gas and water requirements were satisfied by the plants in the two phytotrons. Waste wash water from the living compartment was delivered to the wheat nutrient solution, while the liquid and solid crew wastes were removed from the system. Food for the crew was provided by the higher plants in the form of vegetables and bread, and any remaining nutritional requirements were satisfied by stored freeze-dried supplies.

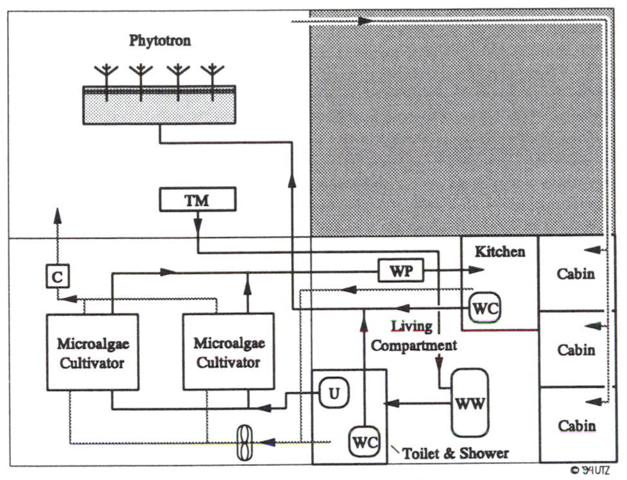

Figure IV.27: **Second Stage of BIOS-3 [48]**

For the second stage, one phytotron was disconnected from the system and replaced with a compartment of *Chlorella* cultivators. Gas and water exchange requirements for the crew were satisfied by the activity of the algae cultivators and the remaining phytotron. Liquid crew wastes were consumed by *Chlorella,* and the solid crew wastes were dried to remove water and return it to the system. Food was supplied from the single phytotron and stored supplies.

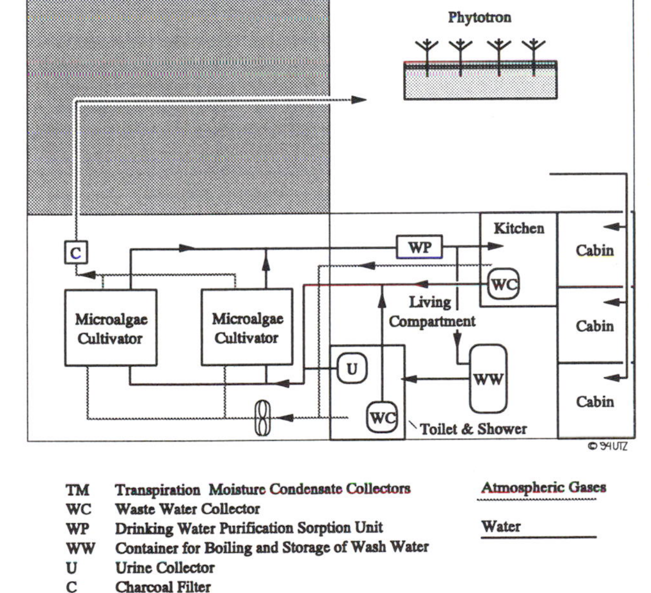

Figure IV.28: **Third Stage of BIOS-3 [48]**

For the third stage of the experiment, the phytotron containing wheat and vegetable plants was replaced with a phytotron containing only vegetable crops. Waste wash water was exchanged between the crew and algae compartments. However, unlike the previous two stages of the experiment, none of this water was mixed with the phytotron nutrient solution.

McDonnell Douglas Long-Duration Life Support Testing Program

From 1965 to 1970 three regenerative life support tests with human test subjects were conducted in a space station simulator chamber by a team at McDonnell Douglas in Santa Monica and Huntington Beach, CA. The testing program consisted of three phases :

-	Phase 1	1965	30-day manned test
-	Phase 2	1968	60-day manned test
-	Phase 3	1970	90-day manned test

The hardware and procedure were upgraded after each phase. The objectives for this simulator test of a Regenerative Life Support System (RLSS) included :

- Operation of the RLSS without resupply
- Evaluation of the advanced life support systems
- Determination of the crew's ability to operate, maintain, and repair equipment
- Achievement of an equilibrium in a closed environment
- Determination of physiological and psychological effects of long-duration confinement
- Determination of the role of humans in performing in-flight experiments
- Determination of material balance, thermal balance, and power requirements

The tests were conducted in a double-walled, horizontal, cylindrical chamber with altitude capability which prevented any inward leakage to the 68.7 kPa chamber atmosphere. There were glass windows between the chamber interior and the test control area outside the chamber. During the 30- and 60-day tests, the equipment was on the left of the chamber with the galley and sleeping quarters on the right. This was found to be not a good arrangement. Thus, for the 90-day test, the equipment was rearranged from an acoustics, living viewpoint. Equipment and living area were separated by an acoustic bulkhead made of fiberglass.

The 30- and 60-day tests were conducted at Santa Monica. The four crewmen for the tests were college student volunteers. During the 60-day test, two of them drank distilled water, while the other two drank reclaimed water. It was the first time for humans to consume their own recovered waste water. Two systems were used for the reclamation of water from urine - a Vacuum Distillation Vapor Filtration (VDVF) system that was heated by radioisotopes, and a wick air evaporation system as backup. Also, a waste water electrolysis unit was onboard. The carbon dioxide from the atmosphere was collected with a solid amine system, with a molecular sieve as backup, and the oxygen was recovered with a Sabatier unit. A toxins burner with a catalyst controlled the CO and other atmospheric contaminants. Two different types of food were used during the tests - frozen and freeze-dried. The basic energy intake was 12000 kJ per day. The waste management system included commode, urinal, canner and baler, and sink. An overboard dump tank was used for reagents used in onboard tests and similar wastes. The baler was to pack and compress dry trash.

For the 90-day test the chamber was moved to Huntington Beach to have it closer to the shops and analysis labs. The spares and backup systems got the crew through the 90 days. Spares could not keep all systems running at all times, but but the test could be continued by using the backup systems.

An interesting aspect of the tests in the the space station simulator chamber was the fact that the crew test volunteers were not locked in the chamber. If they had left, the test would have been compromised - the test conductors had to entice them to stay in.

The CELSS Breadboard Project

In 1986, the Breadboard Project, NASA's higher plant-based CELSS program, was begun at Kennedy Space Center. The Breadboard Project has as its goal the demonstration of the scaling-up from previous laboratory-sized research studies into the production of food for human life support, water recycling, and atmospheric gas control in its Biomass Production Chamber (BPC). The BPC is a renovated cylindrical steel hyperbaric facility, approximately 3.5 m in diameter and 7.5 m high, and with a total internal volume of 113 m^3 (see figure IV.29).

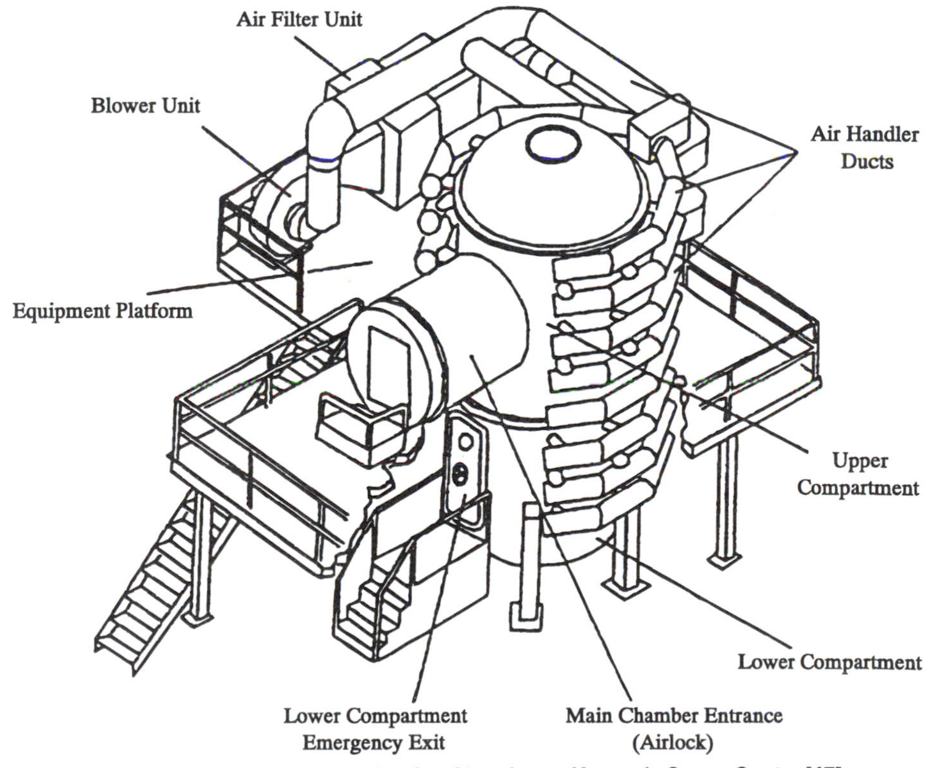

Figure IV.29: **Biomass Production Chamber at Kennedy Space Center [47]**

Originally used in the Mercury program, it has been modified for plant growth by the creation of two floors with eight plant racks and the installation of High Pressure Sodium Lamps or Metal Halide (MH). The configuration of growing area yields a total plant area of 20 m². Ventilation of the chamber is accomplished by ducts which lead to an external air-handling system including filters. Air circulation is provided by two 30-kW blowers, providing an air flow near 400 m³/min with velocities at the plant level of about 0.5 to 1 m/s. Temperature and humidity are controlled by a chilled water system through atomized water injection. A compressed gas delivery system is used in the manipulation of atmospheric carbon dioxide and oxygen. The leak rate of the BPC is about 5 % of its volume per day. Air turnover is about three times a minute, with ventilation air being ducted at a rate of 0.5 m³/s into the chamber between lights and growing trays. The BPC is divided into two compartments (upper and lower levels). Each compartment contains two plant growing levels with a growth area of about 5 m² per level. Lighting is provided by 96 400-W HPS or MH lamps. With HPS lamps, photosynthetic photon flux (PPF) levels average about 650 µmol/m²/s, and with the MH lamps about 500 µmol/m²/s at 0.6 m below the barrier. A cross-section of the BPC is given in figure IV.30. [27, 32, 47]

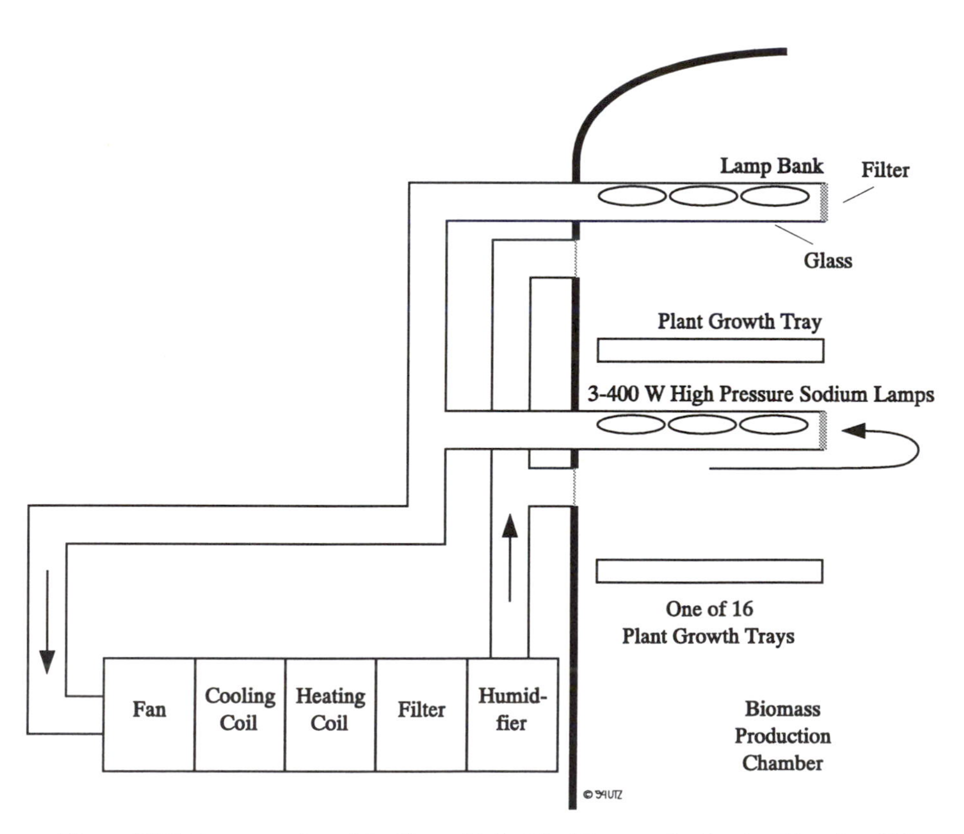

Figure IV.30: **Cross-section of the Upper Half of the Biomass Production Chamber [47]**

Crop	Date	Average PPF	Photoperiod	Daily PPF	Length of Study
		[μmol/(m²·s)]	(hrs)	[μmol/(m²·s)]	(days)
Wheat	5/88	666	24	57.7	68-86
	1/89	535	20	38.5	86
	5/89	691	20	49.7	85
Soybean	11/89	815	12	35.2	90
	5/90	477	12	20.6	97
	11/90	644	10	23.2	97
Lettuce	3/90	290	16	16.7	28
	9/90	280	16	16.1	28
	9/91	293	16	16.9	28

Table IV.42: **Irradiance Levels for Crop Tests in NASA's Biomass Production Chamber [47]**

Crop	Date	Edible Yield		Total Biomass	
		[kg / m²]	[g/(m²·day)]	[kg / m²]	[g/(m²·day)]
Wheat	5/88	1.16	15.0	2.88	37.4
	1/89	0.67	8.0	2.36	27.4
	5/89	0.82	9.6	2.76	32.5
Soybean	11/89	0.54	6.0	1.66	18.5
	5/90	0.4	4.1	1.18	12.2
	11/90	0.49	5.0	1.3	13.4
Lettuce	3/90	0.16	5.7	0.17	6.0
	9/90	0.16	5.8	0.18	6.3
	9/91	0.2	7.2	0.22	7.9

Table IV.43: **Yields (Dry Matter) of Crops Grown in the Biomass Production Chamber [47]**

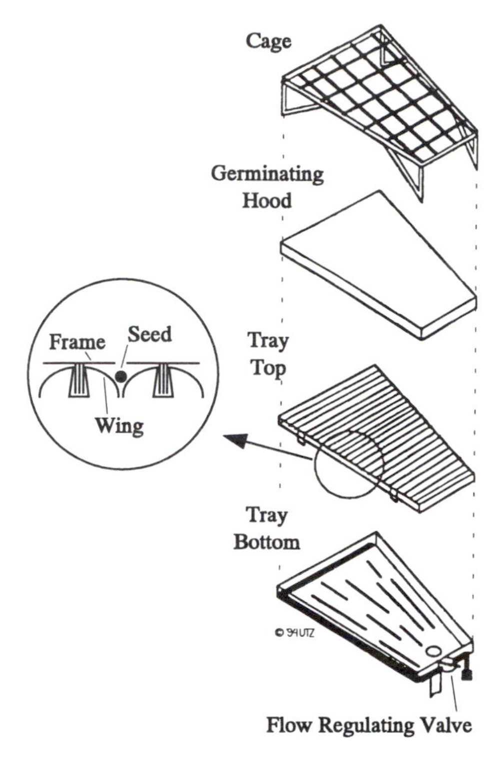

Figure IV.31: **Thin Film Tray System for Wheat Growth [32]**

The initial crop tested was wheat, grown in nutrient film, with plant supports holding the canopy about 50 mm above the nutrient level (see figure IV.31). In the following, studies of soybean, potato, and multiple crops in continuous production are planned. The first phase of Breadboard, scheduled through 1993, calls for integrating and demonstrating three major components of a CELSS: biomass production, biomass processing, and waste conversion. Some results of tests conducted from 1988 to 1991 are given in tables IV.42 and IV.43.

The Biosphere 2 experiment in Arizona is similar to BIOS-3 in that it is a sealed system in which, in that case eight, crew members work to maintain ecological balances and, thus, their life support. It greatly differs from BIOS-3 in that it is much larger, depends exclusively on sunlight to drive photosynthesis, although external power is used for air conditioning, and relies on more biological feedback for basic recycling. Biosphere 2 is described in great detail in chapter VII. Also, the BioHome, developed at Stinnes Space Center, is discussed later in chapter IX.

The Human-Rated Test Facility

The Closed Ecological Life Support System Human-Rated Test Facility (HRTF) is currently being developed at the NASA Johnson Space Center in Houston, TX. The HRTF is a human-rated, ground-based CELSS facility where full-scale, integrated, long-duration testing of the CELSS concept can be conducted. The overarching objective of the HRTF is to aquire the information and operational experience necessary to define performance and design requirements for eventual flight systems. Extended-duration, full scale, integrated testing performed in this facility under closed, controlled conditions will meet this objective. Especially, this facility will provide NASA with the capability of integrating advanced, regenerative physico-chemical life support subsystems with biological systems on a scale large enough to effectively demonstrate staged evolution of planetary base life support systems.

The ability to accomodate testing pertinent to disciplines other than advanced life support is considered an additional benefit of the HRTF. For example, human factors are also planned to play a significant role with respect to maintaining test crew productivity and ensuring their overall comfort. Potentially, medical disciplines like human physiology, psychology and sociology, will also benefit from cooperative efforts in support of HRTF human testing.

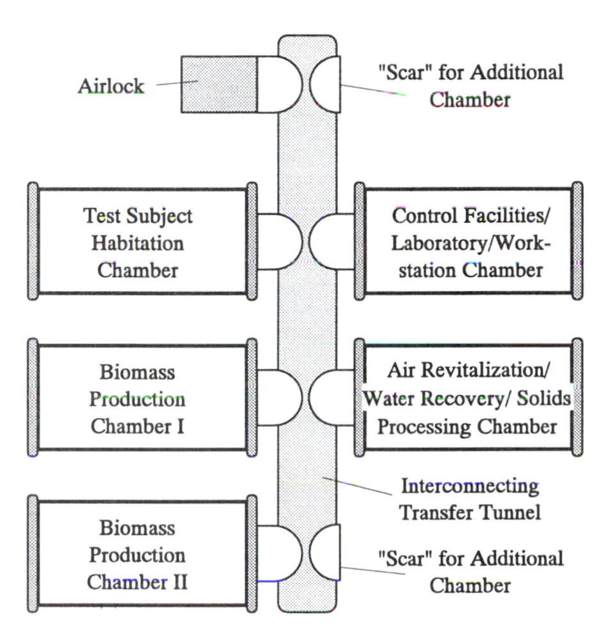

Figure IV.32: **HRTF Conceptual Reference Configuration**

The HRTF multichamber reference facility is currently (1995) under construction in Building 241 at the Johnson Space Center. A schematic of this configuration with the five functionally distinct chambers, the interconnecting transfer tunnel and the airlock is given in figure IV.32. Also shown are "scars" which permit the growth of an additional two chambers to the reference facility. The five chambers, each measuring 4.6 m in diameter by 11.3 m in length, have identical basic interior structures consisting of two decks, support beams and stairway and ladder accesses.

The HRTF project will be comprised of three distinct segments which follow parallel design, developement, integration and testing paths:

- Multichamber Facility
  It will provide the basic structure in which to conduct closed, controlled life support testing on a large scale. Chamber internal utilities will provide distribution of essential utilities, like power, data and fluid distribution,from chamber systems to each other and to and from chamber external utilities, like control and data acquisition systems. Located externally to the HRTF chambers will be facilities which provide support to the operation and testing conducted in the chambers. These support facilities will include analytical laboratories and a medical support facility.

- Life Support Systems
  The life support provisions will accomodate testing of biological and physico-chemical life support systems in the areas of air revitalization, water recovery, food production, solid waste processing and thermal management.

- Crew Accomodations
  The crew accomodations, including the personal, work, galley and hygiene areas, will provide a comfortable, productive environment for the test crews while occupying the HRTF during the planned long-duration missions. Additionally, there will be essential interfaces between the crew and the life support systems, most notably in the hygiene area.

To provide a basis from which a set of system requirements can be derived, a reference scenario has been established for the HRTF. The HRTF reference mission timeline (see figure IV.32) provides a simple set of facility and operating guidelines, including mission duration, crew size, crew changeout schedule and anticipated off-nominal operating conditions. Testing will be comprised of two distinct phases:

- Physico-Chemical Systems Phase - during which no biological system will be activated.
- Integrated Biological and Physico-Chemical Systems Phase - during which biological systems are brought on-line and become the predominant life support provider.

Test phase durations of sixty days and one year, respectively, will be performed, with periodic crew changeouts. The HRTF systems will be designed to support smooth transitioning between these two test phases. Additionally, to investigate the full range of CELSS performance characteristics, off-nominal conditions will be imposed on the test scenario, e.g., reduction of crew metabolic loading during extended duration "extravehicular activities". [27]

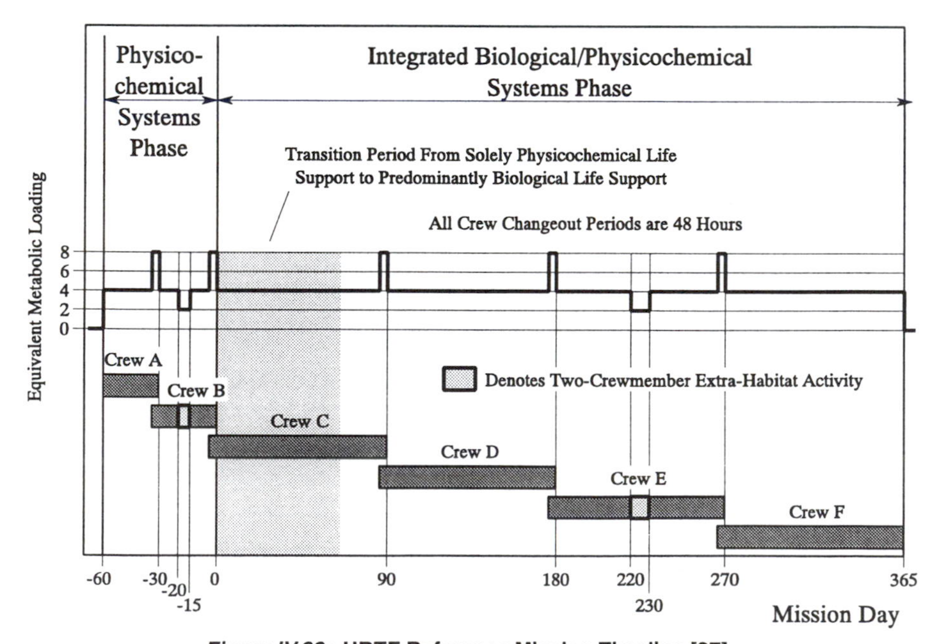

Figure IV.33: **HRTF Reference Mission Timeline [27]**

Japanese and European CELSS Programs

Japanese and European CELSS programs, although smaller, are also underway. They concentrate on gas recycling systems involving oxygen and carbon dioxide separation and concentration, water recycling systems, plant and algae cultivation techniques, as well as animal and fish physiology and breeding. European issues have also included much work on microgravity issues of biological development, essential to the successful translation of ground-based CELSS to space. In Japan, a Closed Ecology Experiment Facility (CEEF), named "Biosphere J", is currently under discussion. Design and planning began in 1990, and the completion is scheduled for 1998. The CEEF is planned to cover 6500 m^2 with an overall volume of 5200 m^3, and is designed to support two people. It will consist of linked modules for agricultural crop cultivation, a human habitat, domestic animals, a land ecosystem, an ocean ecology, and further technology and support modules. The modules are designed to be separated to permit control of carbon dioxide and oxygen. For the agricultural crop module a combination of sunlight and artificial light is being planned. [52]

IV.6 SPACE SUITS AND EXTRA-VEHICULAR ACTIVITIES (EVA)

IV.6.1 Working Outside a Spacecraft or Space Habitat

An Extra-Vehicular Activity (EVA) is defined as any space operation or activity performed outside the protective environment of a spacecraft or space habitat, therefore requiring supplemental or independent life support equipment for an astronaut.

On March 18, 1965, the first EVA egress was performed by Alexej Leonov. He spent ten minutes outside his Voskhod spacecraft. Since then, a considerable amount of EVA experience has been gained. The EVAs, covering research studies and various repair and assembly operations, have proven the advisability and even necessity of using man's capabilities to support activities in free space. A human being can do almost everything during a spacewalk, in free space or on a planetary surface, that can be done on the Earth's surface. However, working in lower or zero gravity in a space suit necessitates certain design compromises to facilitate the productivity of the space walker. Two factors that always have to be considered when designing an EVA are:

- A space suit is stiff, restricts visibility and movement, is fatiguing to work in, and all work is done with gloves that significantly reduce tactility and dexterity.

- Under zero gravity conditions all astronauts and tools have to be tethered to preclude them from drifting away or getting lost.

Overall, EVAs are very demanding activities and there are many aspects to be considered in the design of hardware, tools, and procedures to be used on an EVA mission. EVA equipment and tasks should be designed from the beginning to be EVA compatible, i.e., maximize the probability of success, minimize the expense and prevent the need for inefficient, time-consuming operational workarounds due to bad design concepts. There are three categories of EVA:

- Scheduled EVA — Planned before launch and included in the nominal mission timeline.
- Unscheduled EVA — Not included in the nominal mission timeline but maybe required to achieve payload or mission success.
- Contingency EVA — Required to effect the safe return of crew (and spacecraft).

Unscheduled EVAs were required, e.g., during some Skylab, Salyut, and Space Shuttle missions. About 40 % of the EVAs performed by the Russians were unscheduled and needed for repair works to restore space station performance and provide mission safety. No contingency EVA was ever required by any manned space mission. [23, 35]

The essential prerequisite for EVA effectiveness and safety is proper crew training. The training program on the ground comprises the familiarization with the space suit and its systems, altitude training in vacuum chambers to acquire skills in the space suit and airlock system control, parabolic flights using flight mockups while performing some tasks, and the use of specific simulators such as those designed to acquire skills in Manned Maneuvering Unit (MMU) control. Most importantly, the training in neutral buoyancy facilities, i.e., water tanks, provides excellent simulation capability for EVAs in zero gravity. Astronauts with EVA experience recommend their use early in the design process with mock-ups of the flight hardware to simulate the tasks to be performed. Functional tests of EVA hardware and tools in thermal / vacuum chambers have been proven to be mandatory - many design changes resulted from these tests. Simulations of EVAs under lower gravity conditions, e.g., on the lunar or Martian surface, can also be conducted on parabolic flights, just as the thermal and vacuum environment of a planetary surface can also be simulated in vacuum chambers. [4, 33]

IV.6.2 Extravehicular Mobility Units (EMU) - Requirements, Design, and Operations

An Extravehicular Mobility Unit (EMU) is a closed and isolated environment that protects the astronaut from exposure to the space environment. It consist of a space suit assembly and a portable life support system (PLSS). The space suit has to retain the oxygen pressure required for breathing and ventilation, and provide adequate dexterity, mobility, and visibility for performing tasks in orbit. In U.S. space suits, for example, a suit bladder is protected from external puncture by a bladder restraint fabric and meteoroid protection garment. The bladder seams and flanges are reinforced with a second layer of bladder material for added strength. The probability of a micrometeoroid penetrating the EMU during a single 2-man 6 hour EVA has been calculated to be 0.0006. For radiation protection, the current Space Shuttle EMU provides an approximate aluminum equivalent shielding of 0.5 g/cm^2. For EVAs in a polar or geostationary orbit about 1.62 g/cm^2 would be required due to the different radiation conditions (see section III.1). [5, 46]

The PLSS has to provide the crew member with the necessary life support functions, in particular oxygen supply and purification, temperature and humidity control, metabolic and environmental heat removal, and an emergency and warning system to detect hazardous situations and to inform the crew member about the situation. In general, the oxygen is supplied from tanks, the ventilation circuit is driven by a fan, the carbon dioxide and odors are removed by means of LiOH and an activated charcoal filter. Cooling is provided by a condensing heat exchanger. The power supply is based on batteries. [39]

The most vulnerable part of the space suit assembly are the gloves. Caution must be exercised not to accidentally puncture or damage a glove during an EVA . In general, the glove is the most important part of the suit because of the fundamental importance for dexterity and tactility performances, and is thus custom sized. It includes three gloves worn each on top of each other. The first is the glove bladder for gas retention. The second is the glove restraint for pressure load retention. The third glove is the thermal protective one, based on a multi-layer insulation, felt and rubber layers concepts. Astronauts with EVA experience emphasize that EVA tasks should be designed so that they require only one hand. This is due to considerations of working in the space suit and the frequent requirement to stabilize oneself with the other hand. As a rule of thumb it is also recommended that the maximum force to perform a task should be < 110 N and the task should also be designed to use larger muscle groups whenever possible to reduce crew fatigue. [33, 39]

For a modern space suit visor, the angle of view has to be at least 120° in the vertical plane (55° upward view and 65° downward view), and at least 200° in the horizontal plane (100° both left and right view). As an example, the Space Shuttle EMU is depicted in figure IV.34. [4]

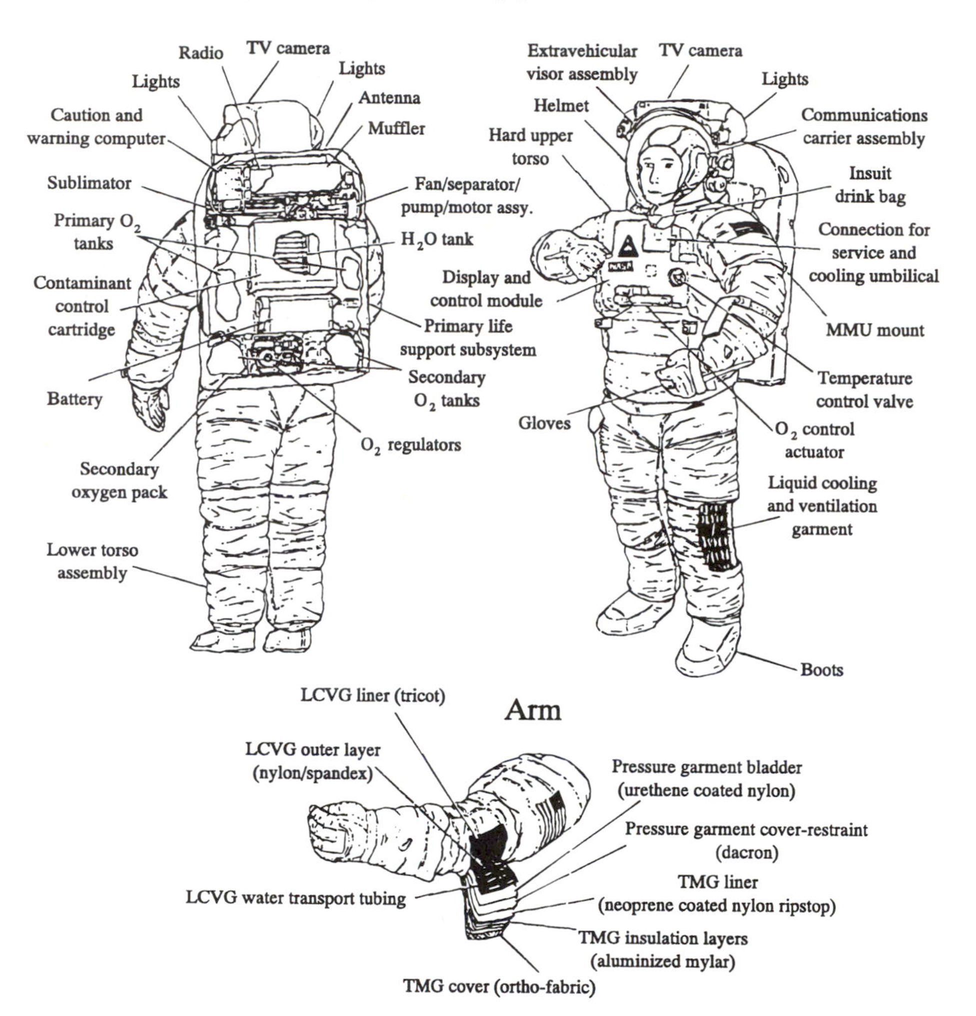

Figure IV.34: **The Space Shuttle Extra-Vehicular Mobility Unit [29]**

The most important EVA suit design issues include flexibility, ability to use existing suits, and operational impact due to prebreathing. Use of lower pressure suits, e.g., 29.6 kPa, has two advantages - current availability and flexibility. However, the use of the low pressure suit will cause the need for extensive prebreathing with higher cabin pressures. Use of high pressure suits has one distinctive advantage: no prebreathing, which is both a safety and operational benefit. (see section IV.6.3) [4]

In general, the operating pressure in a space suit with an oxygen atmosphere (nitrogen content < 5 %) is 30-40 kPa. The optimal selection of space habitat pressure is a function of a safe cabin environment, achieved by avoiding O_2 toxicity and hypoxia, and by numerous safety, mission and vehicle design, science, and EVA suit considerations. The sea level equivalent (SLE) of O_2 and pressure is based on maintaining the same amount of O_2 in the blood at lower pressure as exists at sea level of 101.3 kPa. As the total pressure is decreased in the atmosphere, the percentage of O_2 must be increased in order to maintain the sea level equivalent in the blood. SLE cabin atmosphere can be established in the range of 25.7 kPa and 100 % O_2 to 101.3 kPa and 21 % O_2. At varying SLE cabin atmospheres, evaluation of candidate cabin environments can then be made. (see figure IV.35) Historically, both the U.S. and the Russian space programs have been operated at various combinations of cabin and suit pressures, as shown in table IV.44. Moreover, the Russian suit provides the possibility of switching over to a lower pressure (27.6 kPa) operating mode for 15 minutes which facilitates improved mobility. [3, 26]

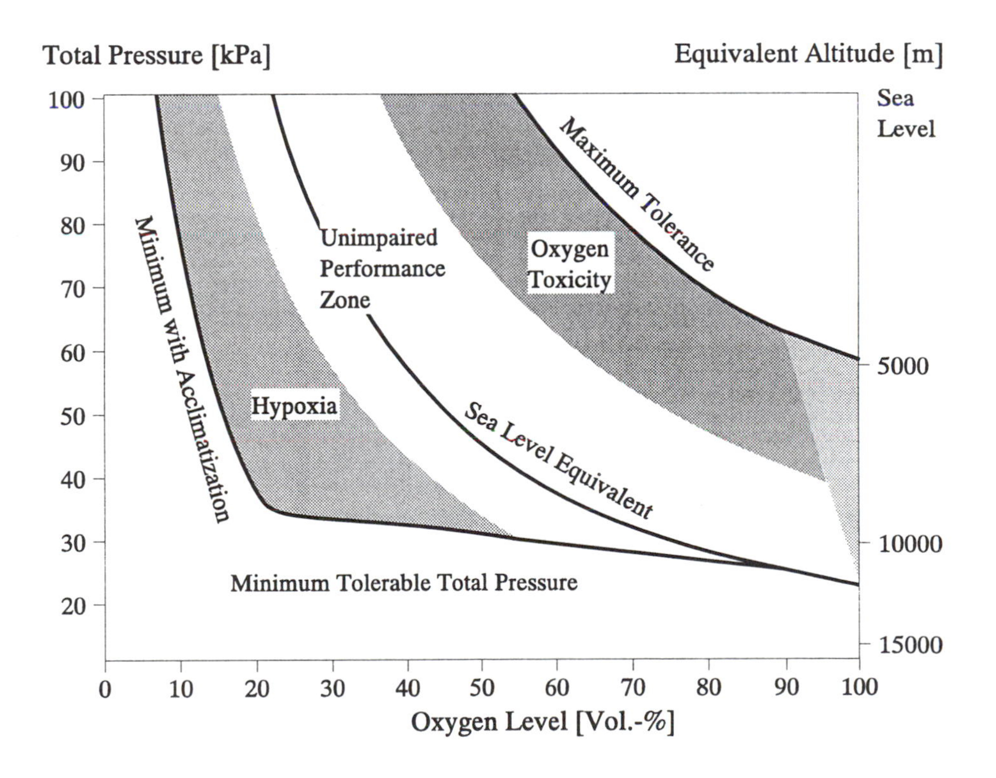

Figure IV.35: **Sea Level Equivalent Cabin Environments [26]**

Space Program	Cabin Pressure [kPa]	Suit Pressure [kPa]	Oxygen Level [%]	
			Cabin	Suit
Mercury	34.5	24.0	100	100
Gemini	34.5	24.0	100	100
Apollo	34.5	24.0	100	100
Skylab	34.5	24.0	70	100
Space Shuttle	101.3 70.0	21 28.5	29.6 29.6	100 100
Russian Programs	101.3	40.6 (27.6)	22	100

Table IV.44: **Historical Space Suit and Cabin Pressure Data [26]**

The CO_2 partial pressure in a space suit does not exceed 2.0 kPa if the absorbing cartridge works properly, thus also meeting physiological requirements. If the CO_2 partial pressure exceeds 2 kPa (Hypercarbia) this may cause annoying symptoms of increased respiratory rate, headaches, cognitive impairment, and, finally, loss of consciousness. [4]

The space suit microclimate indices also include the thermal mode. A sublimative heat exchanger as well as a method of heat removal with a water-cooled garment is used in a suit thermal control system. Such a method of heat removal provides a nominal thermal mode for sustained operation at practically any workload. Some of the primary criteria for the selection of a portable life support system are the requirements of operation (metabolic rate, duration of duty cycle), and minimum weight, volume, and power consumption. [4, 35]

Crew member performance capability is essentially influenced by the space suit mechanical resistance to man's motion activity. There is a clear dependence between positive pressures and forces in a soft space suit enclosure. The value of the space suit absolute pressure should be a compromise between the risk of high altitude decompression sickness and the necessity to maintain the required level of astronaut mobility and working capability. In this respect, the space suit joints corresponding to the main human body joints in terms of their location must be mentioned. There are two major problem areas: technological and biomechanical. The fact that the major human joints provide motions about two or more axes drives the design to provide a decrease in resistance during enclosure bending. The main drawback of those soft joints is the difficulty for the suit to maintain its constant volume. Therefore, additional resistance to the enclosure bending has to be introduced in the joints, leading to a decrease in crew member productivity. To prevent this, attempts have been made to develop rigid space suits. Currently, hybrid semi-rigid suits are used. In the area of shoulder and wrist joints, the suits use both convolutes and pressure bearings possessing a low moment of resistance. The essential advantages of semi-rigid space suits are:

- Minimum overall dimensions of a suit torso in the pressurized state
- Easy and rapid donning/doffing (suit entry).
- Easy handling and improved reliability since the lines connecting the life support system with the suit are arranged inside
- Adequate hatch pressure tightness
- Possibility of using one single standard size suit by crew members with different anthropometric dimensions
- Easy replacement of parts and reparability by arrangement of main life support system components inside the suit in locations easy to access.

Under zero gravity, the average work rates of astronauts are 460 +/- 40 W. If the necessity arises to transport additional mass (up to 150 kg), the work rate increases up to 600 +/- 50 W. The use of a MMU can decrease the average work rates during an EVA to about 170 W. According to measurements during Russian EVAs, when semi-rigid suits are used the hourly average metabolic rate equals 290 W with short-term maximum values of 910 W. Overall, EMUs are to meet numerous, sometimes conflicting requirements. These are summarized in table IV.45. [4, 35]

Since EVAs are performed in extreme environments, they can be classified as highly dangerous operations. That is why the main scope is to make these activities as effective and safe as possible and why EMUs are designed to be fail-safe. This means that safe return to the spacecraft or habitat must be provided following any single failure. Central to the safe return to the spacecraft or habitat is the detection and alarm in case of loss of a safety critical life support function. Failure detection is performed by a caution/warning system and by the crew member. The caution/warning system monitors the oxygen tank pressure, CO_2 level, ventilation flow, and battery voltage and current. [35, 46]

In case of a system failure or an emergency, the EVA crew members return to the airlock or proceed with the sortie under a more careful monitoring and control of the suit status. The airlock is an important element to provide EVA effectiveness and safety. Its dimensions, design and equipment should ensure the performance of all the necessary operations for the spacecraft egress/ingress within the shortest possible time and with minimized workload. The airlock should be equipped with the means to prevent the depletion of the life support consumables during the pre-EVA space suit checks, prebreathing, and airlocking. It is clear from the overview in table IV.46 that the main emphasis in the design phase has to be on protection of the crew members against space suit decompression, in particular, decompression of arms, legs, and gloves, and most critical elements of the space suit. [35]

Space Suit Requirements	Comments
Pressurized volume and respirable atmosphere	Provision of oxygen, carbon dioxide removal, and pressure control
Thermal control and external thermal insulation	—
Physical protection from external objects	Micrometeoroids, sharp edges
Communication	With spacecraft and ground control
Radiation shielding	No shielding from Solar Flares and GCR possible
Proper bioengineering characteristics of the space suit	For crew members of different anthropometry (95 percentile)
Proper physiologic and hygiene conditions	Peak metabolic rate of up to 700 W Management of urine and feces
High reliability	—
Fire safety	In 100 % oxygen atmosphere
Durable lifetime	3- 5 years onboard a spacecraft
Reusability	25-50 sorties with some spares onboard the spacecraft
Universality	Possibility to provide EVAs with different spacecraft Usability in combination with MMU
Minimum weight of consumables	Feed water for coolant Absorbent for carbon dioxide removal
Simplicity in operation and maintainability	Repair onboard the spacecraft possible
Minimum weight and overall dimensions	—
Minimum costs	—

Table IV.45: **Summary of Space Suit Requirements [35]**

Emergency	System Actuation and Crew Action
Loss of Pressure - Space Suit (Q = 0.2-2.0 kg/h) (Q > 2.0 kg/h) - Gloves (Q > 3.0 kg/h) - Outer Bladder (Q > 2.0 kg/h)	- Actuation of injector; proceed with activities according to oxygen reserve - Activation of oxygen emergency supply; return to airlock within 25 minutes - Activation of automated cuff; return to airlock within 15 minutes - Redundant bladder becomes operational automatically; proceed with activities according to oxygen reserve
Primary Oxygen System Failure	- Activation of oxygen reserve; proceed with activities according to oxygen res.
Primary Fan Failure	- Actuation of redundant fan; proceed with activities according to oxygen reserve
Power Supply Malfunction	- Actuation of injector; return to airlock
Malfunction of Carbon Dioxide Absorption Cartridge	- Actuation of injector; return to airlock
Heat exchange Malfunction	- Actuation of injector; return to airlock
Pump Failure	- Actuation of two fans; proceed with activities according to oxygen reserve

Table IV.46: Design Measures to Provide Safety During EVA Activities [26]

IV.6.3 Cabin Pressure vs. Space Suit Pressure

In the design of space vehicles and space suits, an important constraint is the fact that optimization of the cabin pressure is contradictory to the optimization of the space suit pressure:

- It is operationally undesirable to have a cabin pressure of less than one atmosphere (101.3 kPa) as the increasing oxygen concentration results in, for example, an increase of the flammability potential. Also, it may be desirable for the conduction of certain experiments to have one atmosphere pressure provided.

- On the other hand, the space suit pressure should be as low as possible in order to provide high mobility and a maximum dexterity and tactility of the glove.

Resulting from the combination of spacecraft cabin pressure and space suit pressure, the crew member must prebreathe oxygen. Oxygen prebreathe or denitrogenation is conducted by astronauts prior to each EVA to prevent the potential occurrence of incapacitating decompression sickness (DCS) and bends. Prebreathing reduces the nitrogen content in the astronaut's body which prevents the formation of nitrogen bubbles in body tissues, leading to the so-called "bends" when the atmospheric pressure is reduced. However, it is operationally undesirable to prebreathe oxygen due to wasted crew time and therefore reduced crew flexibility. Prebreathing time should thus be kept at a minimum. [18, 24]

The duration of prebreathing primarily depends on the ambient atmosphere composition (cabin atmosphere) and initial and final pressure levels. The required time spent breathing pure oxygen is a function of the ratio of initial and final suit pressure. A safety measure is the R-factor, or bends ratio, an analytical ratio of tissue nitrogen pressure prior to decompression and the final suit pressure:

$$R = \frac{P_{N_{2i}}}{P_{ss}}$$

$P_{N_{2i}}$: Initial absorbed tissue N_2 pressure, i.e., cabin atmosphere N_2 partial pressure
P_{ss}: Final ambient pressure, i.e., space suit pressure

The lower the R-factor, the lower the risk for decompression problems. In the U.S., Russia, and Europe different approaches and R-factors are used or planned:

U.S.: R = 1.4; 2-4 hours prebreathe at a suit pressure of 30 kPa
Russia: R = 1.7; 0.5 hours prebreathe at a suit pressure of 40 kPa
Europe: R = 1.2; 2.5 hours prebreathe at a suit pressure of 50 kPa [18]

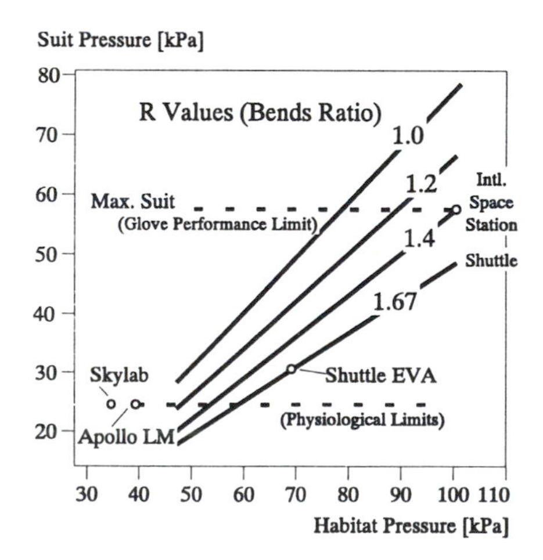

Figure IV.36: **Cabin Pressure vs. Suit Pressure as Influenced by Bends Ratio [20]**

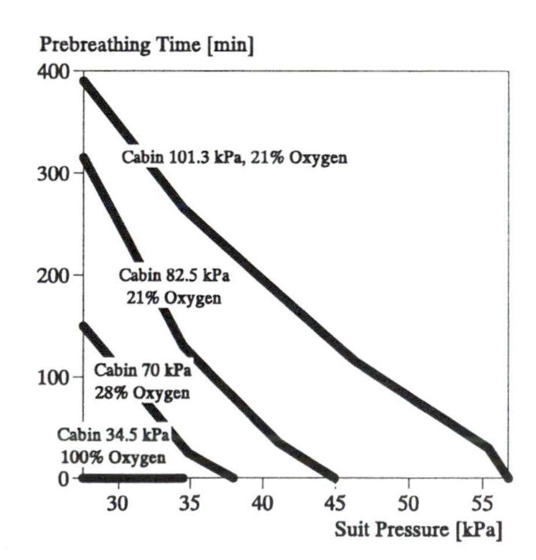

Figure IV.37: **Prebreathing Time as a Function of Cabin and Suit Pressures (R = 1.4) [26]**

Figure IV.36 shows several cabin and suit pressures plotted against various R-Factor curves that express the range of operational combinations that are potentially available. Prebreathing time as a function of cabin pressure and EVA suit pressure is graphically portrayed in figure IV.37.

At high cabin pressure, e.g., 101.3 kPa, there are serious issues related to the specification of the space suit pressure and the need for prebreathing. High pressure suits will be costly but use of the existing, flexible suits requires up to 6 hours of prebreathing. A relatively high operating pressure (40 kPa) of the space suit practically excludes the probability of DCS or bends and does not require durable prebreathing before egress. A cabin pressure of 70 kPa prior to EVAs could also be used in order to adopt, e.g., the existing space suits with only 1 hour of prebreathing. If a 34.5 kPa, 70-100 % O_2 atmosphere is adopted the existing U.S. suit can be used with no prebreathing required. [26, 35]

Onboard the Space Shuttle, prebreathing for scheduled and unscheduled EVAs is normally accomplished from an orbiter cabin pressure which has been reduced to 70.3 kPa. The final prebreathe is an unbroken prebreathe performed in the suit with the time of prebreathe necessary determined by the time at 70.3 kPa. The prebreathing factors in the U.S. and Russia are 1.4-1.6 and 1.5-2.0, respectively. A typical protocol for the prebreathing procedure onboard the Shuttle is given in figure IV.38. An alternate prebreathe protocol which may be used is the in-suit option with a minimum of 4 hours of unbroken prebreathe with > 95 % oxygen at a cabin pressure above 86.2 kPa prior to any cabin or airlock depressurization below 55.2 kPa. [24]

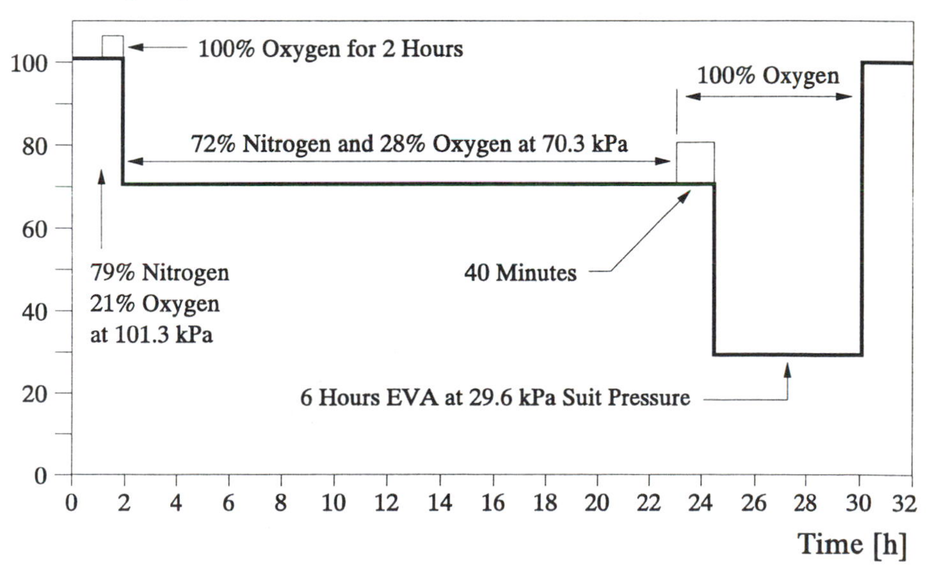

Figure IV.38: Prebreathing Protocol for EVAs from the Space Shuttle Orbiter (Staged Compression) [24]

The Russian EVA suit operates from 27-41 kPa. However, the cosmonauts prefer to work at lower pressures because the higher pressure reduces mobility, increases the hazards due to leaks, and increases physical discomfort during and following EVA tasks. The atmosphere in the Russian suits consist of oxygen. The nitrogen content before prebreathing is not higher than 1-2 %, and the CO_2 content is close to zero. The pressure profile for the Russian prebreathing procedure is given in figure IV.39.[3, 26]

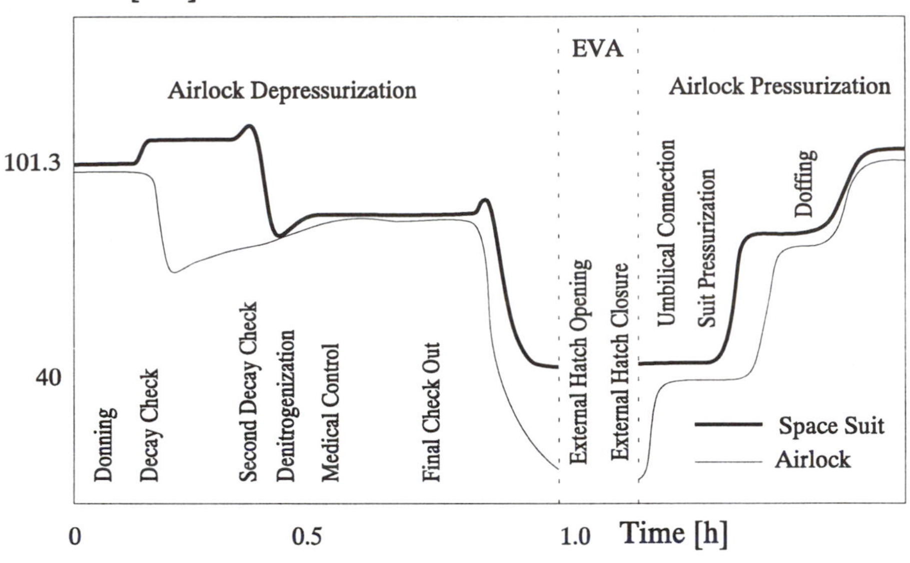

Figure IV.39: Prebreathing Protocol for Russian EVAs [3]

The original European EVA reference concept was based on a zero prebreathe design with an R-factor of 1.2, which represents a conservative, low risk approach, assuming a Hermes cabin pressure level of 70 kPa. Hermes was the planned but now cancelled European manned spacecraft. The R-factors for nominal and emergency modes were later set to 1.22 and 1.6, respectively, resulting in a prebreathing time of 150 minutes in the nominal mode. [39]

IV.6.4 Past and Present Space Suits and EVA Experiences

EVAs have been conducted during the Gemini, Apollo, Skylab, and Space Shuttle programs of the U.S., and the Russian Voskhod, Soyuz, Salyut, and Mir programs. The overview in tables IV.47 and IV.48 shows all EVA mission hours accumulated during all space missions until 1991. Except for the Gemini and Voskhod programs and the EVAs performed by the command module pilots during the Apollo 15 to 17 missions, all EVAs have been conducted by two-crew member teams. It is significant that all Gemini, Voskhod, Soyuz, and Apollo EVAs were open-hatch operations, which means the spacecraft cabins were depressurized and all inside crew members were exposed to vacuum conditions. All non-EVA intravehicular crew members wore protective pressure suits with necessary life support and communication functions provided by spacecraft systems. The major objective of the Voskhod, Gemini, and Salyut EVAs was to demonstrate EVA feasibility and develop and evaluate EVA system capabilities. [23]

	Gemini	Apollo	Skylab	Space Shuttle	Total
Number of EVAs	9	19	10	15	53
Total EVA Mission Duration [h]	12:21	84:05	41:55	78:34	216:55
Total EVA Crew-Time [h]	12:21	160:53	83:15	158:10	414:39
EVA of Max. Length [h]	2:29	7:37	7:03	7:08	-

Table IV.47: **Summary of U.S. EVAs (until 1991) [29, 35]**

	Voskhod	Soyuz	Salyut 6	Salyut 7	Mir	Total
Number of EVAs	1	1	3	13	25	43
Total EVA Mission Duration [h]	0:10	0:37	4:46	48:28	113:04	167:05
Total EVA Crew-Time [h]	0:10	1:14	9:32	96:56	225:42	333:34
EVA of Max. Length [h]	0:10	0:37	2:05	5:00	7:00	-

Table IV.48: **Summary of Russian EVAs (until 1991) [29, 35]**

The Voskhod and Soyuz EVA suits, as well as all American suits which were used prior to the Shuttle program, were of the "soft" design. Their enclosure was made as an overall with a removable helmet. The life support system was located in a removable backpack, fixed on the back or legs. During the Salyut and Mir space station programs, the EVA space suits were of the semi-rigid type with high reliability for multiple free-space EVAs. In Russia, these semi-rigid suits were derived from design concepts for the planned Russian lunar EVAs. The weights of space suits and PLSS have tended to increase significantly from program to program in the U.S. as shown in table IV.49. [35]

	Space Suit Mass [kg]	PLSS Mass [kg]	Total EMU Mass [kg]
Apollo	27.2	61.2	88.4
Space Shuttle	49.9	72.6	122.5
Mir	-	-	70.0
Alpha (US Suit)	~ 90	~ 195	~ 285

Table IV.49: **Masses of Space Suits and PLSS [2, 20]**

The Manned Maneuvering Unit (MMU) of the U.S. was first used during STS-41B in February 1984. First successful tests of the Russian MMU were performed in February 1990 in the vicinity of the Mir space station. During the Apollo program, the astronauts also accomplished various tasks on the lunar surface. Specifically they demonstrated the feasibility of working in a reduced gravity environment. A total of 160 EVA hours were spent on the lunar surface. [6, 35]

IV.6.5 Trends for Future Extravehicular Mobility Units

In the future, there will be a multiplicity of space habitat types at space destinations corresponding to various space environments: on-orbit, on the lunar surface, and on the Martian surface. These habitats will have a variety of functions. Some will be living habitats and might properly be at normal pressure and O_2 conditions, as found here on Earth. Some might effectively be kept at reduced pressures because they are utility habitats or habitats for conducting space experiments and growing food. Overall, each of these habitats will have its own astronaut visitations and work duty cycles, and each would tolerate a variety of pressure / O_2 ratio / suit combinations. The establishment of a lower habitat pressure that would still maintain a zero-prebreathe condition for EVAs, e.g., with R =1.2, allowing for lower EMU operating pressures could significantly reduce the mass of EMUs.

Typical EVA support tasks of the future will include the assembly, operations support, and maintenance for the International Space Station Alpha. Current plans call for using an enhanced capability Shuttle EMU to support the station. The EMU is being tested to determine its capability of performing 25 mission EVAs, lasting 6 h each. [23]

Further EVA system development objectives identified for manned space missions beyond the space station, like a manned lunar base or a flight to Mars, include light weight materials, minimum prebreathe time, mobility improvement for crew member locomotion in reduced gravity conditions, and regenerable individual life support subsystems to reduce expendable transport and logistics penalties. It is important to note that on the long run the use of open-loop systems will be prohibitively expensive. For an 8 hour EVA, one person requires 1.22 kg LiOH, 0.63 kg O_2, and 3.5-5.4 kg water for cooling. On a six months mission with five two-member crews, 8 hour EVAs per week, this would require more than one ton of water alone. [20, 26, 28]

An advantage of a low-gravity field, i.e., a field with less than 1g but not too close to 0g, is that it will give crew members a sense of up and down. On the other hand, when they lose their balance they will fall into the planetary dust. Of prime interest for the development of future planetary space suits will therefore be the incorporation of dust-proof protective measures. At a planetary EVA worksite, any movement by the crew member or work activity involving an interaction with the planetary surface will raise dust. Because of the lower gravity environment on Moon and Mars, the dust will propagate and cover the lower extremities of the space suit. Moreover, the dust contamination will present a hazard to planetary surface operations by causing abrasion of the EMU surface. Also, it is important that the dusty suit surfaces can be cleaned before the habitat is entered again to prevent the risk of contaminating the entire habitat with dust particles. Using U.S. Occupational and Health Administration Standards for loads being carried not to exceed 20.3 kg in the respective gravity field, Earth weight design goals for future planetary EMUs would be 122 kg for a lunar PLSS and 53.3 kg for a Mars PLSS, respectively. [6, 20]

The European EVA space suit design has progressed as part of the Hermes spacecraft development program which has been canceled in 1994. The European EVA space suit system was supposed to consist of an autonomous anthropomorphic system for unassisted donning / doffing through a back entry design, providing an EVA sortie capability of up to 7 hours and an operational life of 15 years or 35 sorties. It was designed for an average metabolic rate of 300 W, with maximum metabolic rates of 580 W for 15 minutes and of 470 W for any given EVA hour, and a minimum metabolic rate of 120 W for 60 minutes. The nominal operating environment of the suit was to be a 50 kPa atmosphere of 95 % pure oxygen, with an emergency operating environment of 27 kPa for a minimum time of 30 minutes. The autonomous suit wet mass was planned to be 125 kg. It is quite unlikely, though, that the European space suit program will be continued in the near future. [39]

IV References

[1] Ahmed S.
An Overview of Soviet Concepts of Food and Nutrition in Spaceflight
International Conference on Life Support and Biospherics, Proceedings, p. 401-410, 1992

[2] Alexandrov A.
The EVA Program of the USSR
Lecture Notes, International Space University, 1992

[3] Barer A., Filipenkov S.
Decompression Safety of EVA: Soviet Protocol
Acta Astronautica, Vol. 32, No. 1, p. 73-74, 1994

[4] Barer A., Filipenkov S.
Suited Crewmember Productivity
Acta Astronautica, Vol. 32, No. 1, p. 51-58, 1994

[5] Barrett M.
Medical Aspects of EVA
Lecture Notes, International Space University, 1992

[6] Brown M., Schentrup S.
Requirements for Extravehicular Activities on the Lunar and Martian Surfaces
20th Intersociety Conference on Environmental Systems, SAE Techn. Paper 901427, 1990

[7] Chu W.
Mass Analysis for the Space Station ECLSS Using the Balance Spreadsheet Method
19th Intersociety Conference on Environmental Systems, SAE Techn. Paper 891502, 1989

[8] Conger B. et al
First Lunar Outpost Extravehicular Life Support System Evaluation
23rd Intersociety Conference on Environmental Systems, SAE Techn. Paper 932188, 1990

[9] Crump W.; Janik D.
Introduction to Life Support Systems
International Space University, Space Life Sciences Textbook, p. 167-181, 1992

[10] Crutcher A.; Perez Vara R.
ECOSIM - Environmental Control & Life Support Simulation
Preparing for the Future, Vol. 2 No. 4, p. 15-16, ESA, 1992

[11] David K.; Preiß H.
DEBLSS - Deutsche Biologische Lebenserhaltungssystemstudie
Dornier GmbH, Bericht: TN-DEBLSS-6000 DO/01, 1989

[12] Diamant B.; Humphries W.
Past and Present Environmental Control and Life Support Systems on Manned Spacecraft
20th Intersociety Conference on Environmental Systems, SAE Technical Paper 901210, 1990

[13] Doll S.; Case C.
Life Support Functions and Technology Analysis for Future Missions
20th Intersociety Conference on Environmental Systems, SAE Techn. Paper 901216, 1990

[14] Doll S.
Life Support Systems for Manned Space Exploration
Lecture Notes, International Space University, Kitakyushu, 1992

[15] Doll S.
Supporting Life in a Space Environment
International Space University, Space Life Sciences Core Curriculum Notes, 1993

[16] Granseur P.
Human Factors for Permanently Manned Systems
Presentation at the Technical University of Munich, 1990

[17] Hienerwandel K.; Kring G.
ECLSS for Pressurized Modules - From Spacelab to Columbus
Space Thermal Control and Life Support Systems, ESA SP-288, p. 45-50, 1988

[18] Hudkins K.
Prebreathe vs. R-Factor - Introduction and Overview of the Problem
Acta Astronautica, Vol. 32, No. 1, p. 71, 1994

[19] Huttenbach R. et al
Physico-Chemical Atmosphere Revitalization: The Qualitative and Quantitative Selection of Regenerative Designs
Space Thermal Control and Life Support Systems, ESA SP-288, p. 57-64, 1988

[20] Kosmo J.
Design Considerations for Future Planetary Space Suits
20th Intersociety Conference on Environmental Systems, SAE Technical Paper 901428, 1990

[21] Leiseifer H. et al
Biological Life Support Systems
Environmental & Thermal Control Systems for Space Vehicles, ESA SP-200, p.281-288, 1983

[22] Malnig H.
Homes, Drink / Food Water Supply Assembly
Space Thermal Control and Life Support Systems, ESA SP-288, p. 367, 1988

[23] McBarron J.
Past, Present and Future: The U.S. EVA Program
Acta Astronautica, Vol. 32, No. 1, p. 5-14, 1994

[24] McBarron J.
U.S. Prebreathe Protocol
Acta Astronautica, Vol. 32, No. 1, p. 75-78, 1994

[25] Miernik J.
Closed Loop Life Support Mass Balance Calculations for Space Station Freedom
International Conference on Life Support and Biospherics, Proceedings, p. 473-484, 1992

[26] Morgenthaler G. et al
An Assessment of Habitat Pressure, Oxygen Fraction and EVA Suit Design for Space Operations
Acta Astronautica, Vol. 32, No. 1, p. 39-50, 1994

[27] Nelson M.
Bioregenerative Life Support for Space Habitation and Extended Planetary Missions
Space Life Sciences, ed. S. Churchill, Chapter 22, Orbit Books, Malabar, Fla., 1993

[28] Newman D.
Life Support and Performance Issues for Extravehicular Activity (EVA)
Lecture Notes, International Space University, 1992

[29] Newman D.
Life Sciences Considerations for Extravehicular Activity (EVA)
Lecture Notes, International Space University, 1992

[30] Phillips R.
Food and Nutrition During Spaceflight
International Space University, Space Life Sciences Textbook, p. 245-271, 1992

[31] Preiß H.
European Life Support Systems for Space Applications
Space Thermal Control and Life Support Systems, ESA SP-288, p. 39-44, 1988

[32] Prince R.; Knott W.
CELSS Breadboard Project at the Kennedy Space Center
Lunar Base Agriculture: Soils for Plant Growth, ASA, p. 155-163, 1989

[33] Ross J.
EVA Design: Lessons Learned
Acta Astronautica, Vol. 32, No. 1, p. 1-4, 1994

[34] Salisbury F.
Report on BIOS-3
Presentation at the NASA Johnson Space Center, 1992

[35] Severin G.
Design to Safety: Experience and Plans of the Russian Space Suit Programme
Acta Astronautica, Vol. 32, No. 1, p. 15-24, 1994

[36] Skoog I.
Lebenserhaltungssysteme
Handout of Presentation at the TU München, 1985

[37] Skoog I. et al
Life Support Systems for Man
Life-Sciences Research in Space, ESA SP-1105, p. 97-108, 1989

[38] Skoog I.; Brose H.
The Complementary Role of Existing and Advanced Environmental, Thermal Control and Life Support Technology for Space Station
Environmental & Thermal Control Systems for Space Vehicles, ESA SP-200, p.281-288, 1983

[39] Skoog I.
The EVA Space Suit Development in Europe
Acta Astronautica, Vol. 32, No. 1, p. 25-38, 1994

[40] Tamponnet C. et al
Implementation of Biological Elements in Life Support Systems: Rationale and Development Milestones
ESA Bulletin 74, p. 71-82, 1993

[41] Tamponnet C. et al
Man in Space - A European Challenge in Biological Life Support
ESA Bulletin 67, p. 38-49, 1991

[42] Tamponnet C.; Binot R.
Microbial and Higher Plant Biomass Selection for Closed Ecological Systems
Acta Astronautica, Vol. 27, p. 219-230, 1992

[43] Thews G.; Vaupel P.
Vegetative Physiologie
Springer-Verlag, Berlin, 1990

[44] Tri T., Henninger D.
Controlled Ecological Life Support Systems Human-Rated Test Facility (HRTF): An Overview
SAE Paper 932241, 1993

[45] von Puttkamer J.
Der Mensch im Weltraum
Umschau Verlag, Frankfurt, 1987

[46] Tremblay P.
EVA Safety Design Guidelines
Acta Astronautica, Vol. 32, No. 1, p. 59-68, 1994

[47] Wheeler R. et al
Crop Tests in NASA's Biomass Production Chamber - A Review of the First Four Years of Operation
International Conference on Life Support and Biospherics, Proceedings, p. 563-573, 1992

[48] Wieland P.
Designing for Human Presence in Space: An Introduction to Environmental Control and Life Support Systems (Draft)
NASA Marshall Spaceflight Center, 1992 (to be published)

[49] Columbus Human Factors Engineering Requirements
COL-RQ-ESA-013, 1989

[50] Space Station Freedom Man-Systems Integration Standards
NASA STD-3000, Volume IV, 1989

[51] Workshop on Long-Duration Life Support System Test with Humans - Minutes
NASA, Johnson Space Center, October 25 & 26, 1994

[52] International Journal of Life Support and Biosphere Science,
Vol. 1, No. 1, 1994

V PHYSICO-CHEMICAL LIFE SUPPORT SUBSYSTEMS

In physico-chemical life support systems man is the only biological component. The basic life support functions that can be fulfilled by physico-chemical subsystems are:

- Atmosphere Management
- Water Management
- Waste Management

Concerning the potential reduction of supply mass, the most important physico-chemical regeneration methods concern the water and atmosphere recycling. The provision of food is only possible with bioregenerative methods. These are further dealt with in chapter VI. This chapter outlines the most important physico-chemical life support subsystems for space applications that are presently under discussion in the United States, Europe and Japan. Information on developments made in the former Soviet Union is given where available. For the selection process of any technology it is, of course, most important to know whether the respective technology is actually readily available or, at least, which is its state of development. In order to judge the technological maturity of a life support subsystem a Technology Readiness Level (TRL) can be measured on a maturity scale as shown in table V.1. Rarely would a technology with a TRL less than 5 be considered viable during a spacecraft design phase due to the time required to bring it to flight status. The TRL are indicated in the following chapter where available for a respective subsystem. [35]

Technology Readiness Level	Description of Technology Maturity
1	Basic principles observed and reported
2	Conceptual design formulated
3	Conceptual design tested analytically or experimentally
4	Critical function demonstrated
5	Components tested in relevant environment
6	Prototype tested in relevant environment
7	Validation model demonstrated in relevant environment
8	Design qualified for flight

Table V.1: **Technology Maturity Scale [35]**

V.1 ATMOSPHERE MANAGEMENT

The function of the atmospheric revitalization and control subsystem is to continuously control and regenerate the module atmosphere as required. It has to provide for CO_2 removal, CO_2 reduction, O_2 generation, and monitoring and removal of harmful trace contaminants that are generated by crew and equipment. Also, the atmosphere temperature and humidity have to be controlled. The several components of the atmosphere management system and their respective functions and requirements are listed below:

- *Atmosphere Control and Supply*
  Atmospheric pressure and composition control functions shall provide a method of regulating and monitoring the partial and total pressure of gases in the space habitat atmosphere, i.e., mainly O_2/ N_2 pressure control vent and relief, storage, and distribution. The controls shall operate autonomously with limited or no crew intervention necessary.

- *Temperature and Humidity Control*
  In any space station or habitat some heat load is generated by the crew itself and cabin located equipment. The crew is also a source of humidity. The conditions to be maintained are a temperature between 18° and 27°C and a relative humidity between 25 and 70% for the nominal case. The space habitat atmosphere temperature and humidity shall be maintained by providing for cabin air temperature and humidity control, air particulate and microbial control, ventilation, avionics air cooling, and thermally conditioned storage.

- *Atmosphere Monitoring*
  Atmosphere monitoring instruments shall require as little crew time as possible for operation and maintenance. The monitoring of volatile organics shall be accomplished. Hydrocarbon analyzers shall provide continuous real time indication of total organic contamination in the air. An additional monitor shall be available to unequivocally identify and quantify each organic compound and take measurements at regularly scheduled intervals. Compound specific monitors shall be located near equipment, chemical operations, and processing activities that are potential sources of chemical contamination of the space habitat. These monitors shall be equipped with audible and visible alarms to alert crew members when contaminant concentrations exceed maximum acceptable levels. Also, atmosphere microbiological contamination, e.g., with bacteria, yeast, and molds, will have to be monitored. Microbial decontamination is required when acceptability levels are exceeding the requirements. Decontamination procedures, anti-microbial agents, and supporting equipment shall be provided to counteract and control all contamination events.

- *Cabin Ventilation*
  The lack of gravity and therefore the lack of natural air convection requires that all inhabited pressurized areas of a space habitat be forced ventilated with fans. By this means the atmosphere stays uniformly mixed avoiding thermal gradients and localized contaminant buildup. Also, cooling of the astronauts is accomplished by this forced convection. Some of the ventilation air must be passed through a heat exchanger that interfaces the waste heat transport loop in order to provide thermal control of the atmosphere. Medical doctors at NASA's Johnson Space Center (JSC) have set face atmosphere velocity requirements for space habitats to a maximum of 0.2 m/s to prevent atmosphere drafts. For control of the CO_2 level of a habitat a lower design requirement of 0.08 m/s is customary.

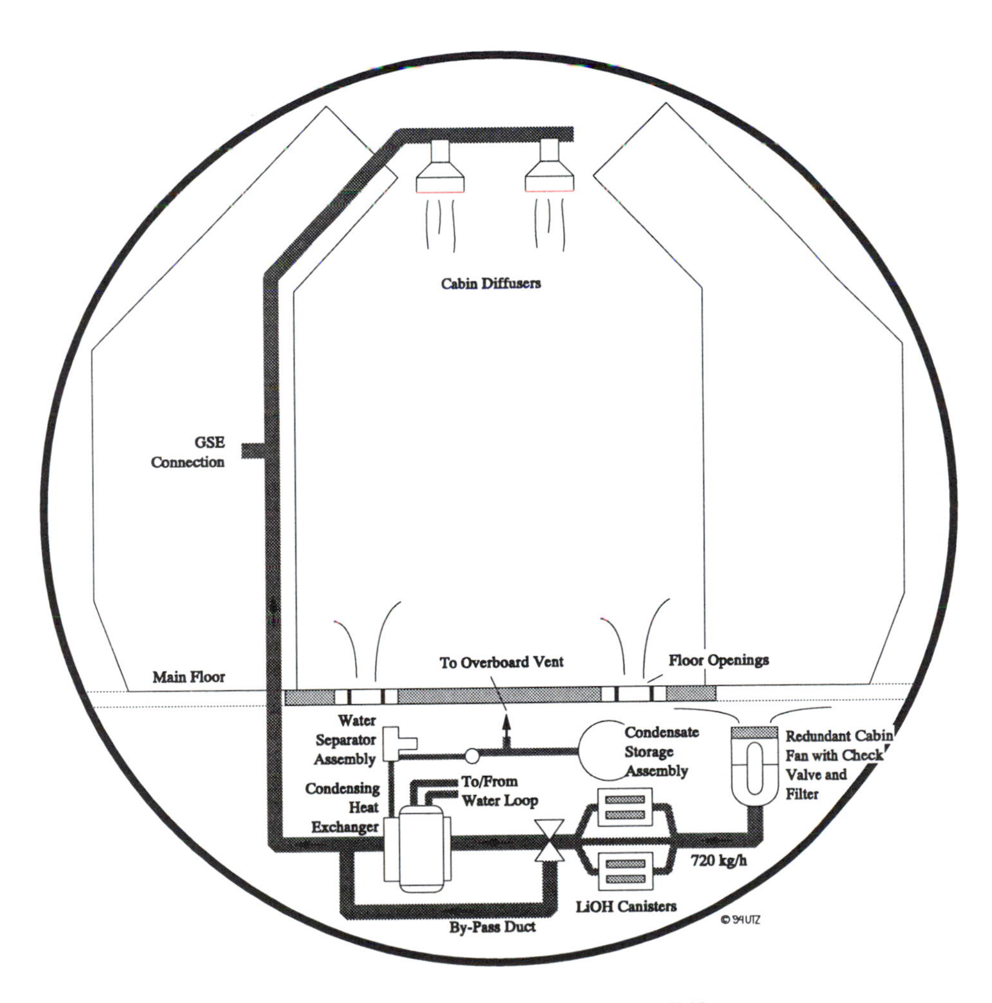

Figure V.1: **Cabin Atmosphere Loop [29]**

Cabin ventilation has to be performed in a way that a minimum air velocity is guaranteed in the cabin. For space station module-type habitats or labs, two concepts are under discussion: a radial flow system where the air is blown through ceiling mounted air diffusors into the cabin and sucked through floor grills back to the fan inlet and, as a second concept, an axial flow through the cabin. In that case, the air is blown into the cabin via a ring shaped diffusor (see figure V.1).

- *Fire Detection and Suppression*
  Fire detection in a pressurized space habitat relies on two different principles. The first one is smoke detection separately performed in each air loop. The second is temperature measurement at the equipments that are identified to be potential fire sources. The fire suppression system should be distributed over the complete habitat. Halon 1301 may be used as suppression agent. Each bottle has to contain a sufficient amount of Halon to guarantee the minimum concentration required to suppress a fire in the related volume. If all bottles are fired, the minimum concentration has to be achieved in the complete module. Nevertheless, the use of Halon 1301 is problematic due to its toxic pyrolyse products.

- *Hyperbaric Treatment*
  Where altitude decompression sickness may occur as a result of operational activity or contingency operation, access to a hyperbaric treatment facility is required.

V.1.1 Air Revitalization Technologies

The three main steps of atmosphere revitalization involve the recovery of oxygen from carbon dioxide, namely:

- CO_2 Concentration
- CO_2 Reduction
- O_2 Generation

The main technologies that may be applied are listed in table V.2. Besides, there are a number of secondary functions which are essential to operation of the atmosphere revitalization subsystem. These include particulate and trace contaminant control, the supply of process consumables, product handling and the disposal of waste materials.

The selection of preferred designs for atmosphere revitalization can begin by defining the boundary limits across which energy and materials flow, and the relationship between the life support system and its power, heat rejection, material supply, and waste disposal systems. Major factors to be considered in this respect which will also impact on capital investment and operating costs, include:

Carbon Dioxide Concentration	- 2-bed Molecular Sieve (2BMS) - 4-bed Molecular Sieve (4BMS) - Electrochemical Depolarization Concentrator (EDC) - Solid Amine Water Desorption (SAWD) - Air Polarized Concentrators (APC) - Lithium Hydroxide (LiOH)
Carbon Dioxide Reduction	- Bosch - Sabatier - Advanced Carbon-Formation Reactor System Carbon - Dioxide Electrolysis - Superoxides - Artificial Gill
Oxygen Generation	- Static Feed Water Electrolysis (SFWE) - Solid Polymer Water Electrolysis (SPWE) - Water Vapor Electrolysis (WVE)

Table V.2: **Atmosphere Revitalization Proposed Subsystem Options**

- Overall thermodynamic efficiency of the life support system
- Achievable degree of material closure
- Efficiency of the power supply system
- Efficiency of the heat rejection system
- Consumption of utilities required to support operation of the life support system

A summary of the most important qualitative and quantitative assessment criteria for the selection of atmosphere revitalization technologies is given in table V.3. [17]

V.1.1.1 Gas Storage Systems

For resupply of a spacecraft atmosphere with oxygen and nitrogen, in the past mainly non-regenerable systems, i.e., storage systems, have been used. For economical storage of these gases aboard a spacecraft or space habitat the following options are available:

- Storage in chemical compounds
- High pressure storage
- Cryogenic storage

QUALITATIVE	CRITERIA
Operational	- Operation at high carbon dioxide concentrations - Operation in degraded or rescue mode - Ease of start-up/shut-down
Mechanical Design / Reliability	- Number of moving parts - Generation of noise - Degree of thermal/mechanical stress
Safety	- Operating pressure and temperature - Presence of hydrogen - Use of exothermic reactions - Nature of side products
Maintenance and Servicing	- Need for scheduled maintenance
Growth Potential	- Spin-offs into other technology areas - Ability to upgrade to higher crew size
PRE-LAUNCH	**QUANTITATIVE CRITERIA**
Technology Status	- Development status
Equivalent Weight	- Launch mass of flight hardware - Power supply mass penalty - Heat rejection mass penalty - Mass of flight spares and consumables - Mass of orbital decay fuel
Launch Volume	- Volume of flight hardware, consumables, spares
Design, Development and Integration Costs	- Design, development, test, and evaluation - Integration into space vehicle
Equipment Cost	- Capital costs: flight hardware, spares, consumables
POST-LAUNCH	**QUANTITATIVE CRITERIA**
Flight Unit Transport to Orbit Cost	- Transport cost of hardware, spares, consumables
Resupply Costs	- Capital cost of resupply spares and consumables
Resupply Mission Costs	- Transport costs of resupply spares, consumables
Maintenance Costs	- Cost required to keep equipment operational over mission life

Table V.3: **Summary of Typical Assessment Criteria**

Although resupply gas storage will lose importance in the future due to the use of regenerable systems, it may still be necessary to provide stored gases, especially oxygen, for emergency backup.

V.1.1.1.1 Chemical Compounds

Numerous chemical compounds include oxygen and nitrogen combined with other elements. A number of oxygen-producing chemicals have been developed for submarine, aircraft and mine applications. The development of high performance chemical suppliers of oxygen and diluent gases is needed for many applications including cabin repressurization, leakage makeup and emergency operation, particularly when long duration storage or stand-by are required. A number of oxygen-producing chemicals which supply the highest possible yield per kg of chemicals are listed in table V.4. The list is limited to those compounds which are not considered toxic and which yield essentially pure oxygen. [15]

Compound	Yield [kg compound / kg oxygen]
H_2O_2	2.1
LiO_2	1.62
NaO_2	2.29
KO_2, K_2O_2	2.96
MgO_4	1.84
CaO_4	2.08
$LiClO_4$	1.66
$NaClO_4$	2.8
$KClO_4$	2.16
$Mg(ClO_4)_2$	1.74

Table V.4: **Yield of Oxygen-Producing Chemical Compounds [15]**

V.1.1.1.2 High Pressure Storage

High pressure storage is used to minimize the tankage volume required for storage of atmospheric constituents at ambient temperature. The use of very high pressures is limited by gas compressibility. Since gases become less compressible at relatively high pressure, an optimum pressure-to-volume ratio exists at a pressure of a few million Pa. [15]

V.1.1.1.3 Cryogenic Storage

Cryogenic or supercritical storage of oxygen and nitrogen is the current state-of-the-art. Supercritical storage of atmospheric gases at cryogenic temperatures provides the means for storing these fluids at relatively high density and operating pressures lower than those used in high pressure gaseous storage. This results in lower specific weight and volume of tankage and enhances crew safety since high operating pressures are avoided. In addition, supercritical storage provides a low-temperature heat sink readily available for thermal integration with other spacecraft subsystems. [15]

V.1.1.2 Carbon Dioxide Removal

Carbon dioxide in a spacecraft atmosphere must be removed and concentrated to prevent it from reaching toxic levels and to provide a concentrated carbon dioxide stream to the CO_2 reduction subsystem for further processing. CO_2 removal units control the amount of CO_2 in the spacecraft atmosphere. They should be capable of keeping the CO_2 concentration at a level of about 0.5 kPa, not allowing it to exceed 1 kPa. CO_2 is usually generated in the cabin atmosphere at an approximate rate of 1 kg/man-day. The CO_2 collected should be of high purity so as not to reduce the efficiency of the CO_2 reduction subsystem. On the Skylab mission a regenerable CO_2 removal subsystem was used for the first time. It employed two canisters of molecular sieve for CO_2 adsorption. Molecular sieves and further promising regenerative techniques are described below. [15, 31]

V.1.1.2.1 Regenerable Processes

V.1.1.2.1.1 Molecular Sieves

Regenerative molecular sieves utilize synthetic zeolites or metal ion alumino-silicates as the basic CO_2 collection material. Two synthetic zeolites are used alternately for adsorption and desorption of CO_2 from the cabin atmosphere. Since moisture in the air is preferentially adsorbed by the zeolites, decreasing their CO_2 adsorption capacity, silica gel beds may be included to remove moisture before the air is circulated through the zeolite canisters. [15]

Thus four-bed molecular sieve (4BMS) is composed of two adsorbing beds, one a desiccant bed for water vapor removal, the other a zeolite molecular sieve for trapping CO_2, operating in parallel with two identical beds in the desorbing mode (see figure V.2). In the process, the wet, carbon dioxide-laden process air first enters the adsorbing desiccant bed, where the water vapor is selectively removed, thereby protecting the downstream carbon dioxide bed from water vapor. The dry air is drawn into a blower, overcoming the system pressure drops. The dry air then passes through a pre-cooler, removing the blower compression and motor heat and the heat of adsorption generated by the water adsorbing on the desiccant. This cool, carbon-dioxide-laden air next passes into the adsorbing carbon dioxide removal bed, which selectively removes the carbon dioxide. Next the air is directed into the desorbing desiccant bed, where the carbon dioxide-free air stream is rehumidified and directed back into the temperature, humidity, and ventilation control system. The parallel bed pairs trade functions when the adsorbing beds reach storage capacity. Adsorption efficiency is highest at low temperatures, requiring that warm cabin air passes through an air-liquid heat exchanger before entering the adsorbing beds. Replacing the Zeolite 5A molecular sieve material with a material that requires less heat for efficient desorption and also reduced residual loading of the sorbent bed would significantly improve the 4BMS. Such a material would reduce the large heater power requirement, while reducing resupply needs by increasing sorbent bed life and efficiency. The 4BMS is a mature technology, already flight-proven on Skylab.

The molecular sieves applied on the Skylab missions employed two canisters for CO_2 adsorption. Each canister contained a predrier section of 13X molecular sieve and a CO_2 adsorption section of 5A molecular sieve. A complete cycle took 30 minutes, i.e., 15 minutes to adsorb and 15 minutes to desorb. The adsorption process was achieved by flowing cabin air through the canisters at cabin pressure and returning the CO_2 depleted air back to the cabin. The desorption cycle worked by merely exposing a canister to space vacuum. At the end of the desorption period, the regenerated canister was ready for readsorbing H_2O and CO_2. Because of an incremental amount of gas remaining on the molecular sieve at the end of each desorption cycle, a periodic bake out was necessary. Bake out was accomplished by merely heating the molecular sieve bed to 478 K while it was exposed to space vacuum. It was found that a critical factor in the performance of the molecular sieve bed was the level of moisture present. Water adsorption on the 5A molecular sieve reduced the rate of CO_2 adsorption. [31]

The 2BMS, similar to the 4BMS, utilizes a carbon molecular sieve (CMS) to remove CO_2. Unlike with zeolites, CO_2 adsorption using a CMS is unaffected by the presence of moisture in the process air stream. The ability of CMS material to preferentially adsorb CO_2 over water vapor eliminates the need for desiccant beds, reducing the 4BMS to a 2BMS. The CMS material also desorbs at a lower temperature than zeolite, decreasing power requirements. The 2BMS is not very mature, yet. An overview of both 4BMS and 2BMS characteristics is given in table V.5. [18, 35]

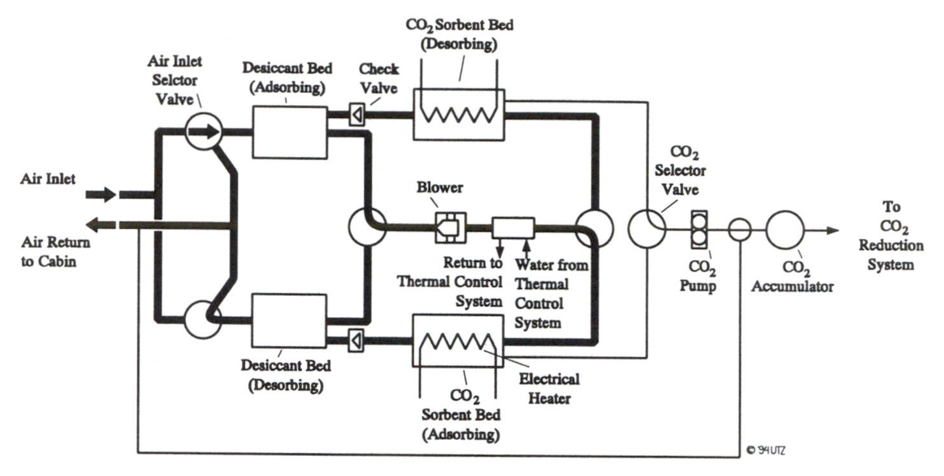

Figure V.2: **4-Bed Molecular Sieve [18]**

V.1.1.2.1.2 SolidAmine Water Desorption (SAWD)

The SAWD operates similarly to the 2BMS, with the main exception that steam-heated solid amine (WA-21) is used to absorb/desorb CO_2. Unlike zeolites, solid amines degrade fairly rapidly with time, increasing bed changeout frequency. SAWD also penalizes the ECLSS by requiring hygiene water, which is eventually vented in vapor form to the THC, increasing the load on the condensing heat exchanger. SAWD has the advantage that desorption takes place at cabin pressure, rather than at near vacuum conditions required by the 4BMS and 2BMS. [34]

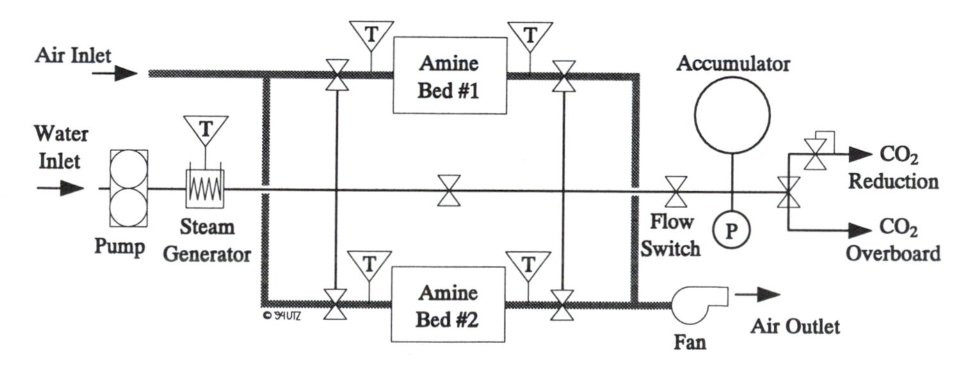

Figure V.3: **2-Bed, Solid-Amine-Resin, SAWD [31]**

Subsystem	**4-BED / 2-BED MOLECULAR SIEVES**

Weight:	87.9 / 48.1 kg	Volume:	0.33 / 0.26 m^3
Heat Generated:	-	Power Required:	0.535 / 0.23 kW

Operating Temperature:	283 / 283 K (min.)	423 / 338 K (max.)
Operating Pressure:	0.395 kPa (min.)	10 kPa (max.)

Operations Mode:	Continuous
Designed Efficiency:	66 / 90 %
Comment:	More effort should be made in sorbent optimization in terms of CO_2 capacity, sorbent characterization to support system design and demonstrate durability, and prototype development
Technology Readiness Index:	8 / 2 - 3
Reliability:	-
Maintainability:	-

Advantages:	- 2-bed approach is about half the weight and power of the 4-bed system - 2-bed working capacity is about twice larger than that of 4-bed - Also no trace contaminant found in 2-bed system
Disadvantages:	- Limited experimental data and testing results in 2-bed approach - No prototype for system design analyses of 2-bed design
Prospective Improvements:	-

Principal Chemical Equation:	-

	Inflow [kg/day]	Inflow Composition	Outflow [kg/day]	Outflow Composition
Gases	-	-	-	-
Liquids	-	-	-	-
Solids	-	-	-	-

Table V.5: **Design Characteristics of 4-Bed Zeolite / 2-Bed Carbon Molecular Sieves for Supporting 3 People [35]**

The main components of the SAWD subsystem are the canisters which contain the resin absorbents and the steam generator (see figure V.2). The resin scrubs CO_2 from the air and the steam is used to desorb CO_2 from the spent resin. The active chemical or resin in the canisters is 0.52-1.13 mm diameter beads of a polystyrene-divinylbenzene copolymer aminated with diethylene-triamine and designated IRA-45. This substrate exposes a large surface area of amine to the space cabin atmosphere for CO_2 removal.

The resin absorbs CO_2 by first combining with water to form a hydrated amine which in turn reacts to form a bicarbonate. These reactions are represented by the following two equations:

$$Amine + H_2O \rightarrow Amine - H_2O$$
$$Amine - H_2O + CO_2 \rightarrow Amine - H_2CO_3$$

The amine is regenerated by applying heat to break the bicarbonate bond, thus, releasing CO_2 by the reaction:

$$Amine - H_2CO_3 + steam\ heat \rightarrow CO_2 + H_2O + Amine$$

Water is important to the absorption process because the dry amine cannot react with CO_2 directly. Between 20-35 wt-% water in the resin bed is required for optimum absorption. This requirement impacts operating ranges by reducing CO_2 absorption capacity at a low relative humidity for the spacecraft cabin.

During desorption, an electrically heated steam generator converts H_2O into steam. The steam heats the bed and pushes residual ullage air out of the canister at a low flow rate. As steam/air reaches the outlet end of the bed, a CO_2 wave evolves of the bed, sharply increasing the flow rate. A flow sensor detects the flow increase and activates a valve, switching the CO_2 flow to a CO_2 accumulator for subsequent use in a CO_2 reduction subsystem. By using steam desorption at 2.110^5 kPa, CO_2 can be compressed in the accumulator. The desorption temperature is controlled by the saturation temperature of steam. The steam used for desorption is recovered by first evaporating it into the process air flow during CO_2 absorption then condensing it in the humidity control heat exchanger of the ECLSS. The major weight penalty associated with the SAWD system is that associated with steam generation and heat rejection. For space missions where power may be limited during specific portions of an orbit, e.g., a solar-cell-powered mission in the dark interval of an orbit, the SAWD subsystem could be desorbed on the light side where power is readily available and absorption could take place on the dark side. A survey of characteristics of a SAWD prototype is given in table V.6. [31]

Subsystem	**SOLID AMINE WATER DESORPTION (SAWD)**

Weight:	51.3 kg	Volume:	0.21 m^3
Heat Generated:	-	Power Required:	0.454 kW

Operating Temperature: Operating Pressure:	292 - 300 K 100 kPa (during adsorption) 30 kPa (during desorption)

Operations Mode:	Continuous; flow alternates between two canisters; 67 minutes adsorption, 67 minutes desorption
Designed Efficiency:	-
Comment:	-
Technology Readiness Index:	6
Reliability:	-
Maintainability:	-

Advantages:	- Regenerable - Lower weight, volume and power requirements than non-regenerable LiOH for longer missions
Disadvantages:	- Amine resin may be degraded by steam during regeneration leading to reduced CO_2 capacity - Amine degradation may yield toxic vapors - Amine groups cannot react with dry CO_2; moisture content of the bed must be optimized and controlled - Excess moisture increases pressure drop and reduces CO_2 capacity
Prospective Improvements:	- Improved method of resin regeneration having a low power consumption - Development of a more stable resin

Principal Chemical Equation:	Primary amines: Secondary amines:	$RNH_2 + H_2O \rightarrow RNH_3 \cdot OH$ $RNH_3 \cdot OH + CO_2 \rightarrow RNH_3 \cdot HCO_3$ $R_2NH + H_2O \rightarrow R_2NH_2 \cdot OH$ $R_2NH_2 \cdot OH + CO_2 \rightarrow R_2NH_2 \cdot HCO_3$

	Inflow [kg/day]	Inflow Composition	Outflow [kg/day]	Outflow Composition
Gases	724	19.6 vol-% O_2 78.6 vol-% N_2 0.5 vol-% CO_2 1.3 vol-% H_2O vapor Trace contaminants	721	19.7 vol-% O_2 79 vol-% N_2 1.3 vol-% H_2O vapor Trace contaminants
Liquids	0	-	0	-
Solids	0	-	0	-

Table V.6: **Design Characteristics of a SAWD Preprototype for 3 people [35]**

For a solid amine system the crucial item is the solid amine material itself. Here, the term "solid amine" is used for resins with amino groups, directly polymerized into a matrix (ion exchange resin). Because the CO_2 molecules form negative ions in the presence of water, anion exchange resins are needed in this case. Anion-exchange resins can be subdivided into weak and strong basic ion exchange resins. For a regenerative CO_2 adsorption process only weak anion exchange resins can be used, because the strong basic resins cannot be desorbed under the conditions applied here. The weak basic resins can be subdivided into primary, secondary, and tertiary amines, where this nomenclature indicates the number of free electron pairs of the nitrogen, where the hydrogen is replaced by other atoms or groups. For a secondary amine, the following reaction, which can be reversed by rising the temperature and/or reducing the pressure, applies for CO_2 adsorption: [21]

$$R_1R_2NH + CO_2 + H_2O \Leftrightarrow R_1R_2NH_2^+ + HCO_3^-$$

V.1.1.2.1.3 Electrochemical Depolarization Concentration (EDC)

The EDC reacts H_2 and O_2 with CO_2 inside an electrochemical cell, generating two outlet streams. The outlet stream from the anode-side cavity of the cell has high concentration of CO_2 along with some H_2, while the outlet stream from the cathode-side cavity contains air with a low CO_2 concentration that is returned to the cabin. In the EDC process, CO_2 is continuously removed from low CO_2 partial pressure in a flowing air stream. The CO_2 exhaust from the EDC is premixed with hydrogen and can be sent directly to a CO_2 reduction subsystem. The EDC also generates direct current (DC) power which can be used by other ECLSS subsystems. The specific electrochemical and chemical reactions which take place within each cell are as follows:

Cathode reactions:

$$\tfrac{1}{2}O_2 + H_2O + 2e^- \rightarrow 2OH^- \quad (1)$$

$$CO_2 + 2OH^- \rightarrow H_2O + CO_3^{2-} \quad (2)$$

Anode Reactions:

$$H_2 + 2OH^- \rightarrow H_2O + 2e^- \quad (3)$$

$$H_2O + CO_3^{2-} \rightarrow CO_2 + 2OH^- \quad (4)$$

Overall Reaction:

$$CO_2 + \tfrac{1}{2}O_2 + H_2 \rightarrow CO_2 + H_2O + \text{electrical energy} + \text{heat} \quad (5)$$

Reactions (1) and (3) are the same electrochemical reactions that occur in a H_2-O_2 fuel cell with an alkaline electrolyte. The products of this electrochemical reaction are H_2O and DC power. With CO_2 present in the cabin air stream, the hydroxyl ions (OH^-) being generated at the cathode react with the CO_2 to give carbonate ions (CO_3^{2-}) and H_2O. The product CO_3^{2-} now takes the place of the OH^- as the primary charge carrier within the cell, migrating toward the anode where a shift in pH results in the CO_2 being released from solution. The overall reaction occurring in an EDC produces DC power.

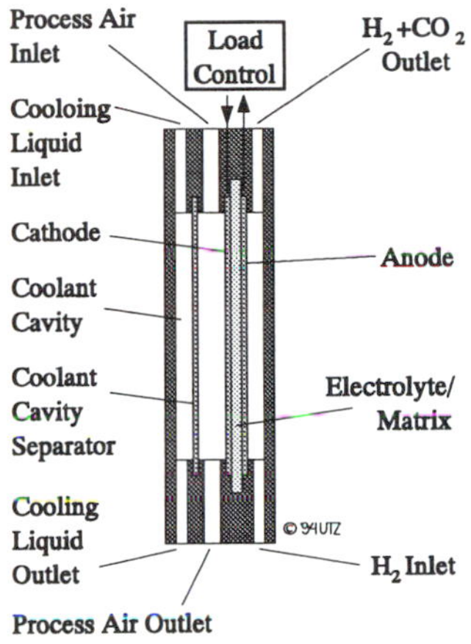

Process Air Inlet
Load Control
H_2+CO_2 Outlet
Cooloing Liquid Inlet
Cathode
Anode
Coolant Cavity
Coolant Cavity Separator
Electrolyte/ Matrix
Cooling Liquid Outlet
© 94UTZ
H_2 Inlet
Process Air Outlet

Figure V.4: **EDC Single Cell [31]**

Figure V.4 illustrates the construction of a typical EDC cell showing an electrode/matrix/ electrode assembly and the various fluid streams. Each cell consists of two electrodes separated by a porous matrix containing an aqueous electrolyte, caesium carbonate (Cs_2CO_3). Approximately 25 % of the heat generated by the reactions occurring in the EDC is removed by the process air and the H_2 and CO_2 streams. The remaining heat is removed by a separate liquid cooling stream. The liquid coolant temperature is controlled to maintain the electrolyte moisture balance within desired limits, allowing the EDC to operate over the relative humidity range of about 20 to 80 %.

The EDC can be operated in a continuous or cyclic mode. The capability of operating in either mode is particularly attractive for day-night cycling to conserve power aboard a solar-cell-powered space habitat. Another advantage of the EDC is that with the same size subsystem, different CO_2 removal rates can be achieved by merely changing operating conditions, specifically the current level. The major weight penalty for the EDC is the indirect penalty for power required by an O_2 generation subsystem to produce the O_2 consumed by the EDC. The process also has the disadvantages of consuming oxygen and producing water vapor, increasing the respective loads on the oxygen generation and humidity control systems. EDC is also a safety hazard due to the potential for hydrogen leakage into the cabin atmosphere, which could cause fire or explosion. [31, 35]

V.1.1.2.1.4 Air Polarized Concentrators (APC)

The ADC is simply an EDC not requiring hydrogen for the CO_2 removal process and including a O_2/CO_2 separator. The APC is safer than the EDC, although H_2 will no longer be available in the outlet stream to facilitate CO_2 reduction. Additionally, without H_2 the EDC becomes a net power consumer. The APC can operate with or without H_2 in order to minimize power requirements while maximizing safety. [35]

Subsystem	ELECTROCHEMICAL DEPOLARIZED CONCENTRATOR

Weight:	44.4 kg	Volume:	0.0713 m^3
Heat Generated:	0.336 kW	Power Required:	0.148 kW (AC) 0.106 kW (DC)

Operating Temperature:	291 K (min.) 297 K (max.)
Operating Pressure:	-

Operations Mode:	Continuous
Designed Efficiency:	-
Comment:	Weight does not include penalty associated with the required increase in size of the O_2 generation system; DC power generated may be used by the O_2 generation system; system may be marginally suitable for cyclic (on-day/off-night) operation.
Technology Readiness Index:	6
Reliability:	-
Maintainability:	Requires N_2 purge periodically

Advantages:	- Regenerable system - CO_2 concentration capacity may be regulated by current adjustment; capacity to handle large CO_2 overload situations - DC power is generated - Good cabin relative humidity tolerance
Disadvantages:	- Heat generated; requires cooling system - Requires supply of H_2 - O_2 is consumed; requires larger O_2 generation system - Potential for H_2 leakage; fire/explosion hazard
Prospective Improvements:	- -

Principal Chemical Equation:	$2CO_2 + 2H_2 + O_2 \rightarrow 2H_2O + 2CO_2 + DCpower + heat$

	Inflow [kg/day]	Inflow Composition	Outflow [kg/day]	Outflow Composition
Gases	2610	99.99 % Cabin air 0.01 % H_2	2610	99.77 % Purified air 0.002 % H_2 0.15 % CO_2 0.08 % H_2O
Liquids	0	-	0	-
Solids	0	-	0	-

Table V.7: **Design Characteristics of an EDC Equipment for 4 people [35]**

V.1.1.2.1.5 Membrane Removal and Other Regenerative Technologies

Osmotic Membranes

The use of membrane technology to effectively concentrate CO_2 would likely result in a simple, light, small, low power CO_2 removal system. However, the gas selectivity of current membrane technology is inadequate. Instead of filtering all other gases besides CO_2 from the process air to produce a stream of high CO_2 concentration, the membranes also filter out some CO_2 and return it to the cabin. This inefficient process requires large membranes to remove relatively small quantities of CO_2.

Electroactive Carriers

CO_2 can also be removed by electroactive carrier molecules that are capable of binding CO_2 when in the reduced form, while releasing CO_2 in the oxidized form. These carriers may be used to develop membranes that selectively and efficiently pump CO_2 when a potential is applied across the membrane. This technology has the potential for being much more efficient than other CO_2 removal technologies, but is currently in the early stages of development.

Metal Oxides

Metal oxides have been considered for use in space habitats and for EVA use since many years. In 1973, a silver dioxide formulation was investigated. It was found to have an absorption capacity of 0.12 kg per kg oxide, at a pCO_2 of 0.4 kPa, and requiring an energy of $1.86 \cdot 10^6$ J per kg CO_2. Due to expansion and contraction during absorption and desorption, the metall oxide pellets have a limited lifetime. Presently, a regeneration temperature of 140° C is required, and the design goal is to achieve 50 to 60 regenerations with cycle times of about 8 hours. For the absorption process water is needed. It enhances sorption capacity, reaction kinetics, and cycle life. Therefore, a moist process atmosphere is an advantage.

Carbonate

Using potassium carbonate, for example, CO_2 is removed from the atmosphere by the following reversible reaction:

$$K_2CO_3\,(s) + CO_2\,(g) + H_2O\,(g) \leftrightarrow 2KHCO_3\,(s)$$

where s refers to solid, and g refers to gas. The desorption process requires a temperature of about 150° C. Since the potassium carbonate is gradually consumed, the process life is limited to about 90 cycles.

Ion-Exchange Electrodialysis

This process uses an ion-exchange resin, reacting with CO_2 to form carbonate ions. The resin can be continously removed by an electrical field imposed perpendicular to the flow path. This field causes the carbonate ions to move from the absorbing cell to the concentrating cell (see also section V.2.2.3). [29]

Other potential CO_2 removal technologies include regenerable adsorbents, e.g., liquid amines, aqueous alkaline electrolytes, and molten carbonate electrolytes, but also biochemical methods, e.g., enzymes, and biological methods, like algae, bacteria, and plants, which can purify cabin air by adsorbing CO_2 and reducing it to oxygen.

V.1.1.2.2 Non-Regenerable Processes

V.1.1.2.2.1 Lithium Hydroxide (LiOH)

In an open ECLSS, CO_2 is removed from the spacecraft cabin atmosphere by flowing CO_2 laden air through a canister containing a packed bed of lithium hydroxide (LiOH) granules. The chemical equation of this process looks as follows:

$$2LiOH + CO_2 \rightarrow Li_2CO_3 + H_2O$$

The spent LiOH is not regenerated and the canisters are returned to Earth for replenishment with fresh absorbent. The amount of LiOH required to remove one person's average daily output of CO_2 is about 2 kg. The theoretical capacity of LiOH for CO_2 is 0.92 kg CO_2 per kg sorbent (see also figure V.5). Carbon dioxide removal using LiOH is acceptable for short, but not long-duration, space missions because of its high weight penalty. CO_2 removal using LiOH has been applied on all manned U.S. space missions since the early 1960, except Skylab. [29,31]

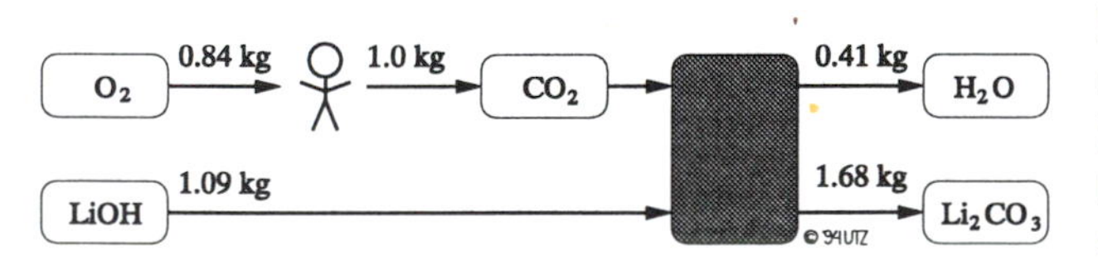

Figure V.5: **LiOH Process**

V.1.1.2.2.2 Sodasorb

Sodasorb is a mixture of $Ca(OH)_2$ (95 % dry-wt.) and sodium-, potassium, and barium hydroxides. For the CO_2 consuming reactions water is necessary and, thus, the mixture also contains 12-19 % H_2O. In a series of reactions CO_2 goes into solution and forms carbonic acid. Next, this carbonic acid reacts with hydroxide to sodium carbonate, thereby regenerating the water consumed earlier. Then, the sodium carbonate reacts with hydrated lime to calcium carbonate, thereby regenerating causic soda and potash. The theoretical capacity of sodasorb for CO_2 is 0.488 kg CO_2 per kg sorbent. [2]

V.1.1.2.2.3 Superoxides

See section V.1.1.3.5.

V.1.1.3 Carbon Dioxide Reduction

To design a physico-chemical ECLSS for near complete closure, CO_2 must not only be collected and desorbed to space, but reduced to useful components inside the system. Thus, in a partially closed or closed physico-chemical ECLSS the output from the CO_2 concentration subsystem is used as the input for the CO_2 reduction subsystem. The main competing regenerative subsystems for CO_2 reduction are currently the Bosch process, and the Sabatier process. For Sabatier all quantitative resources are much lower than those for Bosch, e.g., the expected on-orbit weight of a Bosch reactor for the U.S. Space Station is calculated to be about 1840 kg, while for the Sabatier reactor only a weight of about 500 kg is expected. The Sabatier reactor is considered to be at a high level of technological maturity, while the Bosch reactor would require some fairly major design revisions to reach flight maturity. Also, methane (CH_4) is produced by the Sabatier process, and the choice between these two candidate technologies will be based largely on whether CH_4 is seen as useful or burdensome. Further technologies that are being considered for CO_2 reduction are the Advanced Carbon-Formation Reactor System (ACRS), Electrolysis, and Superoxides. [7, 16, 31]

V.1.1.3.1 Bosch

In the Bosch CO_2 reduction process, carbon dioxide reacts with hydrogen at a high temperature (700-1000 K) in the presence of a catalyst, producing solid waste carbon and water for potable supply and heat:

$$CO_2 + H_2 \rightarrow C + 2H_2O + heat$$

The feed gases, CO_2 and H_2, are compressed and heated before contacting the catalyst bed. The exothermic heat of reaction has to be removed by the thermal control subsystem. Activated steel wool has generally been used as the catalyst. However, the development of better catalysts, such as nickel, nickel/iron or ruthenium-iron alloys that have been used successfully in laboratory studies, may increase Bosch CO_2 conversion efficiency and lower reactor operating temperature. The H_2O vapor exiting the reactor may be condensed and stored for subsequent consumption by the crew, electrolyzed to produce O_2 for respiration or to meet other water needs. A scheme of the Bosch process is given in figure V.6.

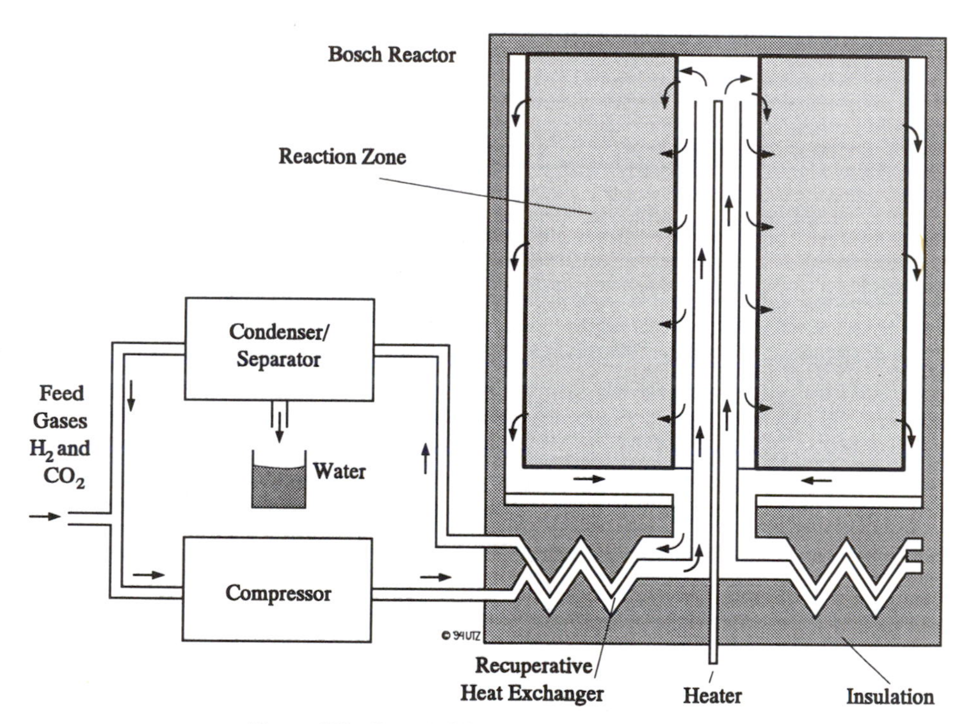

Figure V.6: **Bosch CO_2 Reduction Process [11]**

Single-pass CO_2 reduction efficiencies are generally less than 10 % with the Bosch process. Therefore, to obtain complete reduction with this process, the reactor must be run in a recycle mode. The reactant molar H_2: CO_2-ratio used to achieve 10 % efficiency is 2:1. The recycled gas mixture contains CO_2, CO, CH_4, H_2, and H_2O vapor. The solid carbon deposits on the catalyst, necessitating periodic replacement of the catalyst cartridge. H_2O vapor is continuously condensed and separated from other gases at the reactor outlet.

Subsystem	Bosch CO_2 Reduction Subsystem		
Weight: Heat Generated:	102.1 kg 0.313 kW	Volume: Power Required:	0.57 m^3 0.239 kW

Operating Temperature: Operating Pressure:	800 K (min.)	922 K (nom.) 130 kPa (nom.)	1000 K (max.)

Operations Mode:	Semibatch mode used in continuous steady-state or unsteady-state (20 man-days per cartridge)
Designed Efficiency:	6 % per pass; 100 % conversion with recycle
Comment:	-
Technology Readiness Index:	6
Reliability:	10^5 hours
Maintainability:	-

Advantages:	- No overboard venting of gases - 100 % conversion efficiency
Disadvantages:	- High maintenance (replacement of catalyst beds) - Semi-batch operation of catalyst beds - High operating temperature
Prospective Improvements:	- Structural integrity of ceramic heat exchanger - Recyclability of catalyst beds - Cold seal reactor volume reduction

Principal Chemical Equation:	$CO_2 + 2H_2 \rightarrow C + 2H_2O + heat$

	Inflow [kg/day]	Inflow Composition	Outflow [kg/day]	Outflow Composition
Gases	0.145	54.5 % H_2 45.4 % CO_2 0.1 % N	0.009	6.3 % N
Liquids	0	-	0.102	70.3 % H_2O
Solids	0	-	0.034	23.4 % C

Table V.8: **Design Characteristics of a Bosch Reactor (B-CRS) [35]**

The chemical reactions that occur in the Bosch process can be separated into two groups: low temperature and high temperature. Some reactions proceed best at low temperatures, while others work better at high temperatures. Splitting Bosch into two reactors, one for each type of reaction, will increase system efficiency, decreasing power requirements. Bosch efficiency can also be improved by operating the carbon cartridges over a range of temperatures, instead of a single temperature, since the optimum operating temperature changes as carbon is deposited during cartridge lifetime. Another possible improvement to Bosch could involve use of a CO_2 laser to help break CO_2 molecular bonds, exciting the reduction reaction forward. A laser would send the reaction entirely into the gas phase, eliminating the need for a catalyst. Bosch could ultimately become an extremely useful technology if diamond films could be produced from waste carbon for use in advanced integrated circuits. The activity or adsorption capacity of the C produced in a Bosch reactor has also been studied. These studies revealed that Bosch C had an adsorption capacity about a factor of 50 lower than commercial gas-phase-activated C adsorbents. High-C activity is desirable if the carbon is used for space habitat atmospheric or waste water contaminant removal. [7, 31, 35]

V.1.1.3.2 Sabatier

In the Sabatier process, carbon dioxide is reacted with hydrogen at a high temperature (450-800 K) and in the presence of a ruthenium catalyst on a granular substrate, producing methane and water. The reaction process consist of two steps. The first involves the hydrogenation of carbon dioxide to carbon monoxide. Carbon monoxide is hydrogenated to methane in a second step. The two reactions are:

$$CO_2 + 4H_2 \rightarrow CO + H_2O$$
$$CO + 3H_2 \rightarrow CH_4 + H_2O$$

Thus, the overall reaction is as follows:

$$CO_2 + 4H_2 \rightarrow CH_4 + 2H_2O + heat$$

The heat of this reaction must be managed by the thermal control system. H_2 could be obtained from H_2O electrolysis. Water vapor exiting the Sabatier reactor could be condensed and stored for subsequent consumption by the crew, used as further input to a hydrolysis unit to produce O_2 for respiration or to meet other water needs. Methane (CH_4) could be used for propulsion, e.g., for space station attitude control resistojets, vented to space or decomposed to reclaim H_2. Incomplete CO_2 conversion does occur (98%), with a net loss of usable O_2. This, combined with the potential H_2 loss if methane is vented, results in a small but significant resupply penalty. Venting of CH_4 to space is undesirable if CH_4 interferes with astronomical instrumentation observations.

Subsystem	**Sabatier CO_2 Reduction Subsystem**

Weight:	17.9 kg	Volume:	-
Heat Generated:	0.268 kW	Power Required:	0.05 kW

Operating Temperature:	366.5 K (min.) 700 K (nom.) 811 K (max.)
Operating Pressure:	-

Operations Mode:	Continuous steady-state
Designed Efficiency:	99 % single-pass efficiency
Comment:	-
Technology Readiness Index:	5
Reliability:	-
Maintainability:	1.7 man-hours per year; catalyst susceptible to poisoning by solid amine vapor

Advantages:	- Design of major components, catalyst, and subsystem configuration are at mature level - Reliable operation - Significant savings in weight power, volume and resupply compared to Bosch - Short start-up time - Single pass efficiency > 99 %
Disadvantages:	- Recovered water contains dissolved gases (CO_2, CH_4) - Catalyst susceptible to poisoning by solid amine vapors - N_2 will be vented with CH_4 - Overboard venting or storage of CH_4 required
Prospective Improvements:	- Development of a posttreatment process for the formation of dense carbon by the pyrolysis of methane

Principal Chemical Equation:	$CO_2 + 4H_2 \rightarrow CH_4 + 2H_2O + heat$

	Inflow [kg/day]	Inflow Composition	Outflow [kg/day]	Outflow Composition
Gases	0.196	84.6 % H_2 15.4 % CO_2	0.06	30.8 % CH_4
Liquids	0	-	0.136	69.2 % H_2O
Solids	0	-	0	-

Table V.9: **Design Characteristics of a Sabatier Reactor [35]**

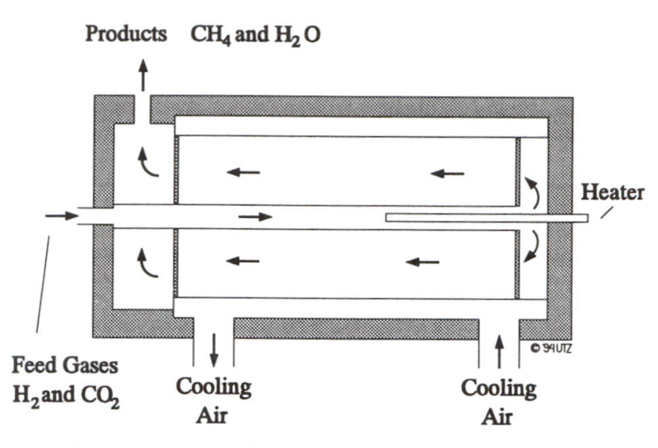

Figure V.7: **Sabatier CO_2 Reduction Process [31]**

Sabatier has the advantage of being smaller and requiring much less power than Bosch. A scheme of the Sabatier process is given in figure V.7. Hydrogen and carbon dioxide, free of contaminant gases, flow into the reactor where CO_2 is catalytically converted to CH_4 and water vapor. The nominal CO_2 flow rate into the reactor is 41.8 g/person-hour. An effective catalyst for the reduction reaction is 20 wt-% ruthenium supported on alumina. Because of the high activity of this catalyst, reduction begins at 450 K and from then on the reaction is self-sustaining, i.e., no external heat is required, a particularly attractive feature for most space habitats where power is limited. Above 866 K the reverse endothermic reaction occurs which prevents overheating. Typically, the operating temperature range of a Sabatier reactor is 450-800 K. Single-pass conversion efficiencies of the lean component are reported to be in excess of 98 % for molar ratios of H_2/CO_2 ranging from 1.8 to 5. The lean component is H_2 for molar ratios of H_2/CO_2 from 1.8 to 4, and CO_2 for molar ratios from 4 to 5. Any excess H_2 is recycled to the inlet of the reactor. By-products such as carbon or carbon monoxide are minimized when H_2/CO_2 feed ratios slightly exceed stochiometric values, i.e., 4:1. [7, 15, 28, 31, 35]

V.1.1.3.3 Advanced Carbon-Formation Reactor System (ACRS)

The ACRS consists of a Sabatier reactor, a gas/liquid separator to remove product water from methane, and a carbon formation reactor (CFR) to reduce methane to carbon and hydrogen. The CFR packs carbon better than the Bosch, reducing resupply. However, its operating temperature (both catalytic and pyrolytic CFRs operate above 1144 K) must be reduced before the ACRS can become feasible. [35]

V.1.1.3.4 CO_2 Electrolysis

CO_2 electrolysis using a solid oxide electrolyte is another method for CO_2 reduction and O_2 regeneration aboard a space habitat. Since it serves the dual purpose of reducing CO_2 directly from a concentrator as well as producing O_2 it obviates the need for a separate CO_2 reduction technology. CO_2 is taken directly from the

Subsystem	Advanced Carbon-formation Reactor System (ACRS)		
Weight:	180 kg	Volume:	0.3 m^3
Heat Generated:	- 0.15 kW	Power Required:	0.4 kW

Operating Temperature:	1400 K (min.) 1600 K (max.)
Operating Pressure:	Vacuum required for insulation between foils

Operations Mode:	Continuous
Designed Efficiency:	90 % single pass
Comment:	Prototype only
Technology Readiness Index:	-
Reliability:	-
Maintainability:	Packaging replacement every 20 days

Advantages:	- Consumables very small
Disadvantages:	- Heat loss - Forced air cooling of CFR tube and seals - Multifoil insulation requires vacuum to minimize heat loss
Prospective Improvements:	- Reliable CFR end seals - Control to maintain H_2/CO_2 ratio

Principal Chemical Equation:	$CH_4 \rightarrow C_S + 2H_2$

	Inflow [kg/day]	Inflow Composition	Outflow [kg/day]	Outflow Composition
Gases	0.1	-	-	-
Liquids	-	-	-	-
Solids	-	-	0.232	-

Table V.10: **Design Characteristics of a Projected ACRS for 3 people [35]**

CO_2 concentrator and electrolyzed to generate O_2. The solid electrolyte subsystem can electrolyze both CO_2 and H_2O vapor to continuously generate enough O_2 to meet a person's metabolic requirement and make up for spacecraft cabin leakage, thereby eliminating the need for an additional water electrolyzer. The basic reaction is:

$$CO_2 + H_2O \rightarrow CO + O_2 + H_2$$

CO is then catalytically decomposed into solid carbon and CO_2, and this CO_2 is recycled to the electrolysis unit. Technological problems, such as ineffective high-temperature (above 1140 K) ceramic-to-ceramic seals, have prevented the solid electrolyte subsystem from reaching the same level of development as the SPWE and SFWE.

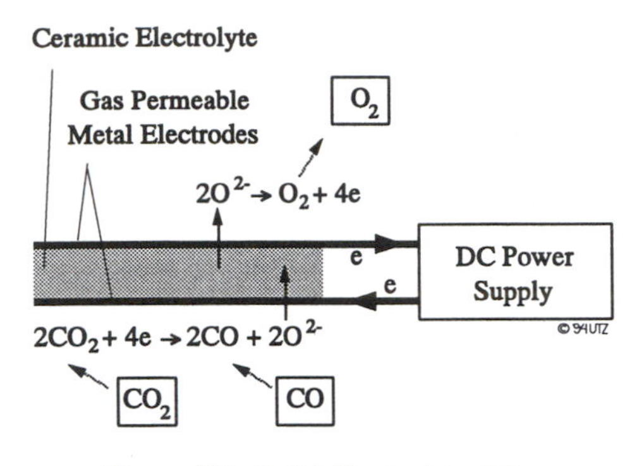

Figure V.8: **Solid Electrolyte CO_2 Electrolysis Cell [31]**

In the solid electrolyte process, as shown in figure V.8, advantage is taken of the fact that only oxide ions (O^{2-}) can migrate through the solid electrolyte at high temperatures when driven by DC voltage. Consequently, separation of O_2 from CO_2 and other product gases is excellent. The process gases, CO_2 or CO_2 and water vapor, are heated to about 1200 K and fed to the cathode side of the solid electrolyte. Both sides of the electrolyte are coated with a porous metal catalyst-electrode, such as platinum. The electrodes are connected to a DC power supply and the following reduction reaction occurs at the cathode:

$$CO_2 + 2e^- \rightarrow CO + O^{2-}$$

Oxide ions diffuse across the electrolyte by ion conduction. At the anode, this oxidation reaction takes place:

$$O^{2-} \rightarrow \tfrac{1}{2}O_2 + 2e^-$$

In a space habitat, the O_2 would be returned to the spacecraft cabin for respiration. Any unreacted CO_2 and CO that leave the cell are cooled to about 800 K. The CO flows to another reactor where most of it reacts by disproportionation as follows:

$$2CO \rightarrow C + CO_2$$

Solid C deposits on the disproportionation reactor catalysts and the CO_2 formed is recycled to the electrolysis cell. In the CO reactor, nickel, iron, or cobalt catalysts are ordinarily used to accelerate the disproportionation reaction. No problems are anticipated in operating a solid electrolyte subsystem in zero gravity since no liquid phases are present in any step process. Since this is an electrolytic process, it is capable of operating at several times of its design capacity. Increased O_2 output is achieved by merely increasing the DC voltage applied to the electrolyzer. [7, 31, 35]

V.1.1.3.5 Superoxides

Alkali and alkaline earth metal superoxides are solid chemicals which serve the dual purpose of providing O_2 and scrubbing CO_2 in a life support system. The usefulness of these active chemicals arises from the reactions that occur when these compounds are exposed to water vapor such as that present in human breath. In the case of superoxides, stored O_2 is released according to the reaction:

$$2MO_2\,(s) + H_2O\,(v) \rightarrow 2MOH\,(s) + {}^3/_2O_2\,(g)$$

where *s*, *v*, and *g* refer to solid, vapor, and gas, respectively. Carbon dioxide is removed from the atmosphere by reacting with the product hydroxide (MOH) causing formation of carbonate (M_2CO_3) and $MHCO_3$:

$$2MOH\,(s) + CO_2\,(g) \rightarrow M_2CO_3\,(s) + H_2O\,(l)$$
$$2MOH\,(s) + 2CO_2\,(g) \rightarrow 2MHCO_3\,(s)$$

Here, *l* refers to liquid. The only superoxide produced on an industrial scale is potassium superoxide (KO_2). Therefore, it is the only active chemical that has been used extensively in breathing applications, especially on the manned missions of the former Soviet Union. KO_2 canisters used in self-contained breathing apparatus for fire fighting and mine rescue work have exhibited poor utilization efficiency (50-80 %) and some overheating problems because of the high exothermic heats of reaction. The low efficiency was attributed to the formation of a hydrous coating of reaction products on the surface of unreacted KO_2. This problem led to recent research on other superoxides, particularly calcium superoxide ($Ca(O_2)_2$) and mixtures of $Ca(O_2)_2$ and KO_2. $Ca(O_2)_2$ also has a higher available O_2 concentration and a greater CO_2 scrubbing capacity per unit weight of chemical than KO_2. Mixtures of $Ca(O_2)_2$ and KO_2 show improved utilization efficiencies when compared to KO_2. Despite these problems KO_2 has been sucessfully used on many missions of the former Soviet Union. The theoretical capacity of KO_2 is 0.309 kg CO_2 per kg sorbent. At the same time 0.38 O_2 per kg sorbent are produced. [29,31]

V.1.1.4 Oxygen Generation

In an open ECLSS, the O_2 required to meet a person's metabolic requirement and to make up for spacecraft cabin leakage is obtained from a stored cryogenic O_2 source and scavenged from spent O_2 fuel tanks. In a regenerative ECLSS, particularly for the first generation of very long duration manned mission, it is generally agreed that water electrolysis will be used to regenerate O_2 for respiration. The basic equation of the electrolysis is:

$$2H_2O + Energy \rightarrow 2H_2 + O_2 + heat$$

The value of this process is not only that it produces precious O_2 for crew use but also H_2, which could fuel a H_2-O_2 fuel cell to meet onboard power needs and could also be used to regenerate H_2O by reduction of expired CO_2. Supply H_2O for the electrolyzer would come from a storage tank which contains H_2O from CO_2 reduction, thus completing the cycle of O_2 recovery from CO_2.

The recovery of oxygen from CO_2 will yield a maximum of 0.74 kg O_2/ man-day from an average amount of 1 kg CO_2 exhaled per man-day. This amount of exhaled CO_2 is based on an average oxygen consumption of 0.9 kg/man-day. The balance of the oxygen is consumed in metabolic processes. Recovery of oxygen from CO_2 thus closes a large part of the oxygen cycle. Only the oxygen needed for the balance of crew respiratory quotient and the leakage losses must then be stored onboard the spacecraft. The major criteria to be considered when evaluating oxygen recovery units are reliability, weight, power, expendable requirements, purity of recovered oxygen, and ease of integration with other life support subsystems. The major competing regenerative subsystems for O_2 generation are currently the Static Feed Water Electrolysis (SFWE) and the Solid Polymer Water Electrolysis (SPWE) process, but also Water Vapor Electrolysis (WVE), CO_2 Electrolysis, and Superoxides are considered. A system for water electrolysis has been in operation aboard the space station MIR since December 1989 and produced several hundreds m^3 of oxygen functioning at intervals from several minutes up to tens of days. [4, 7, 15, 16, 31]

V.1.1.4.1 Static Feed Water Electrolysis (SFWE)

SFWE electrolyzes hygiene water to produce oxygen. Water resting statically in a feed compartment diffuses as vapor through a water feed membrane and into an aqueous KOH electrolyte. Oxygen gas is produced at the anode of the electrolysis cell, while hydrogen gas is produced at the cathode. Product hydrogen is furnished to the CO_2 reduction system with any excess H_2 requiring venting or storage. SFWE has the capability of circulating feed water to provide any needed cooling. An SFWE cell functional scheme and the accompanying electrochemical reactions are shown in figure V.9.

Power for electrolysis is provided by a DC power supply and waste heat is removed by a circulating liquid coolant located adjacent to the O_2 generation cavity. Water for electrolysis is provided by the feed water matrix. Water evaporates from the feed matrix and condenses on the cell matrix where it is subsequently electrolyzed. Both the water feed and cell matrices consist of thin asbestos sheets saturated with a hygroscopic aqueous potassium hydroxide (KOH) solution. Poisoning of the catalyst on the metal electrodes of the electrolysis cell is prevented by not allowing liquid feed water which may contain catalyst poisons from contacting the electrodes.

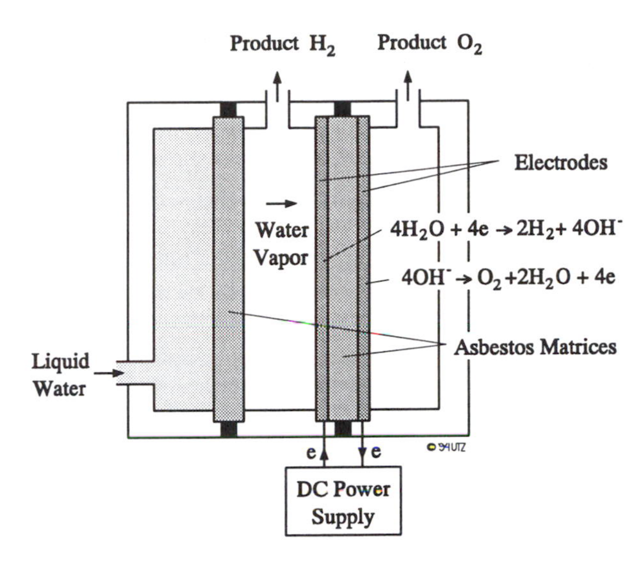

Figure V.9: **SFWE Cell [30]**

At the start of electrolysis, both the water feed and cell matrices have equal KOH concentrations. As DC power is supplied to the electrode, water in the cell matrix is electrolyzed, thereby increasing the KOH electrolyte concentration. The KOH concentration gradient which then exists between the cell and water feed matrices causes water vapor to diffuse from the feed to the cell matrix. Water removed from the water feed compartment is statically replenished from an external water supply tank. If coupled to a water vapor electrolysis dehumidifier module, water vapor can be removed from the effluent O_2 line of a SFWE system, thereby providing additional oxygen and eliminating the need for a water vapor / O_2 separator. Like the SPWE, the SFWE can be operated continuously or cyclically and if desired at high pressures ($7 \cdot 10^6$ Pa). Improvements of the SFWE could include better anode/cathode material to increase cell life and reliability, and an improved electrolyte to reduce cell resistance and power requirements. Such improvements are presently superfluous, given the high level of maturity reached by the SFWE. [31, 35]

V.1.1.4.2 Solid Polymer Water Electrolysis (SPWE)

Similar in concept to SFWE, the SPWE electrolyzes water using a solid polymer electrolyte. The electrolyte in the SPWE is a solid plastic sheet or membrane of perfluorinated sulfonic acid polymer about 0.3 mm thick. When saturated with H_2O, this polymer is an excellent ionic conductor (15 Ωcm resistivity) and is the only electrolyte required in the system. An SPWE cell is shown in figure V.10 with its associated electrode reactions.

Catalyzed electrodes for improving power efficiency are placed in intimate contact with both sides of the membrane. SPWE requires that feed water be in direct contact with the cell anode to provide cooling. The cell membrane prevents intermixing of O_2 and H_2 gases and also eliminates the need for an H_2O-O_2 separator on the O_2 outlet side of the cell. The deionized process H_2O used in the SPWE subsystem acts as both a reactant and coolant and is fed to the

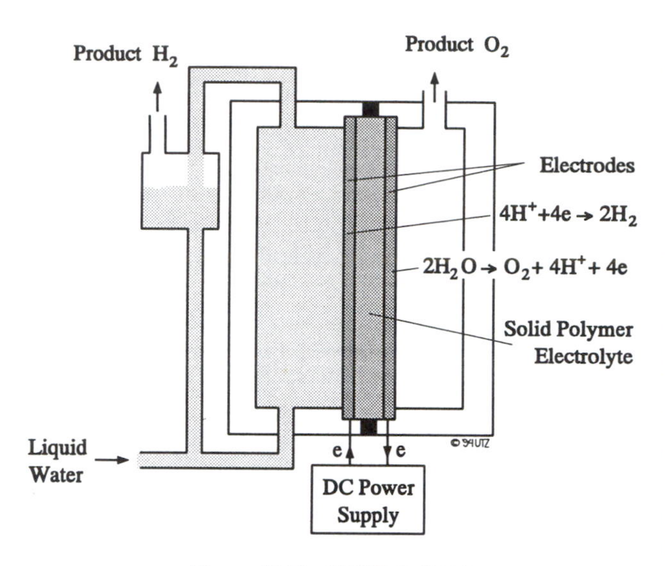

Figure V.10: **SPWE Cell [31]**

H_2 electrode side of the cell. The SPWE cell membrane is sufficiently permeable to allow H_2O to diffuse from the electrode or cathode to the O_2 electrode or anode where it is electrochemically decomposed to provide O_2, hydrogen ions (H^+), and electrons. The hydrogen ions move to the hydrogen evolving electrode by migrating through the membrane. The electrons pass through the external electrical DC circuit to reach the hydrogen electrode. At the hydrogen electrode, the H^+ and electrons recombine electrochemically to produce H_2 gas. Hydrogen gas is separated from the water feed by a dynamic 0-g phase separator pump located external to the electrolyzer. The SPWE subsystem can deliver O_2 or H_2 at any pressure required by simply back pressuring the corresponding gas side. However, for high-pressure operation (790 kPa), the cell is enclosed in a pressure vessel and the differential pressure across the SPWE is controlled by the subsystem gas generator. The SPWE can be operated either continuously or cyclically. Cyclic operation is beneficial for reducing power consumption aboard a solar-cell-powered space habitat when it is shielded from the Sun. [31, 35]

V.1.1.4.3 Water Vapor Electrolysis (WVE)

WVE electrolyzes water vapor directly from cabin air, producing oxygen and hydrogen. Moist air is fed to the anode side compartment of the SFWE-style electrolysis cell, producing an O_2-enriched stream at the anode and hydrogen at the cathode. The membrane prevents mixing of the O_2 and H_2. The process operates continously, with O_2 returning to the habitat and H_2 being

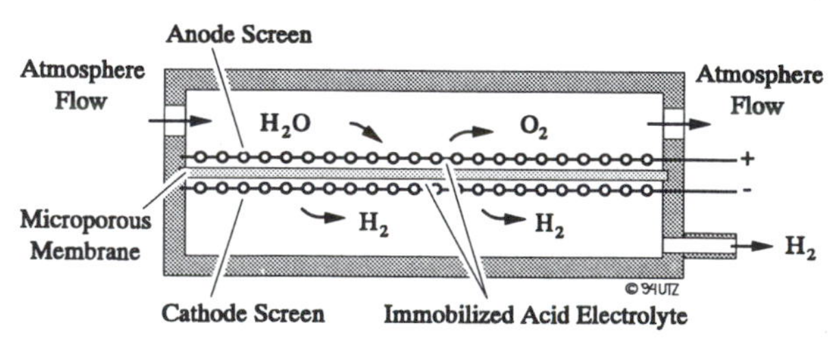

Figure V.11: **Water Vapor Electrolysis [29]**

separated for use of disposal. WVE helps to control cabin humidity, and, having few interfaces, lends itself as a portable technology for providing additional O_2 should the crew gather for significant periods of time. A portable WVE is also a potential candidate for providing a safe haven atmosphere regeneration system. A scheme of WVE is given in figure V.11.

V.1.1.4.4 CO_2 Electrolysis

See section V.1.1.3.4.

V.1.1.4.5 Superoxides

See section V.1.1.3.5.

V.1.1.4.6 Artificial Gill

With an artificial gill, oxygen is extracted directly from air using regenerative oxygen binding organometallic compounds such as hemoglobin. The artificial gill will be ideal for collecting O_2 from a plant growth chamber or CELSS module where plants generate oxygen in low concentrations. This technology is in the early stages of development. It may also be used for the separation of oxygen from the Martian atmosphere. [35]

V.1.1.5 Nitrogen Generation

Because of leakage and air losses the resupply of nitrogen which comprises about 80 % of the loss has an appreciable mass impact on the life support system. The nitrogen content of the food consumed by the crew is only about 5 % of the expected daily loss. Therefore, nitrogen is resupplied from high pressure gas bottles for short-duration missions. For longer duration missions consideration will be given to using other N_2 supply methods to reduce the weight of this resupply subsystem. Cryogenic storage of nitrogen is subject to boil-off losses as well as about 0.5 kg penalty for tank mass and ullage for each kg of recoverable N_2. The tank mass penalty for hydrazine (N_2H_4) a well-behaved liquid with density comparable to water, is only approximately 0.2 kg per kg useable N_2H_4. Therefore, nitrogen can be stored at less mass in the form of hydrazine than it can as a cryogen.

One method of nitrogen generation is based on the thermal catalytic dissociation of hydrazine. Hydrazine is a commodity aboard most space habitats for rocket engine propulsion and refueling of satellites which use N_2H_4 thrusters. Thermal catalytic dissociation of N_2H_4 at 1005 K and 1.7 MPa followed by catalytic ammonia (NH_3) decomposition involves the following equilibria:

$$3N_2H_4 \rightarrow N_2 + 4NH_3 + heat$$

$$4NH_3 + heat \rightarrow 2N_2 + 6H_2$$

$$3N_2H_4 \rightarrow 3N_2 + 6H_2 + heat$$

The N_2-from-hydrazine subsystem consists of a catalytic hydrazine decomposition reactor followed by a hydrogen separator and catalytic oxidizer. Since the hydrazine decomposition reaction is exothermic, three heat exchangers are required to dump waste heat. The separator removes hydrogen from the nitrogen product stream. The catalytic oxidizer oxidizes any remaining ammonia and hydrogen in the product stream to nitrogen and water.

A N_2-from-ammonia subsystem is also under study. It is very similar to the N_2-from-hydrazine system. The major difference between the two systems is the fact that the hydrazine reaction is exothermic while the ammonia reaction is endothermic. Therefore, the ammonia dissociation reactor requires a preheater as the reaction is not self-propagating. Catalytic ammonia decomposition and hydrogen separation are carried out in several stages following N_2H_4 dissociation. Hydrogen is separated from N_2 using palladium-silver alloy tubes and is then available for CO_2 reduction, attitude control or other laboratory uses. Test results have shown that the N_2H_4 dissociation/separation concept is feasible. [6, 31]

V.1.1.6 Trace Contaminant Control (TCC)

The trace contaminant management system has to protect the crew from hazardous contamination through airborne biological and chemical contaminants, including particulates, in the closed environment of a space habitat. A set of hardware has to be provided which is able to detect hazardous trace gases in the atmosphere and to control them below their Space Maximum Allowable Concentrations (SMAC) that have been established for many individual and classes of contaminants (see section IV.3.1.1). The most desirable way to remove trace gases from the atmosphere is to convert them into harmless ones, e.g., CO_2 and H_2O, by non-consuming means.

Trace gases may originate from various sources, e.g., metabolism of the crew members and out- and off-gassing of cabin materials such as plastics, insulation, adhesives, and paints. In addition, there are a number of minor sources, e.g., thermal degradation, leakage, or spills, which will also contribute to the overall contaminant load in the atmosphere. Although the quantities of those contaminants and the likelihood of their occurrence may be small, they cannot be neglected, because these contaminants are hazardous for the crew and form a problem when defining a load model for design purposes. To reduce the burden on the removal subsystem, precautions have to be taken in the design and assembly of space habitats to use materials which minimize the release of contaminants to the atmosphere. In the past mainly expendable charcoal beds have been used for trace contaminant control. Now, regenerable systems are under consideration to decrease the large resupply penalties also because, as crew size and mission duration increase, contaminant control aboard manned space habitats will become increasingly more important and the removal subsystem will become more complex. Typical trace contaminant removal units and their functions are:

- *Particulate filters:* To separate dusts and aerosols
- *Activated charcoals:* To separate high molecular weight contaminants
- *Chemisorbant beds:* To remove those gases which cannot be absorbed by activated charcoal or oxidized to CO_2 and H_2O. These are primarily nitrogen and sulfur compounds, halogens and metal hybrids.
- *Catalytic burners:* To oxidize those contaminants which cannot be readily absorbed [1, 7, 15, 31]

Charcoal (carbon) adsorption beds coupled with platinum catalytic oxidizers remove trace contaminants, such as freons and aromatics, from cabin air. The charcoal beds are expendable, requiring periodic changeout and resupply. Better adsorption materials and catalytic oxidizers could increase trace contaminant removal efficiency, but the most likely improvement will be a completely regenerative system, eliminating resupply needs. The design characteristics of a conceptual TCC system are outlined in table V.11. A safe, practical method for handling contaminant waste after regenerating a charcoal bed has not yet been determined. [35]

In figure V.12 a contaminant control subsystem is delineated. The nonregenerable activated charcoal bed, impregnated with phosphoric acid, is designed for control of well-adsorbed contaminants, ammonia and water-soluble contaminants. A filter located downstream of the fixed bed prevents particulates from entering other parts of the system. About 10 % of the effluent from the fixed bed enters the smaller regenerable activated charcoal bed while the remaining air is ducted to the cabin. The low air-flow rate (long residence time) through the regenerable

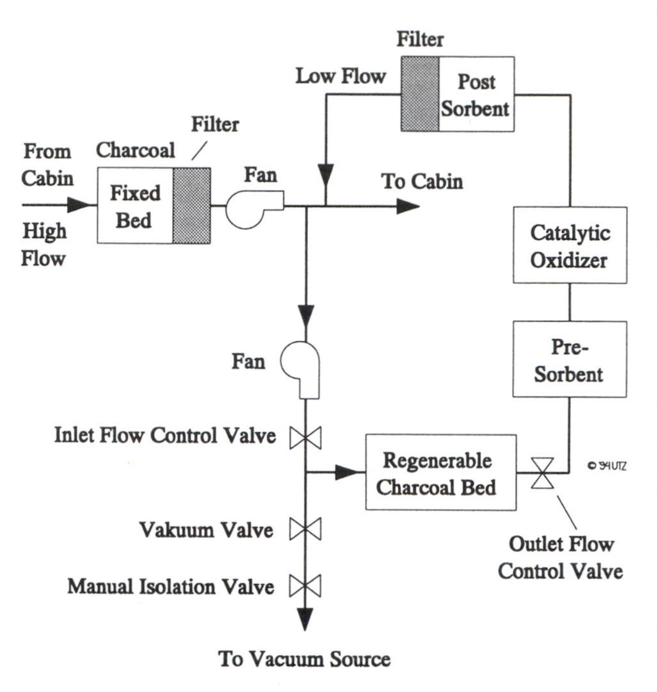

Figure V.12: **Trace Contaminant Control Subsystem [31]**

bed relative to the nonregenerable bed, aids in the removal of contaminants that are poorly adsorbed by charcoal. The use of the combined nonregenerable and regenerable charcoal beds reduces the weight of charcoal needed to handle the load model. The regenerable bed is reactivated by using heat and vacuum desorption. At the end of a desorption cycle the charcoal bed temperature is 366 K and the pressure inside the charcoal canister is $6.7 \cdot 10^{-3}$ Pa. The time required for regeneration is 195 minutes. A nonregenerable LiOH presorbant bed is located upstream of the catalytic oxidizer to prevent acid gases, e.g., HCl and SO_2, from entering the oxidizer and poisoning the catalyst. The high temperature catalytic oxidizer contains an oxidation catalyst, such as palladium on alumina, and operates at 711 K. The primary function of the catalytic oxidizer is to oxidize hydrocarbons not adsorbed in the charcoal beds. The effluent from the oxidizer flows to the nonregenerable postsorbant LiOH bed which removes any undesirable acidic products of oxidation. Halogenated compounds if allowed to reach the oxidizer could produce acidic compounds.

An interesting new TCC technology under consideration is Reactive Bed Plasma (RBP), which can oxidize contaminants at low temperatures (394 K). Contaminants are decomposed by RBP using a synergistic combination of plasma and catalyst. Process air passes through an annular reactor filled with alumina catalyst, where it is partially ionized into plasma. The main function of the catalyst is to increase the time contaminant modules spend in the active plasma region, where plasma-generated high-energy electrons and subsequently produced species (mainly active oxygen) oxidize toxic materials. In addition, RBP can perform as a highly efficient electrostatic precipitator to collect and deactivate hazardous particulate material. The RBP system also includes regenerable posttreatment of toxic reaction products, particularly nitrous oxides. Despite the high power requirements and possible decrements in performance

Subsystem	**Trace Contaminant Control (TCC)**

Weight:	<100 kg	Volume:	< 0.3 m^3
Heat Generated:	-	Power Required:	150 W

Operating Temperature:	723 K
Operating Pressure:	100 kPa

Operations Mode:	Continuous
Designed Efficiency:	-
Comment:	Temperature for catalytic oxidizer; activated charcoal and other beds operate at ambient pressure;
Technology Readiness Index:	5

Advantages:	- Versatile or non-specific - can control many different organic and inorganic airborne trace contaminants - Provides for a country-fresh air environment
Disadvantages:	- Data for load model is often qualitative; quantitative data on material off-gassing needed - SMAC levels are often based on qualitative considerations - Catalyst poisoning must be taken into account - Regenerable charcoal bed may be needed for missions to Moon or Mars; effective method for bed regeneration - Long-term testing of an integrated TCC needed
Prospective Improvements:	- Development of a method for activated charcoal regeneration without decrease of charcoal capacity - Reliability and modes of failure of a TCC subsystem are unknown - Quantitative data (GC-MS) on the outlet of a TCC subsystem during long-term operation

Principal Chemical Equation:	-

	Inflow [kg/day]	Inflow Composition	Outflow [kg/day]	Outflow Composition
Gases	0.0154 (contaminants only)	20 vol.-% O_2 80 vol.-% N_2 < 1 % Contaminants	-	20 vol.-% O_2 80 vol.-% N_2 < 1 % Traces of oxidation products
Liquids	-	-	-	
Solids	-	-	-	

Table V.11: **Design Characteristics of a TCC Conceptual Design [35]**

caused by surrounding electromagnetic interference that are current issues in RBP testing, RBP has the potential for operating as an efficient, low temperature, long life, universal contamination control device. [7, 35]

Another alternative for trace contaminant control is the Super Critical Water Oxidation (SCWO). Since one input to it is cabin air, the SCWO reactor can potentially oxidize all trace contaminants in the process air, eliminating the need for a stand-alone trace contaminant control system. SCWO is currently in the early stages of development (see section V.3.2.1). [35]

V.1.2 Atmosphere Monitoring and Control

The main purpose of the atmosphere monitoring and control system are temperature, humidity, trace contaminant and microbial contaminant monitoring and control. For a space habitat the capability to detect, differentiate, quantitate, and warn the crew shall be provided to maintain crew health for all chemical and selected particulate contaminants by real-time or near-real-time monitoring. Also, the capability to disinfect/sanitize contaminated areas/substances with an approved nontoxic agent which can be accommodated by the ECLSS shall be provided. The trace gas monitoring equipment has to be capable of continuous sensing and display of CO_2, CO and total hydrocarbons with inputs to the subsystems which control the level of these contaminants (see section V.1.1.6). The trace gas monitoring system has to consist of three main parts:

- *Mass spectrometer supplied with air samples via a gas chromatograph*
- *IR-detector*
- *Special sensors*

Three systems are necessary since none of them is sufficient to detect all important traces with reasonable effort. The Gas Chromatograph/Mass Spectrometer (GC/MS) assembly is not capable to distinguish directly substances of identical molecular weight. The IR spectroscopy is not feasible for non IR-active substances like O_2. Moreover the detection of the O_2 partial pressure has to be provided with two failure tolerance due to its importance for life/health of the crew. Therefore, O_2 detection is done via special sensors. For the dimensioning of the contamination control system, for an individual compound *i*, the necessary flow rate through the assembly is calculated as:

$$F_i = \frac{G_i}{SMAC_i \cdot \varepsilon_i}$$

Generation rate of compound i: G_i in [mg/h]

Flow rate of compound i: F_i in [m^3/h]

Efficiency of removal process for compound i: ε_i $(0 < \varepsilon < 1)$

Space maximum allowable concentration of compound i: $SMAC_i$

The parameter selected for the monitoring of atmosphere microbial monitoring is Colony Forming Unit (CFU). As indicated by the name it has to be monitored how many vital microorganisms able to form colonies are in the habitat air. Since sedimentation like on Earth is not possible, this is done by blowing air onto an Agar substrate in an incubator and counting the colonies after 24 hours. The problem of this method is that the number of microorganisms is drastically increased by the incubation. Therefore, special precautions are necessary to avoid contamination of the habitat. The evaluation of the incubated Agar sample is done on ground using a video picture. Atmosphere microbial contamination control is done in combination with trace gas contamination control (see section V.1.1.6). The charcoal filters are impregnated with a bactericidal substance. The filter design has to guarantee that for all nominal conditions "grow through" of the filter is avoided since otherwise contamination of the module with a high number of microorganisms occurs.

V.1.2.1 Temperature and Humidity Control (THC)

The THC system of a space habitat has to control temperature and humidity levels such that a comfortable shirt-sleeve atmosphere for the crew is provided continuously. THC is also important because if the relative humidity in the atmosphere of a space habitat is too high, water may condensate and influence or damage instruments aboard. Condensing heat exchangers (CHX) effect both the thermal and the humidity control. CHX have been used since the beginning of manned spaceflight by both the United States and the former Soviet Union. Thermal control is achieved by transferring internal heat loads and external heat fluxes to a water coolant loop. Removal of heat loads, generated by the crew, electronic equipment and other sources, is effected by transfer from the air loop to one or two coolant loops. For minimum weight and volume at maximum effectiveness and reasonable pressure drops multipass cross-counterflow, plate fin heat exchangers have proven to be superior. This is achieved by the extremely large specific heat transfer areas of this type of heat exchanger.

Humidity, generated by the crew and/or the CO_2 control unit, is removed by decreasing the air temperature below its dew point and by separating the condensed water from the air flow. Removal of the moisture is achieved in three steps:

1. *Decreasing the air temperature below the dew point temperature to condense water*

2. *Formation of a condensate film*

3. *Separation of the condensate film from the air stream*

The separation of condensate film and air stream is realized by including a so-called slurper into the air discharge face of the CHX. A functional scheme of a slurper is given in figure V.13. The condensate film is transported by the air flow to the slurper holes where it is drained off by the negative pressure. The quantities of water and air separated from the main air stream is determined by the size and number of slurper holes. [20]

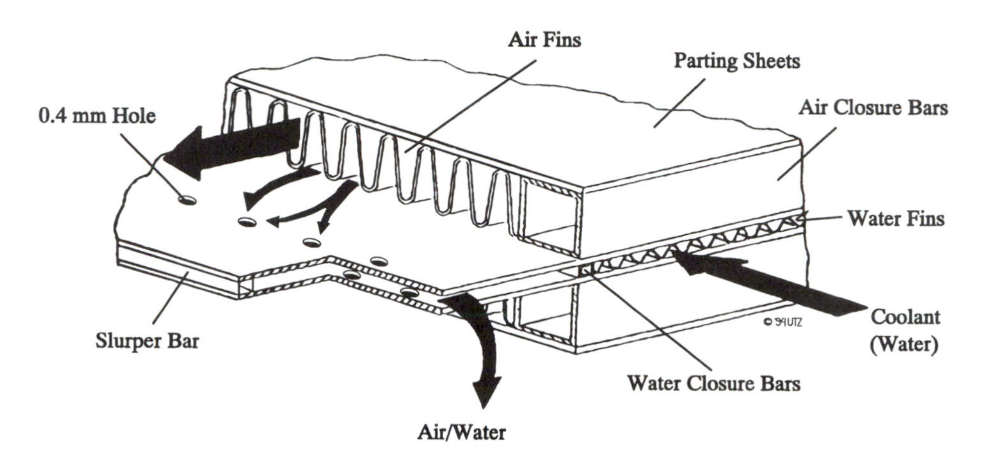

Figure V.13: **The Function of a CHX Slurper [20]**

V.1.2.2 Contaminant Monitoring

A prerequisite to controlling atmospheric contaminants is being able to monitor both the types and concentrations of contaminants. Monitoring is particularly difficult because of the variety of chemical species to be monitored and the high sensitivity required. Often sensitivity is required in the parts per billion range, especially for the more toxic species of contaminants. Current mass spectroscopy (MS) technology coupled to an inlet system for concentrating contaminants can perform this function, and has been proven in microgravity conditions. Basic parameters which determine the selection of trace gas monitoring strategies are seen in:

- *Probability of the occurrence of the compound*
- *Toxicity of the contaminants*
- *Quantities or production rates*
- *Removal efficiency by the ECLSS*
- *Capacity to degrade the ECLSS*

The general difficulty in the application of terrestrial trace gas monitoring strategies, e.g., in buildings and industrial plants, lies in their possibility to introduce fresh and clean air to dilute the contamination of the atmosphere. The strategies onboard submarines mainly cover the field of monitoring the atmosphere composition, rather than monitoring trace gases. Thus, for the numerical analysis and simulation of trace gas generation and removal processes within a space habitat, trace gas contamination control analysis software has been developed. [7, 13, 19, 31]

V.1.2.2.1 Gas Chromatograph / Mass Spectrometer (GC/MS)

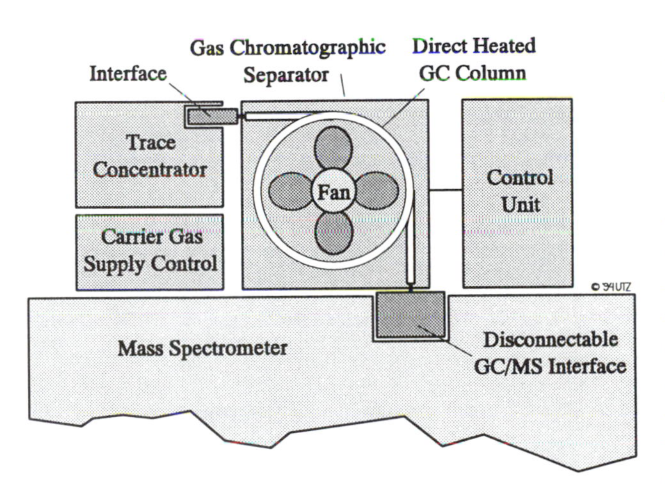

Figure V.14: **Gas Chromatograph / Mass Spectrometer Unit [9]**

The GC/MS separates atmospheric constituents for analysis during a two stage process. The first stage is a gas chromatograph that uses molecular diffusion to separate consti-tuents, and the second stage is a mass spectrometer employing electromagnetic separation. The GC/MS could be improved by reducing interference caused by the normal constituents of air (O_2, N_2, H_2O), which currently precludes the reliable analysis of contaminants weighing less than 50 amu. Increasing detector sensitivity below the present 1 mg/m³ would be another important improvement. GC/MS is flight-proven technology, having flown on the Mars lander *Viking*.

Gas chromatography is a technique for separating chemical substances in order to obtain pure components of a mixture. It is based on the repetitive distribution of compounds between two phases during a continuous flow. Components are detected by means of a thermal conductivity sensor or a flame ionization detector. The identity of components is determined by their retention time in a column. The concentration of components is determined by integrating the detector output. A gas chromatograph (GC) consists of a separation column, an injection port or an inlet interface at one end of the column, in order to insert the mixtures, and a detection device at the other end, in order to monitor the separated compounds exiting the column, and, finally, a gas supply equipment to operate the system.

A carrier gas flow in the GC column carries the compounds through its entire length. The carrier gas forms the mobile phase while the non-moving phase in the column is called the stationary phase. Compounds flowing through the column undergo the repetitive cyclic process of getting dissolved in the stationary phase and then reentering the gas (mobile) phase. The extent of this repetitive partitioning and the dynamic equilibrium of each compound between mobile and stationary phases determines its substance-specific duration time in the gas chromatographic column. This leads to a separation of the components during their transport in the column. A scheme of the GC process is shown in figure V.15.

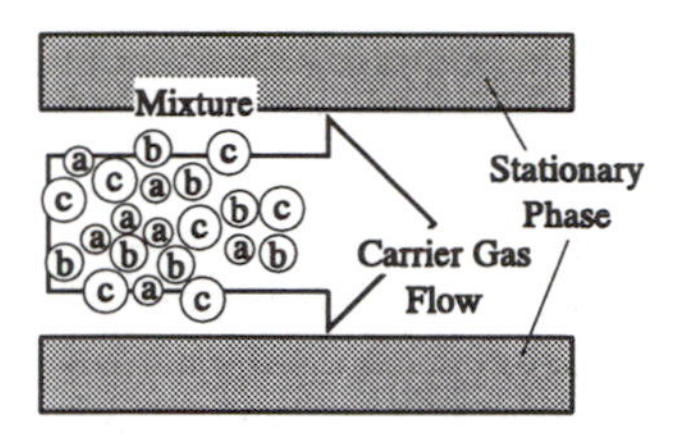

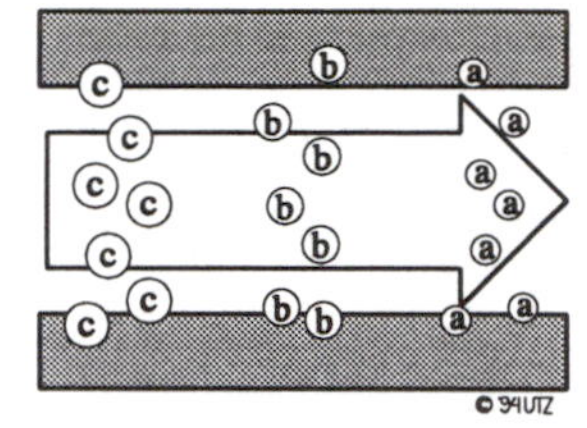

Figure V.15: **Gas Chromatographic Separation [9]**

The time between the injection of the mixture and the breakthrough of one of the components out of the column is the retention time t_r of this particular compound for this particular column under the given conditions. A detection device sensing the eluation of the components can produce a signal proportional to the amount of each compound exiting the column. Together with the specific retention time, this signal is used for identification and quantification (see figure V.16).

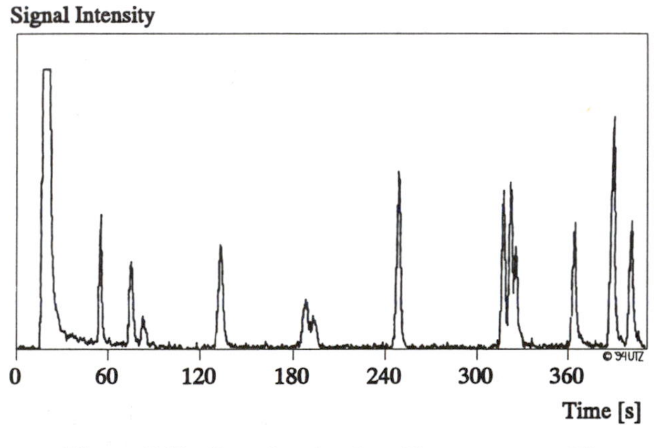

Figure V.16: **Sample of a Gas Chromatogram [9]**

In simple GC equipments, separated compounds are detected by a non-identifying device, which only indicates whether or not a compound is there (the amplitude of the signal can be, of course, used for quantification). The flame ionization detector is an example for these detectors where an identification of compounds is only possible by comparison of retention times under the same GC compositions. In a GC/MS, the detector of the gas chromatograph is a mass spectrometer, an identifying detector. Every retention signal of GC/MS corresponds to series of mass spectra, by which each eluating compound can be recognized. In addition to retention times, this is a new dimension of information. Another

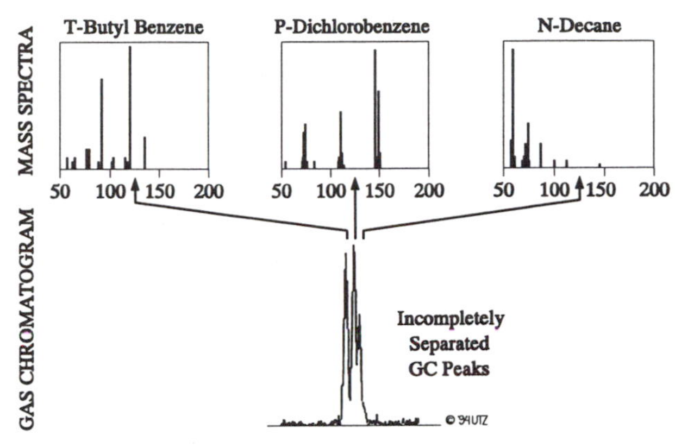

Figure V.17: **Mass Spectroscopy Applied to a Incompletely Separated Gas Chromatograph Peak Group [9]**

advantage of monitoring the GC-separated compounds using a mass spectrometer is that poorly separated GC peaks can still be identified (see figure V.17).

Basically, repetitive analysis can also be achieved by a mass spectrometer (MS). It is one of the most developed monitoring instruments for space habitat usage. The MS ionizes sample gases by means of a heated filament and separates the ionized particles in a magnetic field. The instrument scans the range of mass-to-charge ratios by varying the field so that only one mass of ions strikes the target at any one time. In general, a MS is coupled to an inlet system for concentrating contaminants on selected sorbants. When coupled with such an inlet system, e.g., a chromatograph, for concentrating contaminants the MS has the required sensitivity for detecting low concentrations and provides directly relatable data for compound identification. [9, 15, 31, 35]

V.1.2.2.2 Fourier Transform Infrared Spectroscopy (FTIR)

FTIR passes infrared radiation through a sample of cabin air, creating an IR adsorption spectrum with an interferometer. The IR spectrum is a signature of the constituents in the air sample. One limitation of FTIR is its inability to discriminate between related compounds. Compounds with similar structures may have similar absorbances in their IR spectra. Additionally, some compounds are undetectable by FTIR since they do not absorb IR radiation, e.g., hydrogen and chlorine. [35]

V.1.2.2.3 Ion Trap Mass Spectroscopy / Mass Spectroscopy (MS/MS)

MS/MS uses mass spectroscopy to perform both the separation and identification of analytes. Once in the ion trap, the cabin air sample is ionized, and the resulting ions are separated on the basis of molecular weight and analyzed by an ion detector. This sequence of operations simulates tandem mass spectroscopy. Based on the most recent mass spectral technology, MS/MS offers the fastest analysis time, potentially the lowest weight, and the greatest flexibility of any existing technology. The ion trap MS/MS is still under development. [35]

V.1.2.3 Fire Detection and Suppression

Fire is one of the most feared events onboard any space habitat. Although usually materials and operations are selected that should preclude the start of fire, the probability of a fire increases as missions become longer and more routine. Thus, fire detection devices, different from those that have mainly been used in the past ("crew senses"), become more important in future projects. Since different types of fire produce different characteristics, detection systems must be carefully designed for each application. These different characteristics include small particles produced due to thermal breakdown during the incipient stages of a fire, particulate (smoke) or gaseous products of combustion, energy emissions from the fire (radiation), and elevated temperatures. In any case, false alarms should be avoided. Once the fire has been detected it may be suppressed, e.g., by using CO_2, N_2, or Halon.

Many aspects of the design of a fire detection system for a space habitat are affected by the microgravity environment. For example, in normal gravity, the large amounts of heat released in a combustion zone result in density gradients that drive buoyant convection flows, but in microgravity buoyancy-driven flows are absent. Basically, there are two different kinds of sensors that have been proposed:

- Smoke detectors
- Flame detectors

Smoke detectors may be located in air cooling ducts and standoff areas. In standoff areas, ventilation fans may provide forced ventilation which may facilitate the transport of smoke to the detectors. In open areas fire detection may be obtained by flame detectors. Smoke detectors monitor the presence of smoke particles in the air. There are two basic classes of smoke detectors. Those using photoelectric effects to detect smoke, like obscuration detectors, scattering detectors, and condensation nuclei counters. The other class comprises the ionization detectors that rely on the interaction of smoke particles with ionized air molecules. An obscuration smoke detector operates on the principle that when smoke passes between a light source (laser diode) and a receiver (photodiode), a portion of the light is obscured and the output of the photocell is reduced, thereby causing an alarm.

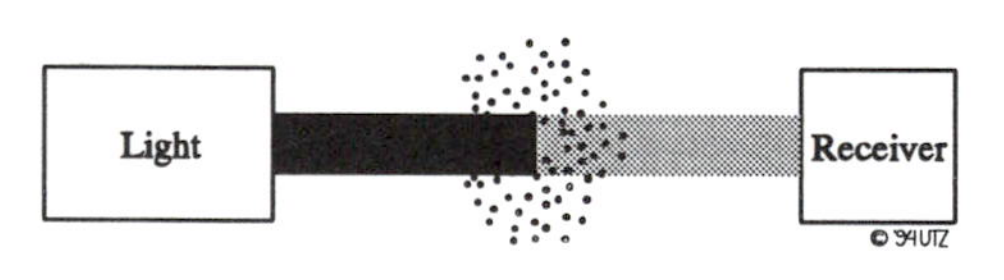

Figure V.18: **Obscuration Smoke Detector [10]**

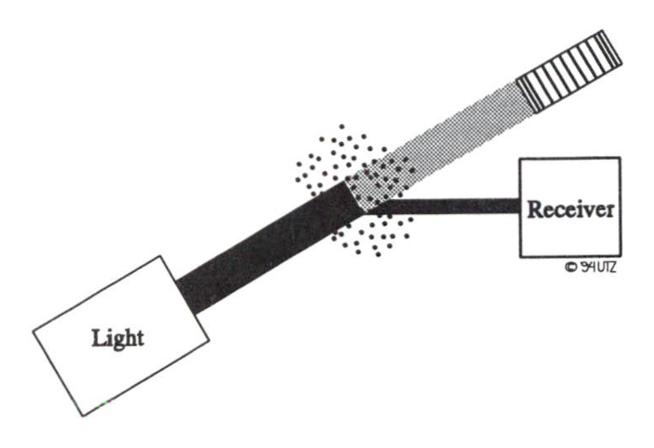

Figure V.19: **Scattering Smoke Detector [10]**

A light scattering smoke detector operates on the principle that when a beam of light passes through a space occupied by smoke, a portion of the incident light will be scattered onto a receiver.

A condensation nuclei counter (CNC) is able to detect smaller particles by increasing their size. Therefore, an air sample is passed through a humidifier where it is brought to 100 % humidity, i.e., water may condense on the smoke particles. These droplets are then large enough to scatter light .

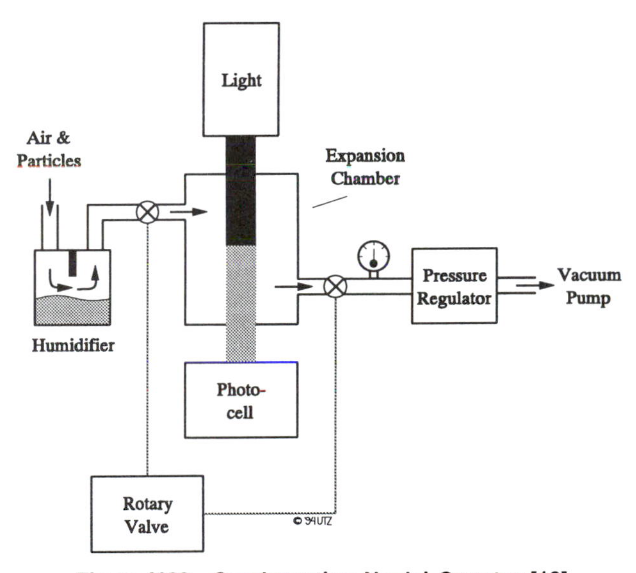

Figure V.20: **Condensation Nuclei Counter [10]**

An ionization smoke detector has a radiation source in a chamber containing positive and negative electrodes. The radiation source ionizes the molecules in the air between the electrodes. When smoke particles are attached to the ions, the increased mass of the particle-ion pair reduces the pair's velocity towards the electrodes. Since the convective velocity remains the same, the pair is carried out of the chamber and the resultant current is reduced, an alarm.

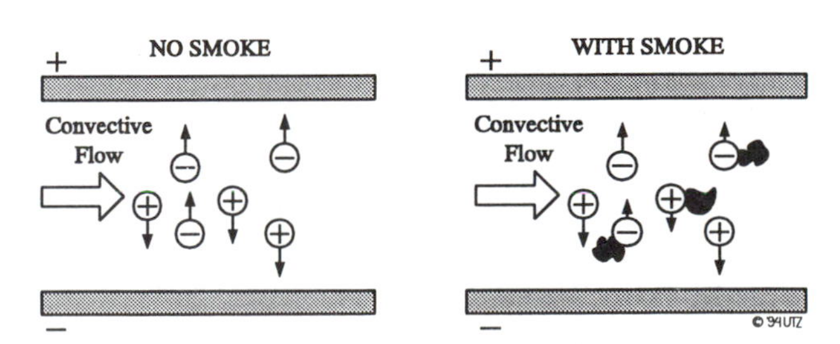

Figure V.21: **Ionization Smoke Detector [10]**

Flame detectors typically monitor one of three bands of the electromagnetic spectrum: UV, visible (VIS), and IR.

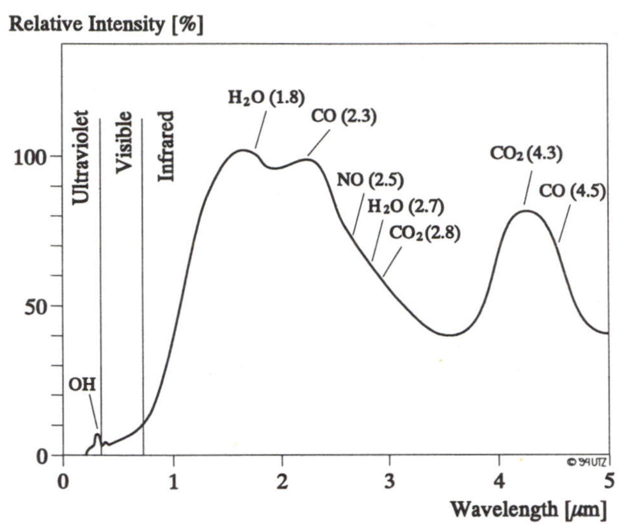

Figure V.22: **Radiant Energy Emitted from a Typical Hydrocarbon Fire [10]**

As an example, the spectrum of a typical hydrocarbon fire is given in figure V.22. Flame detectors may monitor single or multiple wavelengths. Typically, UV detectors monitor wavelengths from 0.18 to 0.26 µm, IR detectors 0.7 to 5 µm, and VIS detectors 0.4 to 0.7 µm. Monitoring different wavelengths reduces or eliminates the risk of false alarms. [10]

V.2 WATER MANAGEMENT

Water is critical to all life support. The water management has to supply the metabolic and wash water requirements of the crew. It functions also to collect atmospheric condensate and waste water. A water requirement of about 3 kg/man-day for drinking and food preparation and about 4.5 kg/man-day for washing and personal hygiene is needed. Waste water includes approximately 1.5 kg/man-day of urine and 2 kg/man-day of respiration and perspiration in addition to waste wash water. Approximately 0.3 kg/man-day of metabolic water is formed from the oxidation of food in the body and constitutes the difference between the water consumption and the body water output. Especially on long duration missions, the large weight penalty for resupply and the need for high purity for human consumption make water management a key enabling technology for manned spaceflight. The basic processes under consideration for water recovery aboard a space habitat can be placed into two categories: distillation and filtration processes. While the distillation methods are mainly considered for urine recovery, the filtration methods will basically process hygiene and potable water. Also, the water recovery from condensate in a space habitat atmosphere is of great importance.

Among the distillation or phase change processes, four different approaches have been pursued:

- *Vapor Compression Distillation (VCD)*
- *Thermoelectric Integrated Membrane Evaporation (TIMES)*
- *Vapor Phase CatalyticAmmonia Removal (VAPCAR)*
- *Air Evaporation*

The three filtration processes receiving the most attention are:

- *Reverse Osmosis (RO)*
- *Multifiltration (MF)*
- *Electrodialysis*

The process selected for water recycling will depend on the source of waste water and the quality of water needed for a given application. It is anticipated that the first generation of long-duration space missions therefore will include two water recycling and storage subsystems. One subsystem will process concentrated feeds, such as urine and flush water, and the second a more dilute feed, such as laundry or shower water. Potable water will probably be recycled using a phase change process while lower quality water will be recovered by filtration. The ultimate goal, however, is to have one space habitat water recovery subsystem.

In addition to the subsystem designed specifically for reclaiming space habitat waste water, by-product water is derived from other space habitat subsystems, such as H_2-O_2 fuel cells, CO_2 reduction and the space habitat condensing heat exchanger used for cabin humidity control. Water from CO_2 reduction is of high quality since it was derived from a high-temperature process which destroys harmful bacteria. Also, the feed gases used for CO_2 reduction are clean. Water derived from fuel cells and the condensing heat exchanger may require posttreatment to remove chemical and biological impurities prior to being reused. [7, 15, 31]

V.2.1 Urine Recovery

Water reclamation from urine in spaceflights is a complicated research and engineering task. The following factors should be considered in the development of both the process and the system:

- Urine is a complex aqueous solution containing more than 100 various organic and inorganic substances. Among the major contaminants are urea (13-20 g/l), sodium chloride (8-12 g/l), and various acids (up to 3 g/l). The total content of the contaminants in the water amounts up to 5 %. The distinguishing feature of urine is low heat resistance and sensitivity to bacterial decomposition of some substances present in it. At a temperature above 60° C and with the growth of microflora urea decomposition results in ammonia and carbon dioxide formation. Urine gives rise to precipitation when stored up.
- Stringent requirements upon water reclaimed from urine in terms of minimum content of organic and inorganic contaminants are placed.
- System equipment shall be operational under overloads and zero-gravity conditions.

In the former Soviet Union, a system for water reclamation from urine has been developed. In the system, water is recovered by evaporation followed by steam condensation. Condensate is purified by a sorption method. For drinking the purified water is conditioned, i.e., it is saturated with food salts and microelements, and decontaminated. This system has been in operation aboard the space station MIR since January 1990. The reclaimed water is used for generating O_2 from water electrolysis and for flushing the urine reception path. The system is automated and requires minimum maintenance. The operation is monitored by the crew and can be manually controlled from the onboard panel. In a 1.5 year phase of operation more than 1300 l of urine have been processed. Under flight conditions the amount of urine averages 1.2 l/man-day. With 80 % water reclamation from urine the crew is fully provided with oxygen generated from water electrolysis. [24]

In the United States, the major competing regenerative subsystems under development for urine processing are the TIMES and VCD processes. The total system resources for the two technologies are similar, except for a lower power requirement for the VCD. Product water quality of the two subsystems are also comparable. The processing rate for the VCD is much better. The TIMES lacks in its ability to process clean water. [16]

V.2.1.1 Vapor Compression Distillation (VCD)

In the VCD process vapor is compressed to raise its saturation temperature and then condensing the vapor on the surface in direct thermal contact with the evaporator. The resultant heat flux from the condenser to the evaporator, driven by its saturation temperature differential, is sufficient to evaporate a mass of water equal to that being condensed. Thus, the latent heat of condensation is recovered for the evaporation process.

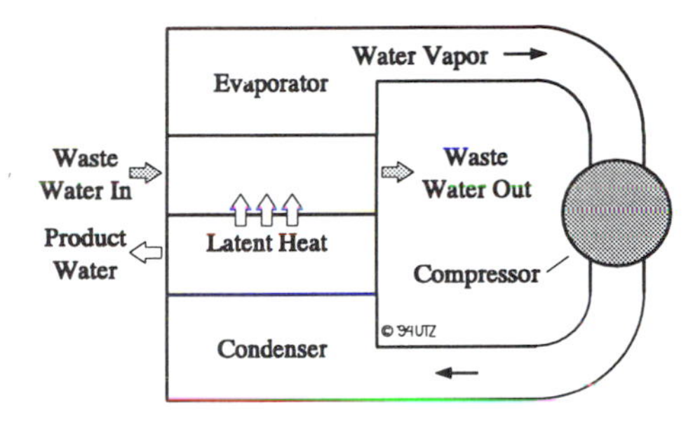

Figure V.23: **VCD Process [12]**

This makes the VCD process a thermally passive process, which requires no active temperature control. Waste heat generated by the process results in an operating temperature slightly above ambient, i.e., between 294 and 308 K. The resulting nominal condenser pressure is 4.8 kPa. The only energy required by the process is that necessary to compress the vapor and overcome thermal and mechanical losses.

The evaporator, condenser, and condensate collector are rotated to provide for zero-gravity phase separation. The VCD process has evolved to enable recovery of more than 96 % of the water contained in urine, concentrating the urine to over 50 % solids. Higher recovery percentages result when processing more dilute water streams. Unevaporated waste water is recirculated until the solids concentration reaches a specified concentration, at which point the solids are removed from the system and stored as brine. The quality of the water recovered from the VCD process will depend on the volatile organics and ammonia that co-condense with the water. Pretreatment of VCD waste water feed, e.g., with an acid, will probably be necessary to stabilize urea and prevent its decomposition to ammonia. Posttreatment of VCD product water may also be necessary before it is reused. Any noncondensible gases, e.g., CO_2, N_2, volatile organics, etc., evolved during distillation by the VCD process cause a pressure build-up and reduced evaporation efficiency. Restoration of efficiency requires periodic evacuation of the evaporator and treatment of the undesirable volatiles. Another disadvantages of the VCD is that its rotating parts can potentially disrupt the microgravity environment of a space station. The design characteristics of a VCD are given in table V.12. [7, 31, 35]

Subsystem	**VAPOR COMPRESSION DISTILLATION (VCD)**

Weight:	101.2 kg	Volume:	0.507 m^3
Heat Generated:	0.115 kW	Power Required:	0.115 kW

Operating Temperature:	289 K (min.)	303 K (nom.)	311 K (max.)
Operating Pressure:	3.6 kPa (min.)	4.8 kPa (nom.)	6.5 kPa (max.)

Operations Mode:	Contin. process operated in a batch mode (750 hours)
Designed Efficiency:	70 %
Comment:	Fundamental P/C process; highly developed; being evaluated against TIMES
Technology Readiness Index:	5
Reliability:	Unknown days
Maintainability:	Life span of the peristaltic fluids pump is the limiting factor

Advantages:	- Designed for application in zero and reduced gravity - Low power consumption due to passive latent heat recovery - Produces slightly higher quality water and processes higher flow rates than TIMES
Disadvantages:	- Processor incorporates rotating components - Generating of gaseous products which must be vented - Processors have not been operated in steady state mode - Feed stream is pretreated - Less than 100 % water recovery necessitates brine storage
Prospective Improvements:	- Adapt processor for steady state operation - Remove pretreatment of inlet stream - Develop modes of operation which minimizes the generation of NH_3

Principal Chemical Equation:	$Urea + H_2O \rightarrow NH_3 + CO_2$

	Inflow [kg/day]	Inflow Composition	Outflow [kg/day]	Outflow Composition
Gases	0	-	0.03	? % Ammonia ? % CO_2
Liquids	32.64	34.8 % Flush water 0.58 % Pretreat water	32.64	34.8 % Flush water 0.58 % Pretreat water
Solids	0.87	0.005 % Ozone 0.01 % Sulfuric acid 64.61 % Urine (= 96 % water)	0.84	0.005 % Ozone 0.01 % Sulfuric acid 64.61 % Urine (= 96 % water)

Table V.12: **Design Characteristics of a VCD Equipment [35]**

V.2.1.2 Vapor Phase Catalytic Ammonia Removal (VAPCAR)

VAPCAR is based on a catalytic chemical process where impurities vaporizing with the process water are oxidized to innocuous gaseous products. It is designed to eliminate the need for expendable chemicals. Unlike VCD and TIMES, VAPCAR is a physico-chemical process that combines vaporization with high-temperature catalytic oxidation of the volatile impurities, such as ammonia or any organic compound that vaporize along with the water.

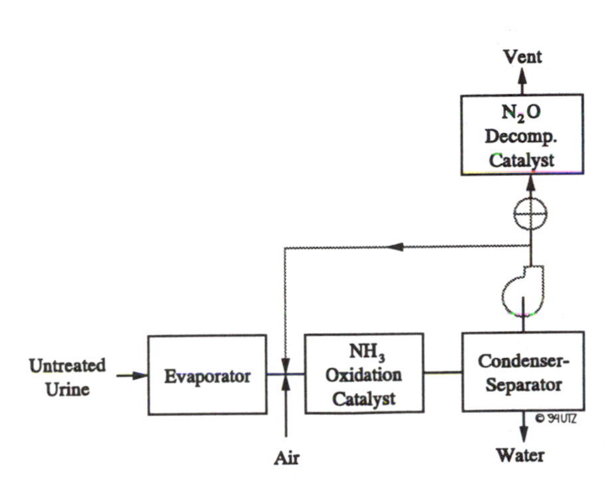

Figure V.24: **VAPCAR Process [31]**

Figure V.24 is a simplified flow diagram of the VAPCAR process using urine as a representative feed. The evaporator contains a bundle of hollow fiber membranes made from a perfluorinated ion-exchange polymer. Waste water is fed to the interior of the fibers and vaporizes from the exterior. The VAPCAR process employs two catalyst beds: in the first bed ammonia is oxidized to a mixture of nitrous oxide (N_2O) and nitrogen (N_2) and volatile hydrocarbons are oxidized to CO_2 and H_2O. In the second bed, N_2O is catalytically decomposed to N_2 and O_2. The N_2 and O_2 by-products from N_2O dissociation can be used to replenish space cabin N_2 and O_2. The ammonia oxidation reactor contains 0.5 % platinum on alumina oxidation catalyst pellets and operates at about 523 K. The N_2O decomposition reactor contains 0.5 % ruthenium on alumina catalyst pellets operating at about 723 K. The urine recycle and vapor loops in the VAPCAR subsystem are maintained above pasteurization temperature (347 K) to maintain water quality by minimizing or eliminating the growth of microorganisms. Unlike dry heat, wet heat as employed in the VAPCAR process is particularly effective in killing microorganisms. VAPCAR requires neither pretreatment nor posttreatment. The recovered water has little ammonia, few hydrocarbons, and low conductivity, requiring only adjustment of its pH to meet potable water standards. VAPCAR employs the same low-reliability membrane technology found in the TIMES, and has the additional disadvantage of operating at high temperatures. Some design characteristics of the process are given in table V.13. [31, 35]

Subsystem	**Vapor Phase Catalytic Ammonia Removal (VAPCAR)**

Weight:	68 kg	Volume:	0.24 m^3
Heat Generated:	0.1 kW	Power Required:	0.1 kW

Operating Temperature:	350 K (nom.)	723 K (max.)	
Operating Pressure:	50 kPa (min.)	80 kPa (nom.)	200 kPa (max.)

Operations Mode:	-
Designed Efficiency:	95 %
Comment:	Max. temperature is operating temperature of one of the catalyst beds; power requirements based on design
Technology Readiness Index:	3
Reliability:	-
Maintainability:	-

Advantages:	- Produces potable water - Catalytic beds decompose hydrocarbons and ammonia - Can operate in microgravity - No pretreatment - Minimum posttreatment
Disadvantages:	- Low technology level - Higher energy usage than VCD - Possible compressor problems
Prospective Improvements:	- Development needed on tube evaporator - Better compressor needed

Principal Chemical Equation:	-

	Inflow [kg/day]	Inflow Composition	Outflow [kg/day]	Outflow Composition
Gases	0.02	-	0.08	-
Liquids	13	-	13	-
Solids	0.45	-	0.45	-

Table V.13: **Design Characteristics of the VAPCAR Equipment [35]**

V.2.1.3 Thermoelectric Integrated Membrane Evaporation System (TIMES)

The TIMES subsystem employs a phase change process which uses a thermoelectric heat pump to transfer heat from a water condenser to an evaporator. A scheme of the TIMES concept is shown in figure V.25. This process recovers only a portion of the latent heat of condensation and transfers this heat to the evaporator via multiple thermoelectric elements. Before entering the TIMES, urine is pretreated with ozone and sulfuric acid to fix free ammonia, inhibit microbial growth, control odor, and reduce foaming. Oxone is an unstable, hazardous chemical in the liquid form, however, dispensing oxone as a solid is difficult in microgravity. Ideally, a relatively benign pretreatment requiring no expendables, such as UV light, will replace oxone. The pretreated urine/flush water entering the TIMES flows through two parallel heat exchangers in contact with the hot side of a thermoelectric heat pump. There the waste water is heated to 339 K and then flows through hollow fiber membranes that make up the evaporation section of the subsystem. The evaporator consists of 600 Nafion tubes assembled into six bundles of 100 tubes each. The exterior of the fibers is exposed to reduced pressure (17.2 kPa) causing water to evaporate from the outer tube surfaces and flowing to the main condenser. There the vapor is condensed on a porous plate surface in contact with the cold junctions of the thermoelectric elements. Partially condensed, the steam flows to an air-cooled heat exchanger, where the condensation process is completed. Latent heat is recovered and reused in the evaporative process. The condensate is tested for water quality, and, if acceptable, gets delivered as product water by a centrifuge-type pump. The pump acts as gas/liquid separator since it removes non-condensable gases entrained in the condensate stream. Unacceptable condensate is reprocessed. The TIMES can attain a 95% water recovery efficiency.

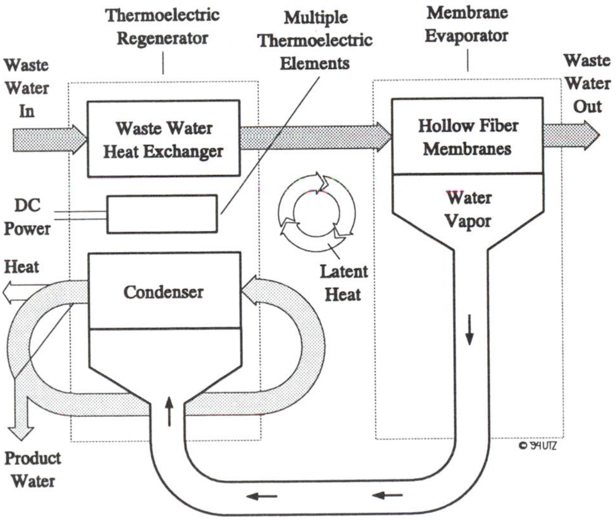

Figure V.25: **TIMES Process [31]**

Thermal control is required by the TIMES process since a major portion of the heat generated by the inefficiencies of the thermoelectric elements is used to partially offset the heat required for vaporization. An equivalent amount of latent heat liberated during water vapor condensation is, therefore, not used and must be removed from the process. Thermal control is maintained by circulating a portion of the condensate through an external heat exchanger and rejecting the heat to the ambient. The solids concentration in the TIMES recycle loop can be increased until 93 % of the water in the water feed is recovered, i.e., until the solids concentration reaches 38 %. The quality of the TIMES water, like VCD water, will depend on the amount of volatile organics and ammonia that co-condenses with water. The TIMES also shares with the VCD the problem of brine storage. Therefore, TIMES feed water may also require acid pretreatment to stabilize urea in urine and prevent breakdown to ammonia, and posttreatment before being reused to decrease the co-condensed organics. The design characteristics for TIMES process, based on a 20 kg/day model are given in table V.14. [7, 31, 35]

V.2.1.4 Air Evaporation Systems (AES)

In the AES pretreated urine is pumped through a particulate filter to a wick package using a pulse feed technique. The pulse and urine flow rate are controlled by a feed pump that responds to the relative humidity of air leaving the wick package and the amount of liquid processed. Sufficient time is provided between pulses to allow the volume of urine to be distributed before the next pulse. A circulating heated air stream evaporates water from the urine, leaving urine solids in the wicks. When sufficient urine solids accumulate in the wicks, the feed pump is stopped and the loaded wicks are dried down and replaced with a new wick package. Humid air leaving the wick package passes through a condensing heat exchanger. A water separator downstream of the condenser removes water from the air stream and pumps it to the product water loop. Acceptable product water passes through a microbial check valve to the posttreatment section, where it is filtered to remove trace contaminants before being stored for crew use. The complete dry-down procedure results in nearly 100 % water recovery from waste water. The characteristics of an AES paper design study for supplying 8 people are given in table V.15.

V.2.1.5 Aqueous Phase Catalytic Oxidation Post-Treatment System (APCOS)

The Aqueous Phase Catalytic Oxidation Post-Treatment System (APCOS) is a waste water posttreatment system. Its function is to remove, via oxidation, organic contaminants from a waste water input. The heart of the APCOS system is a catalytic oxidizer consisting of a catalyst housed within two stainless steel reactors which are plumbed in series. In breadboard tests, the system has successfully processed a contaminant load consisting of organic substances, like acetic acid, ethanol, and phenol, and inorganic salts, like sodium chloride, potassium chloride, and ammonium carbonate.

In the process, gaseous feedwater is delivered to the system, preheated in a stainless steel regenerative heat exchanger, and boosted to to the desired operating temperature of about 140° C with an inline heater. Then oxygen is injected into the feedwater in the reactor at a pressure of about 620 kPa, i.e.,

Subsystem	**THERMOELECTRIC INTEGRATED MEMBRANE EVAPORATION SYSTEM (TIMES)**

Weight:	68 kg	Volume:	0.23 m³
Heat Generated:	0.17 kW	Power Required:	0.17 kW

Operating Temperature:	336 K (min.)	339 K (max.)
Operating Pressure:	17 kPa (min.)	? kPa (max.)

Operations Mode:	Continous / Batch
Designed Efficiency:	91 %
Comment:	-
Technology Readiness Index:	4 - 5
Reliability:	-
Maintainability:	Longest hands-off run was 20 days (1987)

Advantages:	- Non-moving evaporation and condensing surfaces - Temperature differential for latent heat sharing is created by a non-moving device (the thermoelectric elements) - Circulation waste fluid is contained within hollow fiber membranes in the evaporating section
Disadvantages:	- The large number of organic fibers could be attacked by the acid oxidizing environment and fail - Plugging of small diameter tubes can result in a plugging/ maintenance problem - More breakdown of urea due to higher operating temperature causing poorer water quality than VCD - Not as energy efficient as VCD (about twice the energy per kg water recovered) - Requires pretreatment
Prospective Improvements:	- Reliability improvements - Ability to achieve higher water recovery

Principal Chemical Equation:	$Urea \rightarrow NH_3 + CO_2$ (side reaction)

	Inflow [kg/day]	Inflow Composition	Outflow [kg/day]	Outflow Composition
Gases	-	-	0.7	-
Liquids	21.8	-	21.1	-
Solids	0.7	-	0.7	-

Table V.14: **Design Characteristics of a TIMES Equipment Based on a 20 kg/day Model [35]**

Subsystem	AIR EVAPORATION SYSTEM (AES)

Weight:	-	Volume:	-
Heat Generated:	-	Power Required:	0.315 kW

Operating Temperature:	280 K (min.)	333 K (nom.)	333 K (max.)
Operating Pressure:	103.4 kPa (min.)	103.4 kPa (nom.)	103.4 kPa (max.)

Operations Mode:	Continuous - unit turned on until urine/flush water tank is emptied, then shut off.
Designed Efficiency:	100 %
Comment:	Distillation of water from urine by evaporation of water from urine-saturated felt pads, over which air is circulated in a closed loop.
Technology Readiness Index:	5
Reliability:	Relatively high, likely driven by compressor, slurper, fan, and valves (standard space-qualified).
Maintainability:	Wick evaporator design most allow ready access to wick unit for periodic replacement; efficient packaging and sealing of dried, spent wicks for disposal due to foul smell.

Advantages:	- Nearly complete water recovery - Simple device with standard space-qualified mechanical components - Insensitive to feed stocks - Operates at ambient internal pressure (easy sealing) - Operates either intermittently or continuously
Disadvantages:	- Energy intensive - Large logistics resupply requirement (mass and volume of wicks) - Requires pretreatment of urine solution
Prospective Improvements:	- Regenerable wicks - Increasing wicking effectiveness - Increased degree of energy recovery - Building of an engineering development unit

Principal Chemical Equation:	-

	Inflow [kg/day]	Inflow Composition	Outflow [kg/day]	Outflow Composition
Gases	0	-	0	-
Liquids	-	-	-	-
Solids	-	-	0	-

Table V.15: **Design Characteristics of an AES Paper Design Study for Supplying 8 People [35]**

140 kPa above the system pressure. This two-phase mixture passes through the catalyst oxidizing the organic compounds. Reactor effluent is cooled as it passes through the regenerative heat exchanger. Any free gas, i.e., undissolved O_2 and CO_2, can evolve from a recycle reservoir. Thus, the system pressure must always be high enough to assure the water is in the liquid state. [32]

V.2.1.6 Super Critical Water Oxidation (SCWO)

See section V.3.2.1.

V.2.2 Hygiene Recovery and Potable Processing

After use, i.e., washing and showering, water is a detergent solution containing mechanical impurities, e.g., desaquamated epithelium, hair, and dirt, fats, urea, ammonia and other organic and inorganic compounds entering into the composition of man's fat and sweat. The composition of hygiene water depends on many factors, e.g., the grade of detergent, hygienic procedure periods, man's individuality, intensity of physical and emotional loading, and environment parameters. This alone explains the difference in the properties of hygiene water: 450 - 600 mg O_2 / l, pH 4.8 - 7.1, 180 - 800 mg fats/l; 3 - 100 mg urea/l. Aboard a space habitat the amount and the composition of dissolved impurities depends not only on the quantity and the character of substances washed away during hygienic procedures but also on the substances condensed from transport air carrying water to the handwash and shower facilities into the sprayed stream of water, e.g., ammonia, alcohols, acetone etc.

For hygiene water recovery aboard the space station MIR a system for receiving untreated water as a water/air mixture from the handwash and shower facilities, separating the mixture phases, storing untreated water, purifying water from mechanical and dissolved impurities, decontaminating and storing purified water, preheating and feeding purified water to the handwash and shower facilities has been developed. The system is currently in operation onboard MIR providing a 3-man crew with water for hand washing and showering. The system is operated automatically. The operation of the system is monitored by the crew from the system control panel. The operational features of the system are: [23]

- Recovered water capacity: 0.9 l/h
- Energy consumption: 0.55 Wh/l
- Weight in terms of 6600 l service life: 530 kg
- Weight of consumable replaceable blocks: 400 kg

The major competing regenerative subsystems for potable and hygiene water processing of the manned space program of the United States are the Multi-filtration (MF) and Reverse Osmosis (RO) processes. The quantitative resources for the multifiltration and reverse osmosis processes are similar, with MF having slightly lower power and resupply weight and volume. The three major issues of

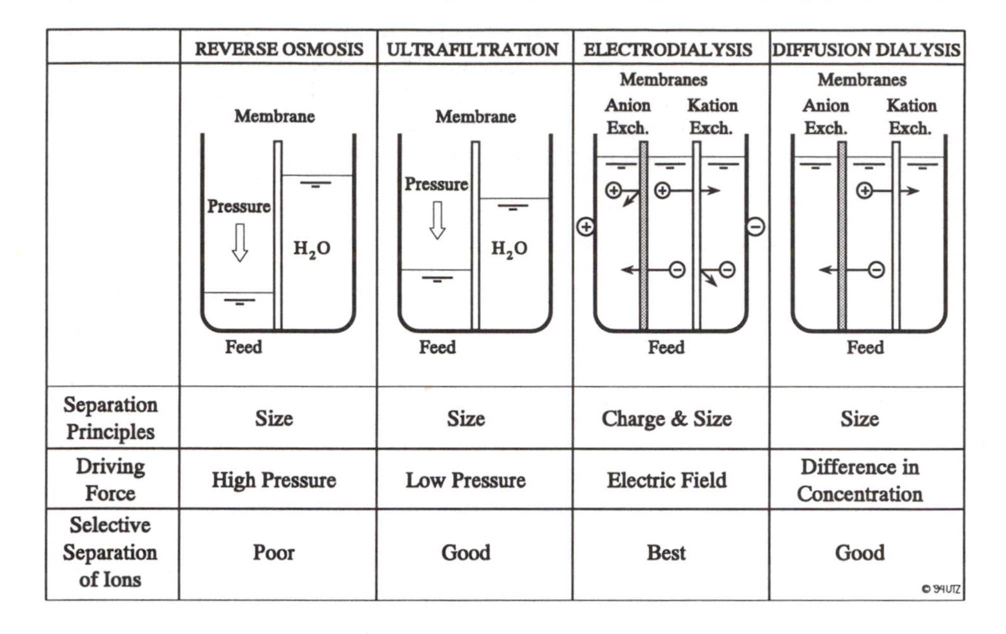

	REVERSE OSMOSIS	ULTRAFILTRATION	ELECTRODIALYSIS	DIFFUSION DIALYSIS
	Membrane Pressure H_2O Feed	Membrane Pressure H_2O Feed	Membranes Anion Exch. Kation Exch. Feed	Membranes Anion Exch. Kation Exch. Feed
Separation Principles	Size	Size	Charge & Size	Size
Driving Force	High Pressure	Low Pressure	Electric Field	Difference in Concentration
Selective Separation of Ions	Poor	Good	Best	Good

© 94UTZ

Figure V.26: **Comparison of Membrane Separation Methods [31]**

reliability, integration, and complexity all favor the MF because of its single pass operation which leads to a less complex, more reliable design. Further principles of water recovery using membranes are the electrodialysis and diffusion dialysis. An overview and comparison of these principle is given in figure V.26. A more detailed description follows below. [16]

V.2.2.1 Reverse Osmosis (RO)

In the standard osmotic process, water moves from a compartment with a less concentrated to a compartment with a more concentrated solution driven by osmotic pressure. In RO, the process is truly reversed: pressure is applied to the waste water until its osmotic pressure is exceeded, forcing water across a semipermeable membrane and leaving most ions and larger organic compounds behind. The RO unit rejects all suspended solids, all macromolecules, and most low molecular weight salts, although current membranes are incapable of removing small organics. The result is a large volume of relatively pure permeate and a small volume of very concentrated fluid, which must be processed by a VCD or stored. Current membranes require pre-treatment by ultrafiltration (UF) to remove suspended solids as well as post-treatment by MF sorbent beds which require expendables. Ultrafiltration is a process that filters most suspended solids and macromolecules, while allowing low molecular weight salts and water to permeate the membrane. As the first stage of the RO system, the primary function of UF is to remove large contaminants that would otherwise foul the RO membranes. Improving membrane performance could eliminate the need for pre- and posttreatment. Improved membrane configurations may also increase system

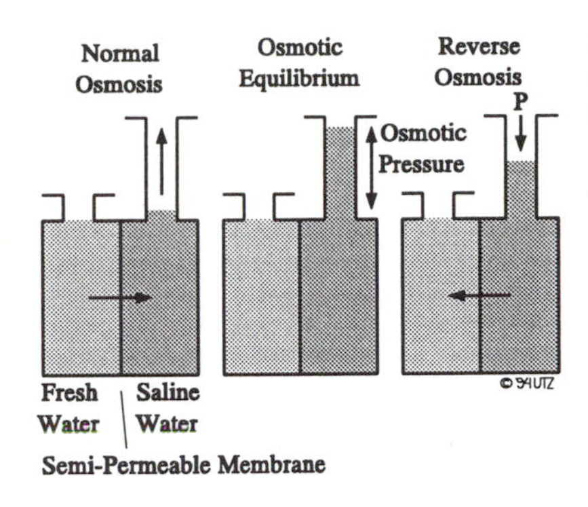

Figure V.27: **Reverse Osmosis [31]**

performance. RO is baselined in a spiral configuration, while UF employs a tube-side feed design. Ideally, the RO system will operate above the waste pasteurization temperature (347 K) to prevent microbial growth. Figure V.27 is a simplified scheme of the reverse osmosis (RO) process.

Two membranes have received the most attention for RO in space, the inside skinned hollow fiber membrane and the dual layer membrane. The hollow fiber membrane consists of a porous polysulfone base with a proprietary solute rejecting thin skin deposited in the fiber interior. The hollow fiber configuration has a high membrane surface area-to-volume ratio which increases module compactness. Waste water is fed to the interior of the hollow fibers to minimize fouling or the collection of contaminants on the mebrane surface. The dual-layer membrane is made from a mixture of zirconium oxide and polyacrylic acid deposited on the interior of a porous metal or ceramic tube. The attractive features of this membrane are its high water flux or throughput ($1.9 \cdot 10^{-5}$ m/s vs. $1.1 \cdot 10^{-5}$ m/s for the hollow fiber membrane at $2.8 \cdot 10^{6}$ Pa) and its stability at pasteurization temperature (347 K). The projected specific energy required for wash water recovery using the hollow fiber and dual layer membranes is about 10 Wh/kg water recovered. The membrane itself is the key element of the process. Total water recovery by RO is impractical because of the high osmotic pressure of the feed solution at high solids concentrations. The most attractive features of RO for space habitat water recovery are low energy consumption and no requirement for a gas-liquid phase separator in zero gravity. Characteristics of a RO system for supplying 4 people is given in table V.16. [7, 31, 35]

V.2.2.2 Multifiltration (MF)

Potable water is obtained through multifiltration of condensate from the temperature and humidity control subsystem and from water formed during the CO_2 reduction process. In the multifiltration process, waste water is purified by flowing it through filters and packed columns connected in series. The only pressure required to sustain water flow through the multifilter is that needed to overcome the small pressure drops across the filters and packed beds. One of the attractive features of multifiltration is that it is a relatively uncomplicated technology which requires very little development for use in space. Disadvantages of multifiltration are the need for expendables to regenerate the ion-exchange beds and a suitable regeneration scheme for activated charcoal.

In multifiltration there are three basic steps. First, particulates are removed by filtration. Second, suspended organic contaminants in waste water are removed by passing water through a bed of activated charcoal, and third, inorganic salts are removed by cation and anion exchange resin beds. MF consists of a

Subsystem	REVERSE OSMOSIS (RO)

Weight:	8.5 kg	Volume:	0.056 m^3
Heat Generated:	0.015 kW	Power Required:	0.02 kW

Operating Temperature:	280 K (min.)	295 K (nom.)	320 K (max.)
Operating Pressure:	310 kPa (min.)	2750 kPa (nom.)	3100 kPa (max.)

Operations Mode:	Continuous
Designed Efficiency:	80 %
Comment:	-
Technology Readiness Index:	5
Reliability:	1000 days
Maintainability:	Flushing may be required occasionally

Advantages:	-
Disadvantages:	-
Prospective :	-
Improvements:	-

Principal Chemical Equation:	-

	Inflow [kg/day]	Inflow Composition	Outflow [kg/day]	Outflow Composition
Gases	-	-	-	-
Liquids	-	-	-	-
Solids	-	-	-	-

Table V.16: **Design Characteristics of a RO Equipment for 4 people [35]**

particulate filter upstream of six unibeds in series. Each unibed is composed of an adsorption bed (activated carbon) and ion exchange resin bed. Over time, the first unibed becomes contaminated sooner than the remaining unibeds downstream. After reaching storage capacity this unibed is removed, and the other five unibeds are moved up to fill the gap. A fresh unibed is placed at the end of the series. Microbial control is achieved by heating the entire subsystem to pasteurization temperature, i.e., 347 K, or chemically treating (iodine injected by a microbial check valve) the process water. Some characteristics of a MF system for supplying 4 people are given in table V.17. [7, 30, 34]

Subsystem	**MULTIFILTRATION (MF) UNIBED**

Weight:	3.9 kg	Volume:	0.00124 m^3
Heat Generated:	0.00038 kW	Power Required:	0.00038 kW

Operating Temperature:	289 K (min.)	328 K (max.)
Operating Pressure:	70 kPa (min.)	210 kPa (max.)

Operations Mode:	Continuous
Designed Efficiency:	99.9 %
Comment:	-
Technology Readiness Index:	5
Reliability:	-
Maintainability:	Typical exchange at 15 days if used for complete treatment

Advantages:	- Contains all polishing elements in one cartridge
Disadvantages:	- Cannot be regenerated practically - Must be replaced and trashed
Prospective Improvements:	- Regeneration technology - Possibly improved sorbants for higher capacity

Principal Chemical Equation:	(sorption)

	Inflow [kg/day]	Inflow Composition	Outflow [kg/day]	Outflow Composition
Gases	0	-	0	-
Liquids	114	100 % Water	114	100 % Water
Solids	0.015	0.01 % Organic carbon 0.01 % Inorganics	0	0.001 % Organic carbon 0.0025 % Inorganics

Table V.17: **Design Characteristics of a MF Equipment for 4 People [35]**

V.2.2.3 Electrodialysis

The electrodialysis or electrodeionization process utilizes ion exchange resins and membranes to deionize contaminated feed water. Ion exchange membranes, used as barriers to bulk water flow, divide the system into three adjacent compartments: a diluting compartment bordered on either side by a concentrating compartment. Feed water enters the diluting compartment, which is filled with mixed-bed ion exchange resin. Ions in the feed solution react with ion exchange resins, transferring through these resins in the direction of an electrical potential gradient applied across the compartments. Due to the semipermeability properties of the ion exchange membranes and the directionality of the potential gradient, ion concentration will decrease in the diluting compartment and increase in the concentrating compartments. The system outputs brine from the concentrating compartments and purified deionized water from the diluting compartment (conductivity about 0.1 microsiemens/cm). Pre- and posttreatment of the feed water is required to extract contaminants not removed by electrodeionization, including non-ionic species, organic foulants, chlorine, and silica. Electrodeionization is capable of deionizing a wide range of feeds, from bulk salt removal to polishing of RO product water.

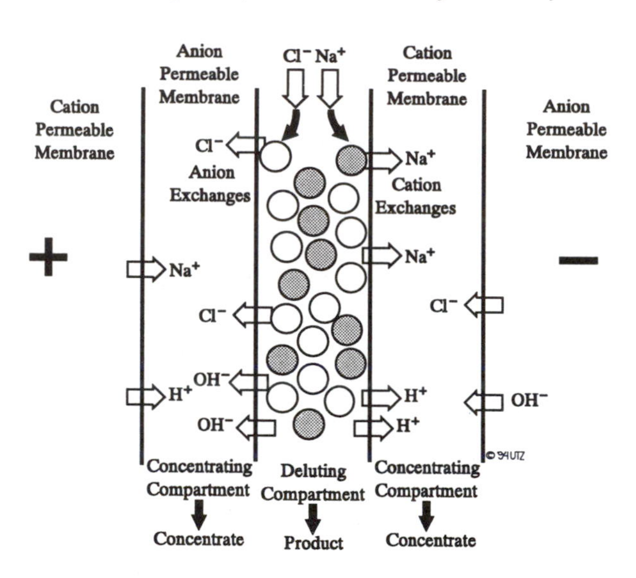

Figure V.28: **Electrodialysis Process [31]**

In the electrodeionization system the ion exchange resin is continually electrically regenerated. Regeneration chemicals are not required, eliminating chemical handling, waste neutralization, and corrosion problems. System disadvantages include high complexity relatively to MF, and logistics penalty of brine storage. [35]

V.2.3 Water Recovery from Condensate

One of the most "pure" water-containing products is condensate of crew respiration products derived from breathing and through the skin. Atmospheric moisture condensate amounts to approximately 45 % (1.1-1.7 kg water/man-day) of the total mass of water collected from a crew. In a water recovery system for

Subsystem	**ELECTRODIALYSIS**		
Weight:	-	Volume:	0.261 m^3
Heat Generated:	0.46 kW	Power Required:	0.46 kW
Operating Temperature:	283 K (min.)	295 K (nom.)	308 K (max.)
Operating Pressure:	140 kPa (min.)	280 kPa (nom.)	420 kPa (max.)
Operations Mode:	Continuous		
Designed Efficiency:	98 %		
Comment:	Pre- or posttreatment may be required		
Technology Readiness Index:	4		
Reliability:	High		
Maintainability:	Cleanable by flushing		
Advantages:	-		
Disadvantages:	-		
Prospective Improvements:	-		
Principal Chemical Equation:	(ion transport)		

	Inflow [kg/day]	Inflow Composition	Outflow [kg/day]	Outflow Composition
Gases	-	-	-	0.0001 % H_2 0.0008 % O_2/Cl_2
Liquids	6000	100 % Feed water	6000	98 % Product water 2 % Waste water
Solids	-	-	-	-

Table V. 18: **Design Characteristics of a Electrodialysis Equipment for 240 People [35]**

condensate, developed in the former Soviet Union, an air/condensate mixture is fed to a gas/liquid mixture filter from the thermal control units and then to the separator of a condensate separation block. After filling the separator liquid cavity a suction pump which feeds separated condensate to a purification unit, is switched on. Air separated from condensate is vented to the cabin. From the purification unit water is fed to the conditioning unit. The water is saturated with mineral salts to the potable water standard and preserved with ionic silver. Mineralized water is then passed on to the potable water container. The operation of the system is almost completely automated and monitored by the crew from the onboard control panel. The system has been in operation since 1975 aboard the space stations SALYUT-4 to -7 and MIR, providing the crew with hot and cooled water for food preparation and showering. The energy requirements for water recovery are 4 kWh/m³ with 100 % water recovered from condensate feed and the quality conforming to the relevant standards. The specific weight of the system equipment amounts to 0.2 kg per kg of recovered water. Based on the performance data of the system aboard the space station, the crew's demands for potable and food preparation water are met not less than by 80 %. The system onboard MIR has recovered about 6 tons of condensate by the end of September 1991. [25]

V.2.4 Water Quality Monitoring

Water quality monitoring is a prerequisite for maintaining a safe space habitat water supply. Space habitat potable and wash water or hygiene water standards have been established (see section IV.3.1.2) and it is the function of the water quality monitoring subsystem to ensure that these standards are continuously met. A monitoring subsystem should be automatic and the important parameters which determine the water quality should be continuously or frequently measured. The parameters to be frequently monitored include pH, ammonia content, total organic carbon, electrical conductivity, and microbial concentration. Commercial sensors are available for making these measurements. Less frequently monitored parameters are color, odor, turbidity, foaming, and heavy metal concentrations. The taste of potable water must also be unobjectionable. Microbial content may be monitored indirectly by measuring iodine levels to ensure that biocide is always present. The principle of a water quality monitoring approach is given in figure V.29. [31]

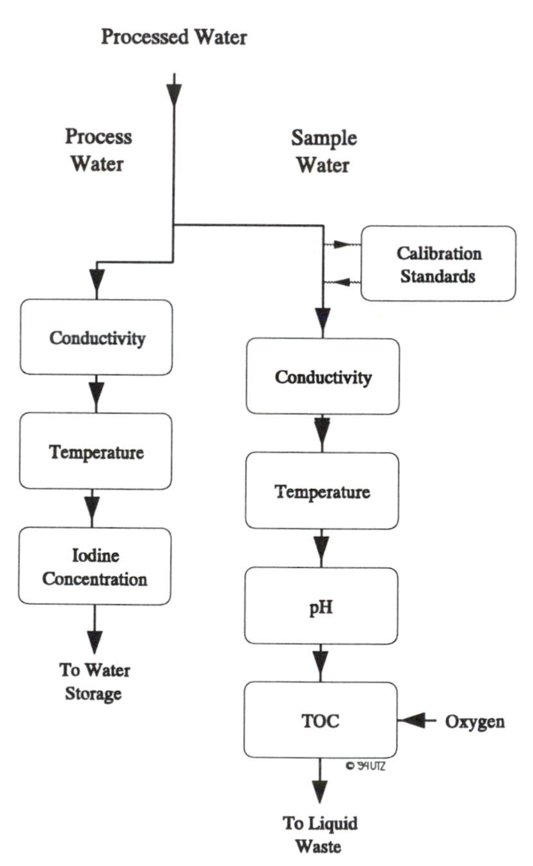

Figure V.29: **Water Quality Monitoring Equipment [29]**

V.3 WASTE MANAGEMENT

The waste management subsystem deals with the collection, treatment and storage of wastes. Waste products, including wet and dry ash, feces, and food preparation wastes aboard crewed space infrastructures must be collected, stabilized to prevent unpleasant or even harmful consequences to humans and/or equipment, and stored. For longer missions, a trade-off between resupply and reprocessing of consumables has to be considered and reprocessing from waste may become an economically justified operation.

The main types of wastes a waste management system must deal with, are wastes produced in semi-solid, solid, or mixed form, waste originating from the water management system and transferred essentially as concentrates, and waste originating from the air regeneration system and transferred as concentrated absorbed or absorbed gases and particles. The sequence of waste management operation is:

1. *Collection and segregation*
2. *Fractionation*
3. *Stabilization or Storage*
4. *Recycling, when appropriate*

For short-term/short-duration missions no recycling is planned. Wastes are simply collected, stabilized as necessary and stored after compaction. For medium duration missions and/or missions close to Earth, with the exception of some water recovered from waste by phase change technology, there will still be no need for extensive waste recycling. Waste storage for longer missions is unacceptable because of the large weight and volume penalty and the potential for space habitat contamination from biological and chemical waste degradation. Nevertheless, for long-range/long-duration space missions the technologies previously developed for the shorter, nearer missions will still be applicable, especially for all non-biodegradable wastes. For the biodegradable fraction the secondary resource concept will progressively drive and permit implementation of new technologies compatible with the later use of such wastes by established planetary bases. For well-established permanent planetary bases and visiting flights, the secondary resource concept will impose strict requirements on the design of consumables, to ensure that they are directly regenerable or suitable for entering various recycling loops, essentially the water and food loops. Furthermore, plants will eventually be grown in space to provide food and at that time large amounts of inedible plant material will need to be processed along with other solid wastes to provide CO_2 and mineral nutrients for the plant growth chamber. Inedible plant material will comprise about 98 % of the total dry solid wastes. An overview of the waste categories produced aboard a space habitat is given in table V.19.

Waste Category	Waste
Biologically decomposable, liquid	Hygiene water, metabolic water, respiration / transpiration water, feces (liquid portion), urine
Biologically decomposable, solid	Feces (solid portion), waste with bound water, solids from urine, sweat and hygiene water, clothes
Metabolic, gaseous	CO_2, trace gases, methane, oxygene
Not decomposable, liquid	Products from experiments, medicine
Not decomposable, solid	Spare parts, plastic, metal
Not metabolic, not decomposable, gaseous	Off-gassing

Table V.19: **Overview of the Different Waste Categories**

The requirements of physico-chemical waste treatment systems depend on the process applied. They differ concerning the O_2 requirement, water content of the waste stream, temperature, pressure and reaction time. The different processes to be considered that are also described in more detail below, involve:

- Super Critical Waste Oxidation (SCWO)
- Wet Oxidation
- Combustion / Incineration
- Electrochemical Oxidation
- Waste Management - Water Systems (WM-WS)

As mentioned earlier, wastes first have to be collected, sorted, and stabilized. Methods for the stabilization of waste are:

- Condensation
- Drying with dry or wet heat
- Osmotic pressure
- Freezing
- Extreme pH levels
- Metallic or organic toxins
- Silver-(II)-oxidation

After stabilization the waste may either be stored or further be decomposed. The steps of the decomposition process are:

- Mechanical processing, e.g., fractionating
- Chemical treatment, e.g., pH assimilation, extraction
- Enzymatic or catalytic treatment, e.g., hydrolysis, desamination
- Fermentation (aerob or anaerob)
- Oxidation, e.g., wet oxidation, super critical water oxidation, combustion
- Post-processing of the output, e.g., demineralization, filtration, reverse osmosis

For physico-chemical waste treatment systems different water contents of the waste is required, depending on the method of waste treatment. In general, a constant amount and a constant composition of the waste makes the processing easier. The waste input can be characterized by:

- The requirement of oxidizable substance
- The water contents
- Particle size and homogeneity

Through the oxidation of organic substances by physico-chemical waste processing methods, like wet oxidation, SCWO or combustion, CO_2 is produced that can be used by plants for photosynthesis or has to be reduced. Because the oxidation rate of processes is very high in comparison with the CO_2 consumption rate some buffer tanks will be required if plants are applied in the system. The O_2, which is then produced by the plants or CO_2 reduction systems, may again be used for the oxidation processes. Because the consumption rate of O_2 of this process is generally very discontinuous, while the plants are producing O_2 quite continuously, again some buffer tanks may be required. Minerals that are produced during the physico-chemical oxidation may be used as nutrients for the plants. Nutrient requirement and production have to be balanced. Waste gases and particles that are produced during these physico-chemical processes have to be filtered from the gas cycle. Altogether, in physico-chemical systems the wastes are almost completely oxidized and these systems are very well controllable. The problem is that they do not yield a balanced mass budget. In this respect, the more promising biological waste processing methods are dealt with in chapter VI. [5, 7, 8, 31]

V.3.1 Feces Collection and Storage

In past manned spaceflight missions, all solid waste was stored and returned to Earth, involving large logistic penalties. Human solid waste was compacted and stored in canisters. When full, these canisters were returned to Earth and resupplied. As described above, the logical improvement to storage is a waste management system capable of processing waste and recovering its useful constituents, an important step in developing a closed ECLSS.

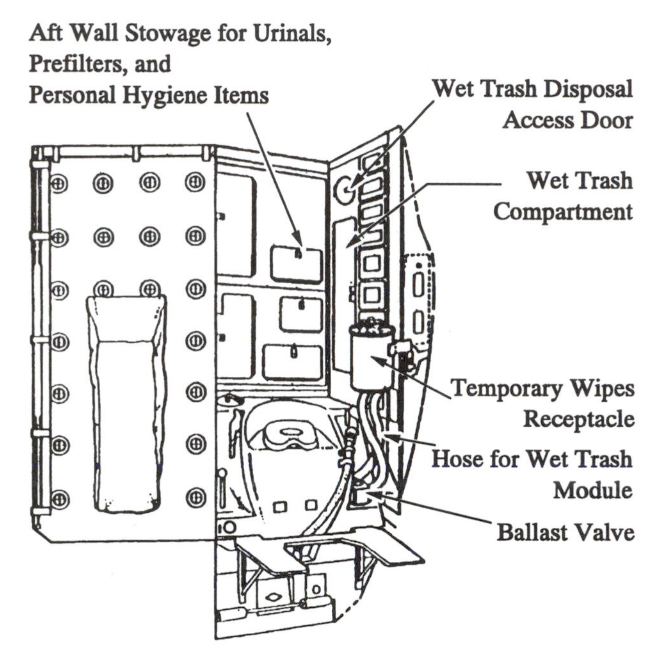

Figure V.30: **The Space Shuttle Waste Management System [27]**

In a typical air flow waste collection system with vacuum dehydration, air is drawn past the user into a collection device. The air flow moves the feces to the bottom of a plastic bag liner. The air is drawn through the device by a centrifugal blower and passed through a filter with activated carbon before being returned to the cabin. Upon completion of the defecation, the user seals the plastic collection bag with a rubber band and turns off the blower. The package of feces is transferred by the user into the drying device. Urine is collected through a tube by means of air flow using the same blower as used for the feces collection. The liquid-air mixture is passed through a liquid gas separator. The liquid phase from the separator is sent to the urine storage tank and the air is drawn through the filter by the blower and returned to the cabin. As an example for a waste collection system, the Space Shuttle Waste Management Compartment is shown in figure V.30. [5, 31, 37]

V.3.2 Solid Waste Treatment

As mentioned before, basically three principal waste treatment methods may be considered:

- Wet Oxidation
- Combustion / Incineration
- BiologicalTreatment

Methods basing on the first two waste treatment principals are further described below, while biological waste treatment methods are dealt with in chapter VI.

V.3.2.1 Super Critical Wet Oxidation (SCWO)

Waste destruction by the SCWO waste treatment process makes use of water in its supercritical state, i.e., above 647 K and $2.21 \cdot 10^7$ Pa, as the process medium for carrying out the destruction of organic compounds by oxidation. The presence of a catalyst is not required for SCWO. Water exhibits unique properties in the supercritical state which makes this process particularly attractive for waste destruction, namely its properties as a solvent change. Organic compounds, normally insoluble in water at standard temperature and pressure, become soluble in supercritical water, as does O_2. This permits oxidation to occur in a single phase, unlike wet oxidation. If sufficient oxygen is available and the reactor temperature and pressure are sufficiently high, i.e., above 922 K and $2.53 \cdot 10^7$ Pa, organic compounds, along with process atmospheric and trace contaminant gases, are completely oxidized to CO_2, H_2, and N_2. Inorganic salts are only sparingly soluble in supercritical water which aids in separating salts from the aqueous product phase, since they precipitate from solution. The temperature and molecular densities allow the oxidation reactions to proceed rapidly and essentially to completion. For example, organic salts can be oxidized with efficiencies greater than 99.99 % for residence times of less than one minute.

With proper reactor design, as in the case of wet oxidation, the heat generated by waste oxidation in supercritical water can be retained in the reactor thereby minimizing the amount of external heat or energy required for the process. SCWO can handle all types of spacecraft waste, in addition to processing condensate, wash, and urine/flush waters. SCWO produces potable water from all input waste waters, creating an entirely potable water supply. Safety concerns associated with the high operating pressures and temperatures, material corrosive problems, and the need for post-treatment to remove a few toxic product gases must be addressed further to SCWO to become an important spacecraft technology. The design characteristics of a SCWO laboratory breadboard model for supplying 4 people are given in table V.20. [7, 31, 35]

Subsystem	SUPERCRITICAL WATER OXIDATION (SCWO)

Weight:	694 kg	Volume:	2.12 m^3
Heat Generated:	0.36 kW	Power Required:	1.44 kW

Operating Temperature:	647 K (min.) 923 K (nom.) 1023 K (max.)
Operating Pressure:	$2.21 \cdot 10^4$ kPa (min.) $2.53 \cdot 10^4$ kPa (nom.)

Operations Mode:	Continuous
Designed Efficiency:	100 %
Comment:	-
Technology Readiness Index:	3
Reliability:	-
Maintainability:	-

Advantages:	- Can combine oxidation of trace organic contaminants in air and air organic contaminants in water in a single unit or reactor - High destruction efficiencies at short residence times (< 5 minutes) - Can be used to separate inorganic salts from water
Disadvantages:	- SCWO reactor operates at high pressures and temperatures - Since oxidation reactions are exothermic, temperature control may be difficult - Metals may corrode and contaminate product water
Prospective Improvements:	- Mass and energy balance on waste streams of interest is poorly defined - Reliability and modes of failure of a corrosion resistant reactor material are desirable - Salt crystal nucleation and growth - Salt separation, particularly in microgravity

Principal Chemical Equation:	$Organic + O_2 \rightarrow CO_2 + H_2O$

	Inflow [kg/day]	Inflow Composition	Outflow [kg/day]	Outflow Composition
Gases	2.59	-	3.77	-
Liquids	25.3	-	25.3	-
Solids	1.27	-	0.098	-

Table V.20: **Design Characteristics of a SCWO Laboratory Breadboard Model for 4 People [35]**

V.3.2.2 Wet Oxidation

Wet oxidation involves the oxidation of either a dilute or concentrated waste slurry (1-10 % solids) at an elevated pressure and temperature. Combustion takes place in air or O_2 at a pressure of about 14 MPa and temperatures ranging from 473 to 573 K, producing largely CO_2 and water. The slurry-oxygen mixture is agitated vigorously for 15-60 minutes. Wet oxidation, unlike incineration, does not require a predried food stock. Dilute waste (5 % solids) can be oxidized using this technique.

The forms of the important biological elements, e.g., C, N, H, O, P, produced by wet oxidation depend on the severity of the temperature and pressure regimes within the process. A detailed knowledge of the inputs to and outputs from the wet oxidation process is lacking. The wet oxidation removes 60-95 % of the initial carbon quantity (COD). Anyway, the percentage of COD removed by the process is not equivalent to the amount of CO_2 it produces because the partial reduction of carbon decreases COD, but not necessarily also CO_2. Since the conditions of wet oxidation are not severe enough to lock dissolved metallic elements into insoluble oxides, these elements will remain in the liquid effluent. If conditions are harsh enough, the equipment might be corroded and metals introduced into the liquid effluent. These hazardous metals might accumulate in the food chain.

The primary advantages of wet oxidation for space habitat applications are the recovery of useful water and the reduction of solid wastes to a very small weight and volume of sterile, non degradable ash. Wet oxidation is particularly attractive for space habitats where plants are grown to provide food. Carbon dioxide for combustion of organics and the mineral nutrients in the aqueous phase and precipitates appearing after wet oxidation can be used as plant nutrients. Nevertheless, the oxygen requirement and potential hazard of high pressure make this process suboptimal for spaceflight. Also, incomplete combustion by wet oxidation may require posttreatment of the product gas and liquid by catalytic oxidation. Experiments were undertaken with catalytic wet oxidation using different catalysts (none, Pd, Ru, Ru+Rh) and with initial pressure and temperature conditions of 25 kgf/cm^2 pO_2, 320° C, and an initial carbon quantity (COD) of 20000 ppm and a nitrogen quantity (kj-N) of 30000 ppm. [7, 8, 31, 34]

V.3.2.3 Combustion / Incineration

Combustion is the rapid exothermic oxidation of combustible elements within a fuel stock. Methods for combustion range from incineration, which is complete combustion in the presence of excess oxygen, to proteolysis, which is destructive distillation, reduction, or thermal cracking and condensation of organic matter under heat or pressure or both in the absence of oxygen. Partial pyrolysis, or starved-air combustion (SAC), is incomplete combustion. It occurs when not enough oxygen is present to satisfy incineration requirements. Biological elements in their highest oxidation state are, as a rule, easiest to reincorporate into the biological cycle. Therefore, only incineration and SAC should be considered based on their potential practicality. The major advantage of incineration over SAC is that it oxidizes most of the important biological elements to their highest oxidation states. [33]

Dry incineration involves the combustion of a concentrated solid waste feed consisting of, e.g., human feces, urine, and non human waste. The waste fed is concentrated by evaporation to a solids content of about 50 % by weight before being heated in air or O_2 near ambient pressure to a temperature of about 813 K. Pure O_2 is the preferred oxidizing agent. The end products of dry incineration are sterile, and consist of water condensate, inorganic ash, and gases (primarily CO_2). A catalytic afterburner may be required when using dry incineration to further the extent of combustion. Disadvantages of dry incineration are incomplete combustion, even with an afterburner. The product water may require further processing before being reused, and an energy intensive evaporation or predrying step is required before combustion can take place. Thus, the process can be described as inefficient, with a catalytic afterburner required to approach acceptable combustion efficiency.

SAC is the alternative combustion for waste treatment. Again, waste water solids are mechanically dewatered before being sent to the combustion furnace. The remaining material which is partly oxidized, exits through the furnace as gaseous effluent and can be fully oxidized in a catalytic oxidation unit. Catalytic oxidation uses relatively small amounts of oxygen. SAC accomplishes the oxidation sequence for sludge with less oxygen than does incineration. Other advantages of the SAC are that it is more stable and easier to direct because it can be controlled by adjusting the rate of gas input, and that it produces fewer toxic emissions and larger particulates (which are easier to remove).

The two combustion methods create different compounds of nitrogen. Incineration yields nitrate (NO^{3-}) and nitrite (NO^{2-}). Although the former is readily useable by plants, the latter is phytotoxic. Thus, effective sinks are required. By contrast, the SAC process converts nitrogen to ammonia, which leaves the furnace in the off-gas and is converted to atmospheric nitrogen (N_2) and water in the afterburner. This form of nitrogen is safe for the life support system atmosphere but unavailable to most plants. [7, 31, 35]

V.3.2.4 Electrochemical Incineration

Electrochemical decomposition is another waste processing technology in early development. It is a non-thermal electrolysis process which degrades organic, solid waste materials waste and urine by oxidation to CO_2, N_2, O_2, and H_2, at the surface of catalytic electrodes. It is anticipated that CO_2 and N_2 will be the products of anodic oxidation, and H_2 will be evolved from the cathode. This process operates at low temperatures (422 K), oxidizes organic waste without the consumption of atmospheric oxygen, has lower power requirements, and does not consume oxygen, making it attractive for further development. [7, 35]

V.3.2.5 Waste Management - Water Systems (WM-WS)

Another system capable of processing solid and liquid waste simultaneously is the WM-WS. This high temperature (920 K) process utilizes catalytic oxidation, and may be more efficient in microbial control than other water recovery processes. The original WM-WS, including a plutonium heat source, was known as the radio isotope thermal energy (RITE) system. [7]

V References

[1] Abele H.; Lawson R.
Trace Gas Contamination Management in the Columbus MTFF
Space Thermal Control and Life Support Systems, ESA SP-324, p. 287-294, 1991

[2] Ammann K.
The Catalytic Oxidizer - Description and First Results of a Breadboard Model for a Component of the Columbus ECLSS
Space Thermal Control and Life Support Systems, ESA SP-288, p.187-192, 1988

[3] Ashida A. et al
Mineral Recovery in CELSS
International Conference on Life Support and Biospherics, Proceedings, p. 275-278, 1992

[4] Belaventsev J. et al
A System for Oxygen Generation from Water Electrolysis aboard the Manned Space Station MIR
Space Thermal Control and Life Support Systems, ESA SP-324, p. 477-479, 1991

[5] Binot R.; Oakley D.
Waste Management
Preparing for the Future, Vol. 3 No. 2, p. 10-11, ESA, 1993

[6] Couch H. et al
Advanced Regenerative Life Support for Space Exploration
Space Thermal Control and Life Support Systems, ESA SP-324, p. 113-120, 1991

[7] Crump W.
The Development of Physico-Chemical Life Support Systems for Manned Spaceflight
International Space University, Space Life Sciences Textbook, p. 481-513, 1992

[8] David K.; Preiß H.
DEBLSS - Deutsche Biologische Lebenserhaltungssystemstudie
Dornier GmbH, Bericht TN-DEBLS-6000 DO/01, 1989

[9] Franzen J. et al
A Gas Chromatic Separator for Columbus Trace Gas Contamination Monitoring Assembly
Space Thermal Control and Life Support Systems, ESA SP-324, p. 301-305, 1991

[10] Fuhs S. et al
Design of the Fire Detection System for Space Station Freedom
International Conference on Life Support and Biospherics, Proceedings, p. 39-50, 1992

[11] Graves R. et al
Technical Breakthroughs in Bosch Carbon Dioxide Reduction Technology
International Conference on Life Support and Biospherics, Proceedings, p. 243-251, 1992

[12] Graves R.; Noble L.
Vapor Compression Distillation Technology for Space Station Freedom
International Conference on Life Support and Biospherics, Proceedings, p. 451-460, 1992

[13] Hafkemeyer H.
Trace Gas Contamination Control (TGCC) Analysis Software for Columbus
Space Thermal Control and Life Support Systems, ESA SP-324, p. 515-520, 1991

[14] Haupt S.
Electrochemical Removal and Concentration of CO_2
Space Thermal Control and Life Support Systems, ESA SP-288, p. 185-186, 1988

[15] Heitchue R.
Space Systems Technology
Reinhold Book Corporation, New York, 1968

[16] Humphries R. et al
Life Support and Internal Thermal Control System Design for the Space Station Freedom
Space Thermal Control and Life Support Systems, ESA SP-324, p. 23-37, 1991

[17] Huttenbach R. et al
Physico-Chemical Atmosphere Revitalization: The Qualitative and Quantitative Selection of Regenerative Designs
Space Thermal Control and Life Support Systems, ESA SP-288, p. 57-61, 1988

[18] Kay R.; Woodward L.
Space Station Freedom Carbon Dioxide Removal Assembly
19th Intersociety Conference on Environmental Systems, SAE Technical Paper 891449, 1989

[19] Klingele S.; Tan G.
Trace Gas Monitoring Strategies for Manned Space Missions
Space Thermal Control and Life Support Systems, ESA SP-324, p. 323-328, 1991

[20] Kuhn P.; Petter F.
Condensing Heat Exchangers for European Spacecraft ECLSS
Space Thermal Control and Life Support Systems, ESA SP-288, p. 193-197, 1988

[21] Preiß H.; Funke H.
Regenerative CO_2-Control
Space Thermal Control and Life Support Systems, ESA SP-288, p. 177-183, 1988

[22] Rotheram M. et al
Space Station Freedom Trace Contaminant Monitor - An Interim Report
International Conference on Life Support and Biospherics, Proceedings, p. 139-148, 1992

[23] Samsonov N. et al
Hygiene Water Recovery Aboard the Space Station
Space Thermal Control and Life Support Systems, ESA SP-324, p. 649-651, 1991

[24] Samsonov N. et al
Water Reclamation from Urine aboard the Space Station
Space Thermal Control and Life Support Systems, ESA SP-324, p. 629-631, 1991

[25] Samsonov N. et al
Water Recovery from Condensate of Crew Respiration Products aboard the Space Station
Space Thermal Control and Life Support Systems, ESA SP-324, p. 625-627, 1991

[26] Skoog I.
Lebenserhaltungssysteme
Handout of Presentation at the TU München, 1985

[27] Skoog A.; Brose H.
The Complementary Role of Existing and Advanced Environmental, Thermal Control and Life Support Technology for Space Station
Environmental & Thermal Control Systems for Space Vehicles, ESA SP-200, p.281-288, 1983

[28] Sperker K.; Tan G.
Carbon Dioxide Reduction System as Part of an Air Revitalization System
Space Thermal Control and Life Support Systems, ESA SP-324, p. 469-472, 1991

[29] Wieland P.
Designing for Human Presence in Space
NASA, RP-1324, 1994

[30] Wood M.
Oxygen Generation by Static Feedwater Electrolysis for Space Station Freedom
International Conference on Life Support and Biospherics, Proceedings, p. 127-137, 1992

[31] Wydeven T.
A Survey of Some Regenerative Physico-Chemical Life Support Technology
NASA TM-101004, 1988

[32] APCOS - Operation and Maintenance Manual
Hamilton Standard Division, United Technologies, 1993

[33] Columbus Human Factors Engineering Requirements
COL-RQ-ESA-013, ESA, 1989

[34] Controlled Ecological Life Support System - Biological Problems
NASA CP-2233, 1982

[35] In-House Life Support Technology Review Databook, FY 91, Volume 5
NASA Ames Research Center, Advanced Life Support Division, Doc. No. 90-SAS-R-003 Rev. 2, 1991

[36] Spacelab Payload Accommodation Handbook - Appendix C
NASA, Marshall Space Flight Center, SLP / 2104-3, 1986

[37] Space Shuttle Waste Management Manual
NASA, Johnson Space Center, 1994

VI BIOREGENERATIVE LIFE SUPPORT CONCEPTS

VI.1 BIOREGENERATIVE LIFE SUPPORT AND CELSS

VI.1.1 Rationale

It was shown in the previous chapters that by means of regenerable physico-chemical subsystems in a life support system the water and oxygen loops can be closed. Future space habitats will require that the carbon loop, the third and final part-loop in the life support system, be closed. Since carrying most of the materials required during the mission at launch, or resupplying them, will become prohibitively expensive as both the number of men and the mission durations increase. Especially, bases on Moon or Mars will have to function with a maximum of regeneration because of the high cost for resupplying such distant settlements. Nevertheless, the closure of the carbon loop will only be practical if advanced life support systems can be developed in which metabolic waste products are regenerated and food is produced. If this is partly or completely fulfilled by regenerative processes, a (partly) closed life support system is obtained. If for the regenerative processes biological systems are used, a biological life support system (BLSS) is obtained. The simplified principle of such a BLSS is shown in figure VI.1.

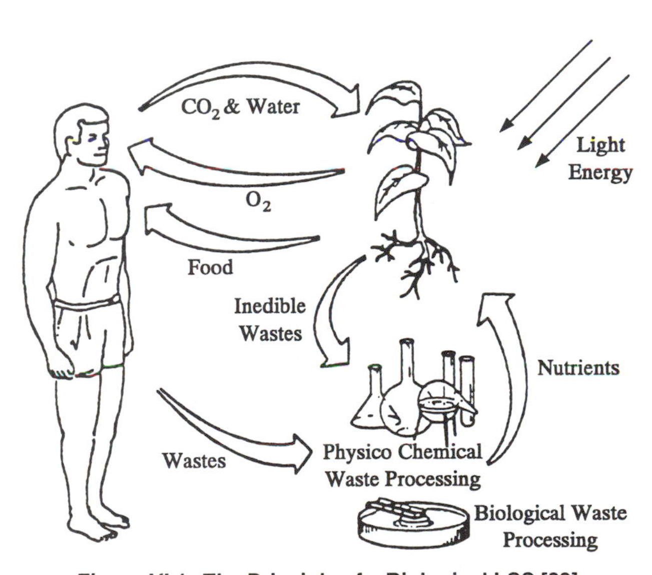

Figure VI.1: **The Principle of a Biological LSS [29]**

A BLSS has to be a balanced ecological system, biotechnical in nature, and consisting of some combination of human beings, animals, plants and microorganisms integrated with mechanical and physico-chemical hardware. Bioregenerative systems can basically be divided into two categories:

- *Chemosynthetic Systems*
  In those systems based on, e.g., hydrogen bacteria, oxygen and hydrogen are electrolytically produced from water. These bacteria contain the stable enzyme hydrogenase. Thus, they are able to produce food containing 70-85 % of proteins from urine, Mg- and Fe-salts, O_2, and CO_2.

- *Photosynthetic Systems*
  An obvious advantage of growing food in space is that plants, in particular, are also capable of atmosphere regeneration. Specifically, autotrophs such as plants, take up carbon dioxide expelled by heterotrophes, including man, and produce the oxygen such organisms consume, e.g., oxygen is produced from carbon dioxide by algae, e.g., *Chlorella*, according to the photosynthesis equation:

  $$CO_2 + H_2O + Light \rightarrow O_2 + Biomass + Heat$$

  A part of the edible biomass is transformed to CO_2 and H_2O again by the human metabolism. The rest of the edible biomass will be partly oxidized and appears in the secretions, i.e., urine, feces, and transpiration. If these products are later completely oxidized, together with the inedible biomass, a completely closed ecosystem, an "artificial biosphere", can be achieved. Only because of the leakage, a constant supply with especially nitrogen and oxygen will be unavoidable.

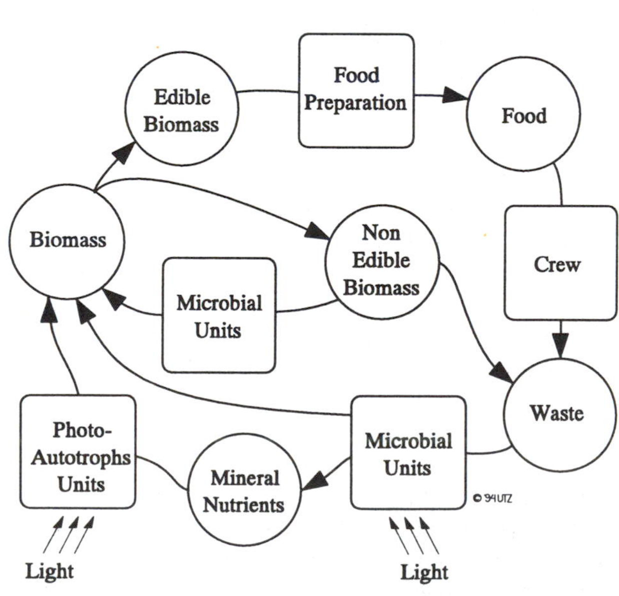

Figure VI.2: **Food Production in a Bioregenerative LSS [64]**

Food production in closed systems will require mineral nutrients from waste regeneration systems. As such waste transformers are of microbial origin, the edible part of their biomass has to be taken into account. Waste mainly consists of organic molecules which are usually degraded into minerals, CO_2, and H_2O. The inedible part of the biomass coming from the photoautotrophs, e.g., higher plants and microalgae, can partly be converted into edible biomass. A global scheme of this scenario is given in figure VI.2. [64]

Thus, the introduction of biological techniques for food production into life support systems produces not only a certain number of problems to be solved, but also opens up a new area of solutions for other life support requirements. A summary of these aspects for the different life support subsystems is given below:

- *Atmosphere management*
  As shown above, oxygenic photoautotrophic organisms, e.g., higher plants, produce food from carbon dioxide and water according to the general photosynthesis equation. But also, the many low-weight, volatile organic compounds which are found as air contaminants in small, closed inhabited volumes and arise mainly from human metabolic processes, equipment off-gassing, leakage from coolant loops, fire control equipment, etc., can be substrates for a variety of microorganisms. Thus, a kind of biological air filter appears as a possible solution for contamination control.

- *Water management*
  The main problem in regenerating water is to eliminate compounds that make it non-potable and/or non-hygienic, which are mainly microorganisms (bacteria, viruses, protozoans, yeasts), and organic and mineral compounds. The solutions that have been developed are either purely physical, e.g., successive membrane filtrations, evaporation and cooling, or chemical, e.g., super critical water oxidation, but are very energy consuming. The solution nature chooses is plant transpiration, e.g., water evaporation from the leaves. This plant water transpiration can easily be recuperated (condensation) and is considered as potable. Another solution would be to adapt the conventional biological treatment processes used on Earth to regenerate used water.

- *Waste management*
  As missions get longer, the volume of waste produced by the crew will increase. A good waste management system will be needed to minimize both the volume and mass of the stored waste, probably relying on a combination of compaction and dehydration. Biological processes can further decrease the mass and volume of the biodegradable elements. Biodegradation can lead to end products such as carbon dioxide and nitrogen, but also some less desirable ones such as hydrogen and hydrogensulphide. [63]

An ideal life support system would mirror the ecological processes on Earth, relying on natural bioregenerative processes to recycle waste and provide, food, air, and water. Consequently, a bioregenerative life support system consists of three functional subsystems that correspond to those of the terrestrial ecosystem:

- Consumers (Humans and animals)
- Producers (Plants and algae)
- Detruents (Microorganisms)

Engineering a scaled-down version of this complex ecological system into a spacecraft or planetary colony is a difficult task. An efficient biological system requires the careful selection of organisms which can perform life support functions while being ecologically compatible with other organisms in the system and with the human crew. In the absence of natural terrestrial forces, maintaining the health and productivity of this system requires stringent control of system processes and interfaces. Without proper control mechanisms the system becomes unstable and unbalanced.

Although such artificial biospheres in space are already theoretically feasible today, they will not become reality in the near future. The reason for this are not only the extreme environmental conditions in space, e.g., microgravity and radiation, but also the high degree of complexity and the manifold feedback processes in such a system. On the other hand it will not be necessary to create a complete copy of Noah's arch if only the provision of life support for humans is required. Therefore, a development program for bioregenerative life support systems for the space applications, called CELSS (Controlled Ecological Life Support Systems), has been established by NASA in the late 1970's. It is directed at studying the feasibility of constructing a bioregenerative life support system for use in extraterrestrial environments. In figure VI.3 is shown, how such improvements in life support technology might be implemented in conjunction with a possible mission growth scenario.

CELSS operation is based on photosynthetic organisms, such as plants or algae, to produce food, oxygen, and potable water, and to remove carbon dioxide exhaled by the crew. Physical subsystems will be required to support these biological functions, including temperature and humidity control hardware, a food processing system to convert biomass into edible food, and a waste processing system to convert waste products, including waste water, into useful resources. Benefits of a CELSS over entirely physico-chemical systems are primarily related to psychological and logistical factors. A CELSS provides a much more Earth-like environment than physico-chemical systems, and is therefore more psychologically satisfying to the crew. In addition, a CELSS produces fresh fruit and vegetables, lending variety and palatability to the crew's diet. The logistical advantage is that a CELSS can produce all the ingredients necessary for human survival with almost no resupply. A small quantity of liquid and solid waste will be non-recyclable, and some organisms that die will have to be replaced, but overall resupply requirements will be significantly lower than for physico-chemical systems. The plant and/or algal growth system will be the largest component of a CELSS. Primarily responsible for food production, the plant growth system also provides for oxygen production, CO_2 removal, and liquid and solid waste treatment.

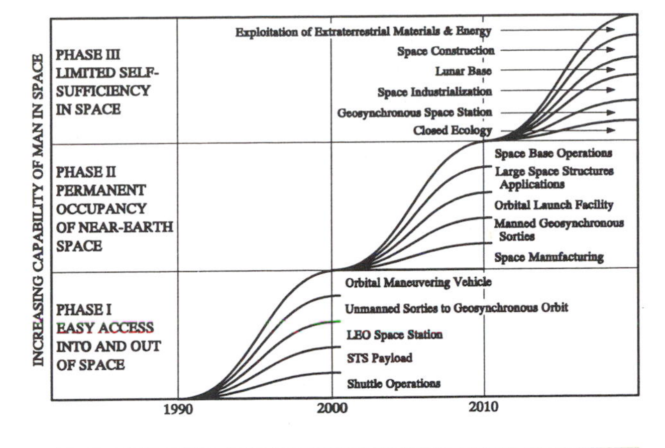

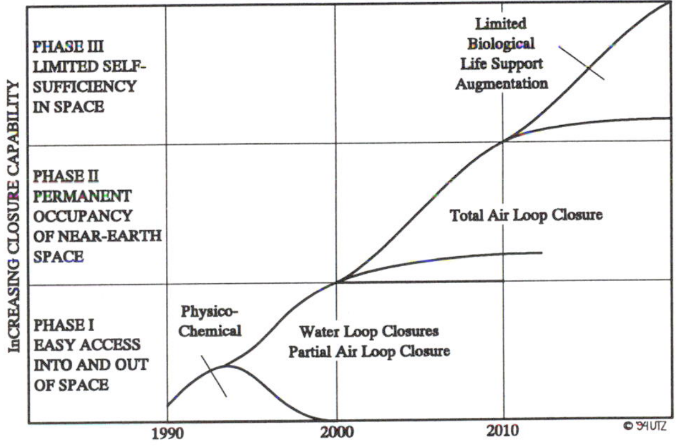

Figure VI.3: **Space Mission Growth Scenario and Prospective Evolution of Life Support [29]**

The major problem of bioregenerative systems is the discrepancy between supply and demand of the several products. The simple closure of a biological system, even if it is exposed to the same amount of light and heat as an open system, results in the eventual death of the system. A fundamental reason concerns the buffering capacity of a system, which is partially provided by the sheer size of the terrestrial ecosystem, e.g., autotrophs in any ecological system require light for photosynthesis. Consequently, oxygen is produced by plants cyclically. Heterotrophs, or oxygen users, similarly undergo cyclic uses of that gas. As a result, CO_2 is produced by each organism in a cyclic fashion depending on its state of maturity, its momentary rate of metabolism, or its supply of food. In a small system these cycles, which are not linked, do not necessarily correspond. Most important, the atmospheric volume in a small system would not be sufficiently large to maintain a steady-state concentration of oxygen, or to absorb CO_2 at a rate fast enough to maintain a steady-state concentration of that gas. Also, not all of the waste can be used by the photosynthetic organism without any preprocessing.

Thus, it will be essential that artificial buffers be created, and that regulation of the system be done through human intervention. In many ways such a system would resemble a typical farm, in that crops must be selected, noncrop plants eliminated, water and nutrients supplied when needed, and food harvested when mature. Atmospheric buffering capacity for a particular gas will have to be supplied by mechanical means. Such mechanical mechanisms must be capable of removing certain atmospheric constituents and supplying others.

The energy necessary for operating both the mechanical and biological systems presumably will come from sunlight, but could also be supplied by other means such as nuclear reactors. Control of the system would no longer be dependent on natural self-regulatory mechanisms, but rather on consistent monitoring by sensing devices and periodic analyses. The behavior of the system would have to conform to an established but variable model of the function of the system. The only system of that kind that has been created is Biosphere 2 (see chapter VII). Finally, it should not be forgotten that, as a kind of spin-off, the experimental study of such biological life support systems with a relatively closed biological cycle may yield basic knowledge and tools needed to cope with, or perhaps even solve, some of the environment-related problems that are of growing concern here on Earth. This subject is further dealt with in chapter IX. [59, 67]

VI.1.2 Modeling and Design

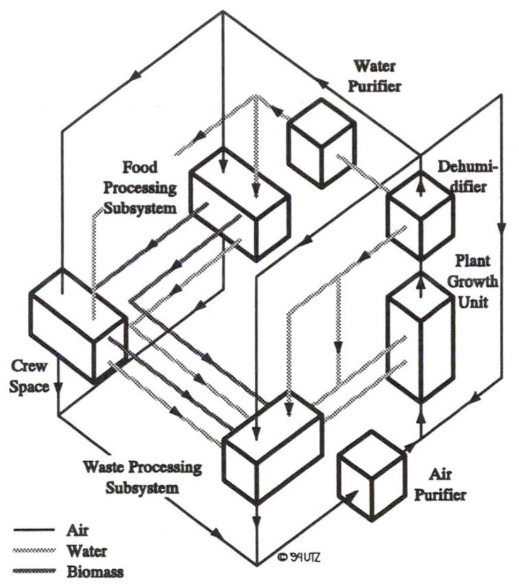

Figure VI.4: **CELSS System Diagram [3]**

The selection of a suitable biological life support system for application in space depends on the mission type and the available technologies. These determine the volume, mass, energy requirement and costs of such a system. In this respect, four characteristic kinds of BLSS or CELSS can be considered:

1. *Earth-based CELSS*
2. *Lunar CELSS (1/6 g, Moon day lasting 29.53 Earth days)*
3. *Martian CELSS (1/3 g, day of 24.7 h)*
4. *μ-g CELSS*

For the application of a CELSS in space, requirements concerning the following aspects have to be taken into account:

- Launch capacity (Low mass and volume)
- Mission requirements (Low energy consumption, high reliability, and easy maintainability)
- Environmental conditions (Functioning under microgravity conditions and radiation)

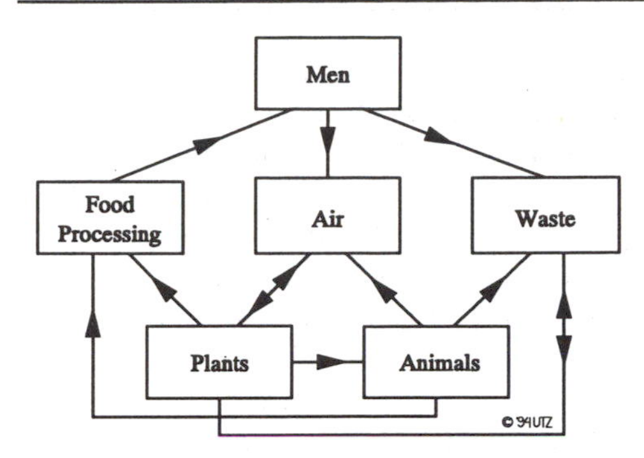

Figure VI.5: **Carbon Mass Flow in a Closed System [29]**

Concerning the costs, a breakeven point will have to be determined in order to obtain an indicator which system is most cost effective for which mission duration. The prime requirement for any life support system is the production of food by biological means. Thus, the carbon cycle has to be closed, as indicated in figure VI.5. It is also important to note that in this case the four components water, atmosphere, food, and waste cannot strictly be separated. In this respect, the principal scheme of a CELSS is given in figure VI.4. [54, 74]

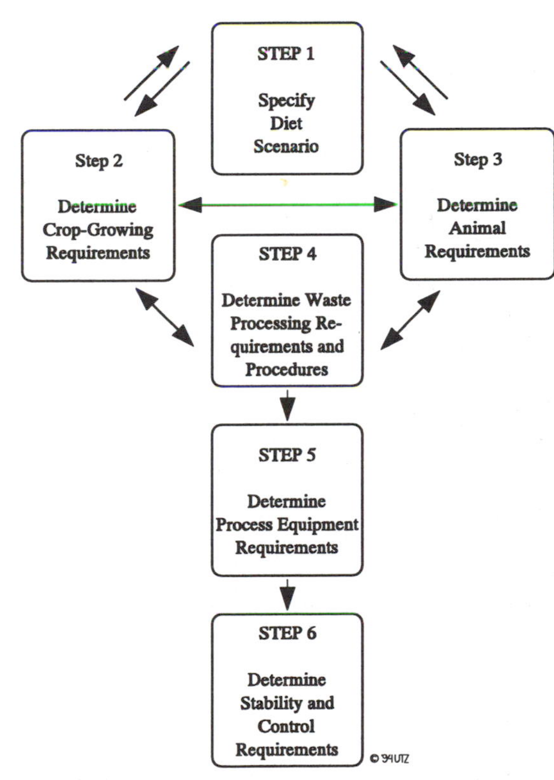

Figure VI.6: **Design Process for a CELSS [29]**

The food production requirement is the design driver for a CELSS. A design program for CELSS, as indicated in figure VI.6, has to start with the specifications of the human diet and the vitamin and trace element requirements. The next step will have to be to select the appropriate plant and animal species. This selection would be reevaluated and retested as the development of CELSS makes progress. In general, the actual use of biological life support techniques must be preceded by three important milestones:

1. The acquisition of complete scientific knowledge about the particular biological process.
2. The development of the required biological technologies based on this scientific knowledge.
3. The integration of these biological technologies into a functioning life support system.

Stage	Description
1	Determination of basic principles and directions of the development
2	Definition of a conceptual design
3	Analytical or experimental testing of the design
4	Breadboard demonstration of critical functions
5	Testing of components in relevant environments
6	Testing of model system in relevant environments
7	Testing of a prototype under space conditions
8	Basic production design

Table VI.1: **Stages of CELSS Development [12]**

Thus, from an engineering point of view, the development process for a CELSS may comprise eight stages as described in table VI.1. [12, 29, 63]

Of course, many single experimental investigations in various disciplines will be necessary for the evaluation of the biological, chemical, and technical basis for these areas before integration into the system. Nevertheless, the development of an operational biological life support system for space requires dual development paths. In parallel to the selection of technologies, plant species, etc., mathematical models will have to be applied to decrease development risks. These should describe the functional coupling between all system components as well as their dynamic behavior. In addition, the models should include the definition of the system stability and eventually form the basis for computerized control and management of the system including problem prediction, trend analysis, crop forecasting, and logistics requirements predictions. Also, the research needs for a CELSS include both ground-based and flight-based studies ranging from subsystem development to total systems integration and behavior.

The current CELSS program schedule as published by NASA is given in figure VI.7. By 1998, NASA wants to demonstrate the feasibility of a CELSS excluding humans. The basic research requirements towards that goal, in the different areas, are summarized and listed in table VI.2. [12, 41]

By 2020, NASA wants a successful demonstration of a human-related CELSS at 1 g completed. Some of the main critical questions to be solved by then include:

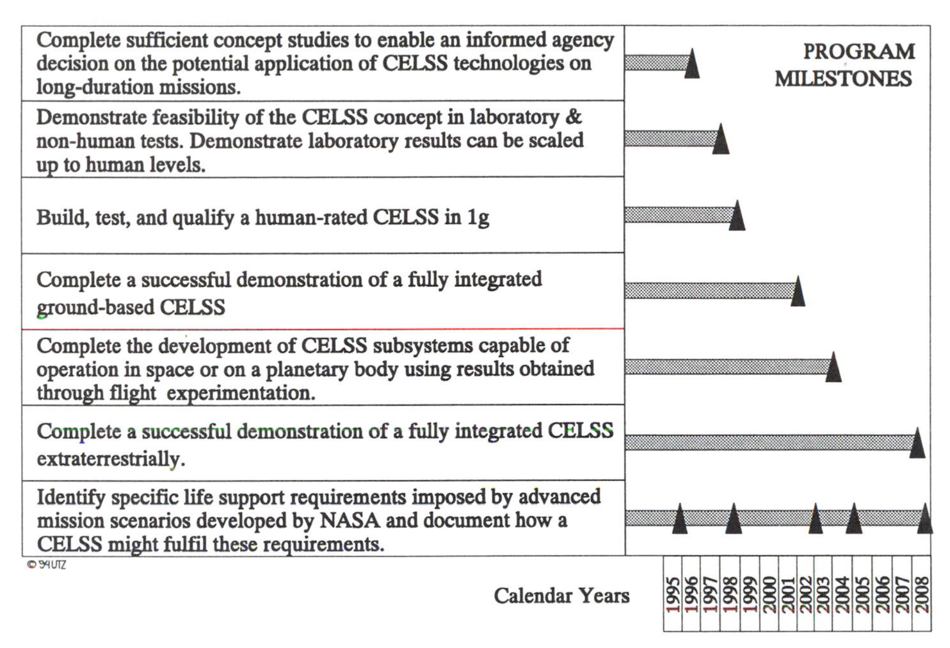

Figure VI.7: **CELSS Program Schedule (Status 1993) [41]**

- Investigation of psychological effects of diet during long-term enclosure
- Determination of health and safety requirements for waste treatment
- Achievement of safe and reliable overall system operation
- Investigation of control systems response to instabilities
- Capability of correction of instabilities by chaos and fuzzy logic

The CELSS program also plans flight experiments on space station, using a small CELSS test facility (CTF), the so-called salad machine, for production of salad vegetables, and miscellaneous hardware developed for specific experiments (see section VI.4.11). This is to be accomplished by 2004 and here the critical questions to be answered pertain to:

- Crop productivity under hypogravity conditions
- Identification of optimizing environments for hypogravity
- Morphology and reproduction
- Countermeasures for anticipated productivity problems
- Questions about the functioning of physico-chemical technologies in space

Thus, by 2008, it should be known how the space environment might perturb an integrated CELSS that functioned smoothly on the ground and how to integrate automation and robotics procedures in the plant culture system. By that time, it

Area	Required Development
Environment	- Materials selection - Atmosphere selection - Gravity selection - Radiation shielding requirements and methodology - Ecosystem tradeoff studies - Chemical analysis and control of contaminants and toxicants - Illumination requirements - Solar reflectors and filters
Management and Control	- Critical biological performance parameters - Biological sensor development - Definition of biological stability criteria - Development of mathematical models for system prediction - CELSS management and control philosophy
Agriculture	- Determining appropriate crop species - Optimum growing and harvesting technologies - Plant culture and physiology in space environments - Equipment concepts for cultivation and harvesting - Radiation effect on genetic drift germination - Plant growth without soil - Forced growth effects on plants - Plant cycle photosynthesis efficiency - Plant hormone activity in microgravity - Plant production of toxic gases - Potential use of animals - Selection of plants that meet water requirements
Aquaculture	- Potential use of fish and/or algae - Food-producing ecologies based on waste-conversion - High yield, high nutrition plant production and harvesting - Photosynthesis process
Food Synthesis	- Acceptable microbiological sources and production methodology - Acceptable chemical synthetic production of protein and carbohydrates
Food Processing	- New concepts for food preparation processing, storage, and distribution to reduce equipment and resource requirements - Improved food preservation and packaging methods
Diet Planning	- Human nutritional requirements - Food and food-source selection criteria - Nutritional equivalency of various food sources - Physiological and psychological acceptability aspects of non-conventional diets and food sources - Definition of crop/plant scenarios - Digestive tract adaptability
Waste Processing	- Physico-chemical processes, e.g., mineral separation and recovery - Characterization of waste streams - Waste treatment technologies - Air treatment technologies - Use of plant transpired water as potable - Conversion of metabolic wastes to plant nutrients - Most complete conversion of waste into usable products - Determination of waste storage requirements - Microbiological processes - Regenerative chemical filters - Chemical separation methods - Auxiliary non-food products from wastes - Plant waste byproduct recycling

Table VI.2: **Required CELSS Scientific and Technology Developments [41]**

might also be possible to make informed decisions regarding the role and future of genetic technology in a CELSS. Ongoing questions to be answered include issues of power consumption and nutritional requirements in space. Also, the big question of how long-term stability of a CELSS can be achieved, will have to be answered. Once all of the above mentioned problems have been solved, a closed, bioregenerative life support system may be obtained that may be similar to the complete reference configuration of CELSS as given in figure VI.8. [41]

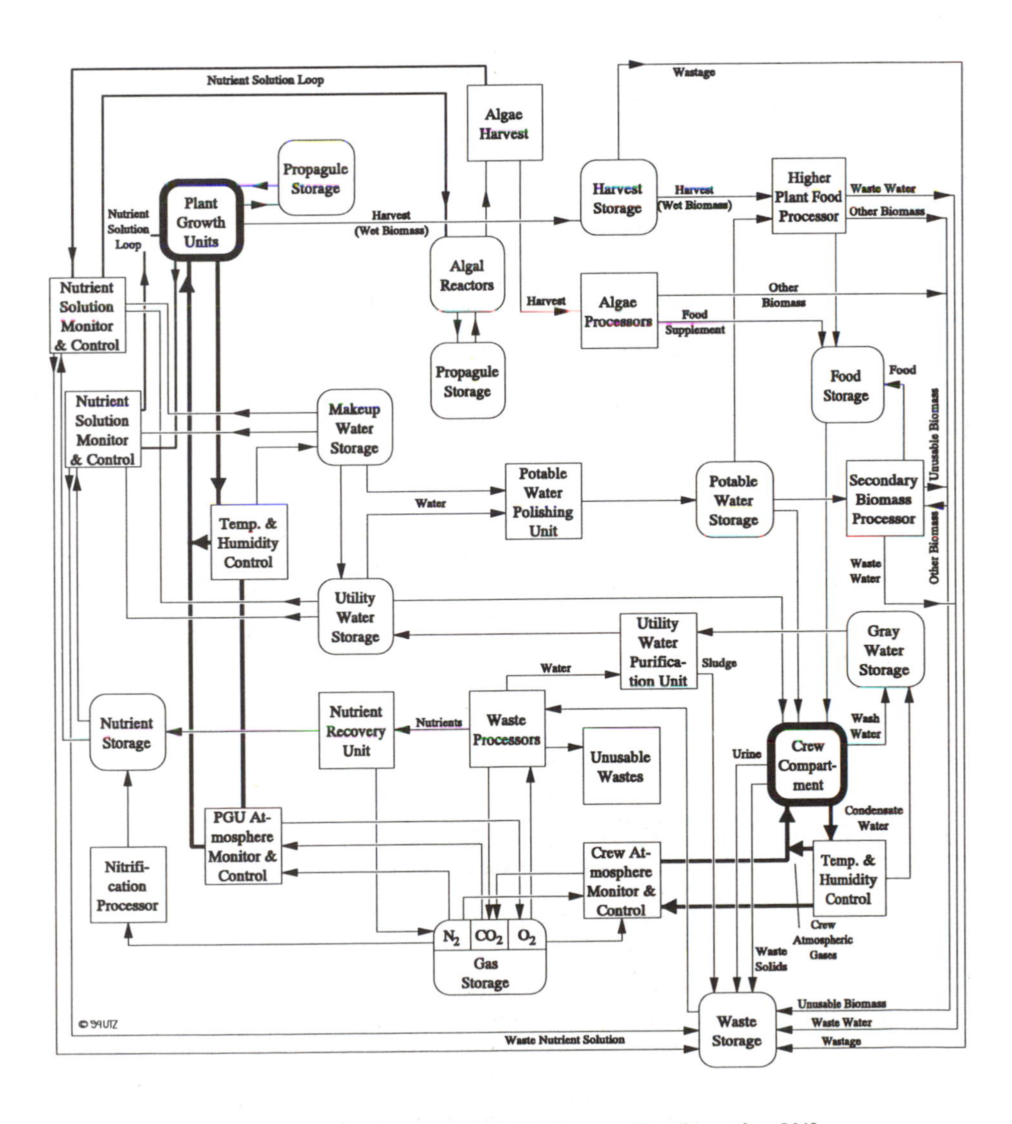

Figure VI.8: **The CELSS Initial Reference Configuration [41]**

Whatever the configuration of a bioregenerative life support system will be like, its mass has to be, of course, as low as possible. The mass of such a system is composed of the different system components. These components can be divided into two groups:

- Dependent on mission duration
- Independent on mission duration

In order to give an idea of which specific components will have to be taken into account in mass calculations, an overview of time-dependents and time-independents of the plant production system is given in table VI.3. Of course, the number of time-independent components is rising with the degree of closure.

Time-independents	Time-dependents
- Plant compartment (structure) - Energy system (solar energy, light) - Structures for plant growth (containers, medium and nutrient solution, pumps, ventilation, light sources, fastenings) - Cooling system (condensators, evaporators, compressors, motors, wires) - Food preparation system (drying, mills, extractors, boiler, conservation, storage) - Tools for maintenance, storage for spare parts, spare parts - Waste treatment system - Control system (sensors, monitors, computers) - Water reservoir including tanks and pipes	- Supply with biomass that gets lost by leakage - Resupply of food that was initially on-board - Spare-parts that are used during the mission - Water of the plant production system (water in growing plants, in stored biomass, in waste, in nutrient solution, in concentrates, in reservoirs, circulating water)

Table VI.3: **Time-Independents and -Dependents of a CELSS [12]**

After all, it shall not be forgotten to mention that CELSS does not only have advantages. This is why there is, e.g., an ongoing debate between specialists, whether physico-chemical or biological life support subsystems are more reliable. An overview of the advantages and disadvantages of the biological agents that in fact play the key role in a CELSS, is given in table VI.4. [12]

Biological Agent	Advantages	Disadvantages
Microorganisms	- Convert organic waste materials to water, CO_2 and useable plant nutrients	- Use oxygen - Slow process - Unknown control mechanisms
Lower plants (Algae)	- Convert CO_2 to oxygen - Simple, stable system	- Unpalatable - Indigestible
Higher plants	- Convert CO_2 to oxygen - Provide food - Process water via transpiration	- Production of biomass - High power and volume - Considered unreliable

Table VI.4: **Advantages and Disadvantages of Biological Systems** [52]

VI.1.3 System Level Problems

When designing an ecological life support system some problems do occur that are not present in terrestrial ecosystems. The problem areas are:

- Harmonization of Mass and Energy Fluxes
- Miniaturization
- Stability

These problems are mainly due to the comparatively small size of such an artificial ecosystem. While on Earth almost "unlimited" buffering capacities exist, these are, of course, not available in artificial systems. Therefore, also the system stability is a problem. Below, these subjects are discussed in greater detail.

Harmonization of Mass and Energy Fluxes

To a certain degree an ecosystem has the ability of selfregulation. The laws of ecosystems are valid for both natural and artificial systems. The energy flux in an ecosystem leads along the food chain, from the primary consumers to the detruents. It starts with the binding of light energy and ends with the complete decomposition of the organic compounds. At each stage energy is released as heat and radiation energy or gets lost as organic substance (see section II.2). In a CELSS man is the only consumer. Thus, the energy flux is directed towards the plant production in order to produce food. For the design of an artificial ecosystem it is necessary to define the turnover of the different subsystems and to adapt it by the environmental conditions. The main factor is the conversion of radiation energy into edible biomass.

Since a space station or habitat will be a closed system, and since it will be necessary to monitor all portions of the system, the technique of mass balance appears best to describe the functional aspects of the station. For the purpose of describing such a model, eight compartments will be considered initially: man (heterotrophs), plants (autotrophs), food, waste material, atmosphere, water, storage (buffer), and physico-chemical processing center. Initially, the flows of the following elements, all common to biological systems, will be monitored: carbon (C), hydrogen (H), oxygen (O), nitrogen (N), sulfur (S), and phosphorous (P). For the most part, methods of rapid and accurate analysis of these elements are available, and flows among the compartments can be followed relatively easily. Through the use of such models of individual metabolic and development needs, instantaneous requirements for the elements CHOPNS can be calculated and the movement of the elements in and out of a compartment can be predicted.

A model of an ecosystem can be considered as representation of a system of living and nonliving components occupying a defined space, through which energy, mass, and information flows. The photosynthetic properties of autotrophs, such as plants, allow them to use the radiant energy of the sunlight to synthesize complex polymers, carbohydrates, lipids, proteins, and nucleic acids from CO_2, NO_2, PO_2^{3-}, and other minerals and gases. The oxidation reactions of heterotrophs ultimately release CO_2 and other minerals required by autotrophs. This cycle of mineral fixation and release is the aspect of an ecosystem that can be simulated using models, because the rates of flow of each mineral constituent in and out of the living organism can be mathematically described as shown below

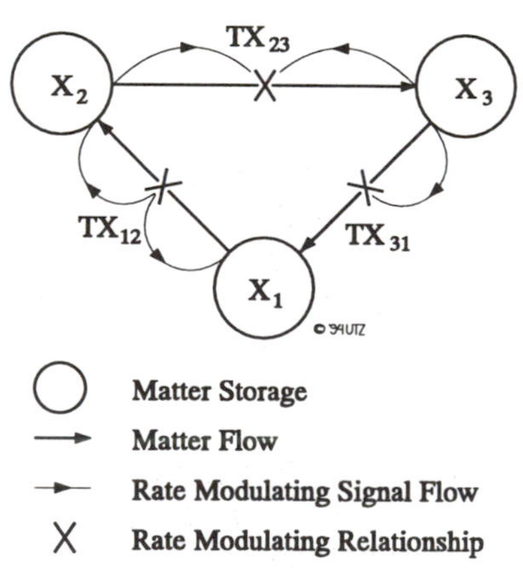

Figure VI.9: **Flow of Carbon in a Small Ecosystem [35]**

Figure VI.9 describes the transport of element X between three storage compartments: an inorganic nutrient storage (X_1), an autotrophic storage (X_2), and a heterotrophic storage (X_3). The autotrophs take up and incorporate the inorganic nutrients into their biomass at a characteristic rate (TX_{12}), which is ingested by the heterotrophs (TX_{23}) and remineralized back into inorganic nutrients (TX_{31}). Thus, in this simple closed loop, the laws of mass conservation dictate that the range of change of mass in each compartment is a function of the rate of mass flow into the compartment minus the rate of mass flow out of the compartment:

$$\Delta X_1 = T_{X31} - T_{X12}$$
$$\Delta X_2 = T_{X12} - T_{X23}$$
$$\Delta X_3 = T_{X23} - T_{X31}$$

The sum of the rates of change of mass in each compartment must, of course, equal zero:

$$\sum_{i=1}^{3} \Delta X_i = 0$$

Or, since mass is neither gained nor lost, it is a constant:

$$X_1 + X_2 + X_3 = M = constant$$

The rates of elemental mass flow between the compartments are a function of the state of the compartments. This is indicated in figure VI.9 by the rates' modulation signal flow. Thus, for this closed elemental cycle, the following functional dependencies can be written:

$$T_{X12} = k_{X12} \left(\frac{X_1}{a_1 + X_1}\right) X_2$$

which describes the observation that the nutrient uptake matter flows go to zero if either the nutrient pool (X_1) or the autotroph population (X_2) goes to zero and saturates in X_1, as X_1 increases to some value. Similarly:

$$T_{X23} = k_{X23} \left(\frac{X_2}{a_2 + X_2}\right) X_3$$

Since the rate of predatory matter flow, TX_{23} will fall to zero if either the predator (X_3) or prey (X_2) population goes to zero, and becomes saturated in X_2, as X_2 increases to some level. Finally, the rate of mineralization can be described by

$$T_{X31} = k_{X31} (X_3)$$

since the rate of mineralization (TX_{31}) will fall to zero when the heterotrophic population (X_3) goes to zero, and is independent of the size of the nutrient pool (X_1) and does not saturate as X_3 increases. The parameters kX_{12}, kX_{23}, kX_{31} (rate constants), and a_1, a_2, a_3 (saturation constants) are determined by a number of variables, e.g., variations in temperature, pressure, biological species, spatial distribution of elements, and light intensity.

These equations then define the behavior in time of a simple closed element cycle model. The major elements in an ecosystem (C, H, N, O, S, P) are all modeled separately, however, and a realistic representation of the flow of elements in any ecosystem must allow for functional couplings such that all the individual elements cycles are integrated into a single dynamic system. This can be done by cross-coupling individual element cycles by single flow-linkages.

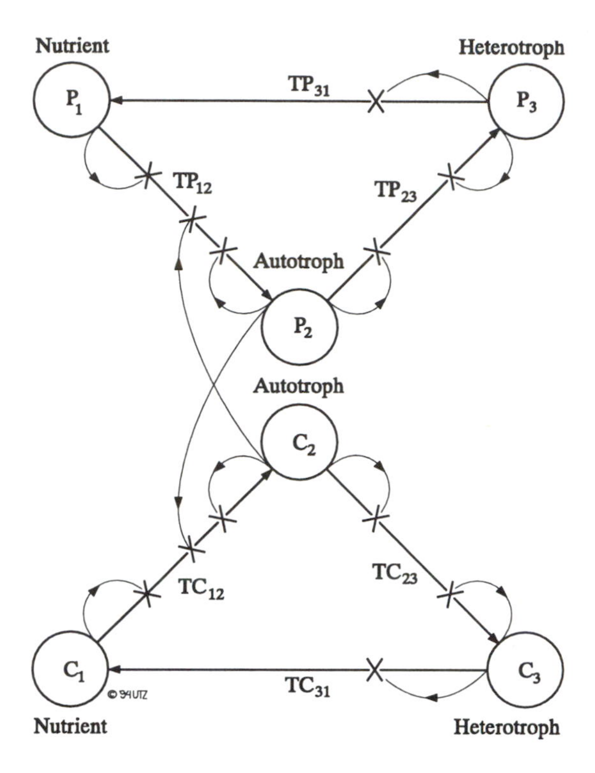

Figure VI.10: **Relations Between the Flows of Two Elements in a Closed System [35]**

These linkages transmit information from one cycle to another such that the behavior of the latter is modulated by the behavior of the former. The example in figure VI.10 depicts the manner in which the flow of one element, e.g., carbon, might modulate the flow of another, e.g., phosphorus, and vice versa in the simple three-compartment closed ecosystem. Real ecosystems are, of course, vastly more complex than this simple two-element, three-compartment cycle model.

Miniaturization

A major problem when designing a CELSS is the miniaturization of the ecosystem. The turnover rates have to be shortened and the masses have to be decreased. Thus, the processes are partly accelerated and the volume is decreased. Because of the miniaturization the buffering capacity is decreased and, thus, the whole system loses stability and self-regulation capacity. One of the problem areas of miniaturization includes the mass flow cycles, i.e., the mass turnover of the subsystems (producers, consumers, detruents) that have to be adjusted because no big reservoirs can be installed. Also, due to the lowered buffering capacity of a small ecosystem the danger of an infection is increasing and the ability of self-regulation is significantly lowered.

Stability of ecosystems

The factors that influence the stability of an ecosystem are as follows:

- Accumulation of substances
- Boundary levels
- Fixation of elements resulting in deficiencies
- Infection through microorganisms

Thus, due to the lack of sufficient buffering capacity, in small ecosystems the equilibrium has to be maintained by monitor and control systems which detect and correct deviations in an early stage. Since a biological system can never be completely sterile, the microorganisms in the system have to be well defined and monitored. Microorganisms can be categorized into accompanying and functional microflora. The accompanying microflora may have positive and negative effects on the overall system. It works as indicator and may protect the system from pathogenic organisms. The microflora can never be completely eliminated but only be stabilized by hygienic measures. In connection with plants the microorganisms in the nutrient solution are decomposing organic substances but there is also a concentration of microorganisms because of the recycling of the nutrient solution

The limited size of the enclosure, the susceptibility of biological systems to environmental perturbation, and the need for control and predictability all suggest that monitoring, sensing and control within an in-space ecosystem be directed through the use of a computerized mathematical model of the system. Such a model could, while operating a life support system, perform a predictive function through simulation. A model developed even before a physical system was constructed, would serve the function of simulating the behavior of a system and could aid in identifying parameters that must be evaluated in advance. The technique of modeling a closed ecological system through mathematical representation of mass flow appears to offer many advantages for understanding these systems. [35]

Besides the mathematical models, in this respect, a Closed Ecological System (CES) testbed for system-level investigations as proposed by ESA is also of special interest. Since the crew is an important biological element, influenced by and acting on the stability of the ecosystem, suitable testing requires such a test facility capable of evolving to accommodate man in the system. Moreover, habitability issues of relevance to these long-duration missions such as natural and induced environments, health management, internal architecture and life management facilities shall be investigated in a habitat called HABLAB, thus calling the whole project CES-HABLAB. It would constitute a dedicated, realistic and constraining environment, created to allow the simulation of virtually all aspects of an extraterrestrial habitat, except for the reduced gravity and radiation.

The CES facility is a stable ecosystem, including humans. It optimizes material recycling, essentially water, air and food from waste streams, and would have the means to simulate on Earth both, life support and habitability aspects, in real conditions of confinement and recycling. Thus, the establishment of such a facility, combining means of studying and testing CELSS and habitability issues, is essential in preparation for future manned missions. The proposed CES-HABLAB is supposed to be evolutionary in nature. It should be able to bridge the gap between existing developments in habitability and biological life support, typified by the use of small-scale mock-ups, computer simulations, microbial systems and very preliminary investigations using higher plants, and the fully-developed center which will be able to support the wide range of activities mentioned earlier. This need for flexibility-for-growth was particularly demanding in the case of the CES facility and has resulted in a concept possessing considerable modularity. Within an overall closed volume, sized according to the estimated needs of the eventual fully-developed CES, many different experiments may be performed, either totally sealed off from one another or interfacing in a controlled manner, e.g., via air connections, water loops etc. Features of the proposed CES include:

- Standard service interfaces for power supply, data exchange, thermal control and fluid exchange, to be located through the floor on a 4x4 m^2 matrix. As a result an extensive service area is located beneath the CES.

- Sunlight will be required within the CES to support the investigations concerning the terrestrial ecosystem and to prepare the technology that will be needed for future Moon or Mars bases.

- A system of "lungs" is required to ensure that the pressure differential between the inside and outside remains within closely controlled units. In principle, individual lungs are required for pressure compensation on each sealed experiment within the CES.

- Air locks for entry and exit of personnel and equipment.

- Docking ports for interfacing with HABLAB elements and habitation modules.

- Standard racks for experiment support and confinement.

- ECLSS equipment (mainly physico-chemical) to ensure a suitable environment in the CES.

- Comprehensive control and data-logging facilities.

The HABLAB part of the center would consist essentially of two areas: One dedicated to architectural and interior design studies, using relatively cheap, quickly reconfigurable wooden mock-ups, and computer simulations, and another dedicated to the exploration of the long-term impact of isolation, internal architecture and different ECLSS configurations. This area should be located adjacent to the CES since it will use many of the same facilities and will eventually need to interface directly with the CES. [1]

VI.2 MICROBIAL SYSTEMS

Microorganisms are a large and diverse group which exist as single cells or cell clusters. Microorganisms include bacteria, algae, protozoa, viruses, and some fungi. Microbial cells are distinct from the cells of animals and higher plants, which are unable to survive on their own but can exist only as part of multicellular organisms. A single microbial cell is generally able to carry out its life processes of growth, energy generation, and reproduction independently of other cells. This is not true for viruses. They are considered to be "acellular", as they lack functional membranes and the ability to replicate outside of a host cell. Microorganisms, without viruses, represent two basic cell types:

- *Eukaryotes*
  They include organisms compartmentalized with distinct nuclei, membrane-bound organelles and 80 S ribosomes. Eukaryotic microorganisms include algae, fungi, and protozoa.

- *Prokaryotes*
  These are organisms in which the cell organelles lack membrane-bound organization, a distinct nucleus, and contain 70 S ribosomes. Prokaryotes are represented by the cyanobacteria (blue-green algae) and the bacteria. [23]

On Earth, bacteria have been found in environments as extreme as volcanic vents on the ocean floor and polar ice caps. This temperature range is even within the temperature ranges that are found on Moon and Mars. Also, some bacteria are resistant to X-rays, γ-rays, and UV radiation in intensities beyond those thought to be present in space. Moreover, bacteria have shown to grow over the entire pH range and, normally, have a rather wide range of tolerable pressure, as long as there are no rapid changes. These abilities of bacteria to withstand very extreme environmental conditions open up many potential applications of microorganisms in space and they may even be used to help "terraforming" Mars, as shown by Lovelock.

The presence of bacteria, fungi, and viruses in manned, self-contained environments is unavoidable. However, the mere presence of microorganisms does not necessarily pose a problem, provided that their numbers are controlled. Just as some microorganisms are detrimental, other microbes may be beneficial to life support and health maintenance. In bioregenerative life support systems, microbial systems may potentially be used for the following three purposes: [34]

- Food Production
- Atmosphere Regeneration
- Waste Treatment

Concerning food production a big advantage of microorganisms is their high growth rate and high harvest index. They are also well known sources of so-called single-cell proteins. The main problems in the use of microorganisms resides in the accompanying nucleic acids production (see table VI.5). According to recommendations, the total nucleic acid intake from all sources must not exceed 4 g per day, and only half of it could come from unconventional sources. Thus, the maximal daily intake of selected microbial biomass is about 50 g of dry weight per day, covering only 30-40 % of the protein space recommended daily allowance.

	Bacteria	**Yeasts**	**Microalgae**	**Higher Plants**
Nucleic Acids [% of dry weight]	10-20	5-10	2-5	1-2

Table VI.5: **Nucleic Acid Content in Various Organisms [64]**

Many microorganisms are able to produce some lipids and, under nitrogen deprivation, carbohydrates. Thus, microorganisms can be envisioned as potential but partial sources of protein and, after further developments, an acceptable source of carbohydrates. Also, a partial completion of daily intakes of specific vitamins and minerals can be obtained. As an example, the nutritional values of *Spirulina* are given in table VI.17 in section VI.4.2.2. [64]

Concerning atmosphere and waste regeneration, two studies have to be mentioned that are currently underway in Europe: the Micro-Ecological Life Support Alternative (MELISSA) and the Biological Air Filter (BAF). MELISSA is a micro-organism-based ecosystem intended as a tool for developing the technology for a further biological life support system of manned space missions. The driving element of MELISSA is the reprocessing of edible biomass from waste, carbon dioxide and minerals, with the use of sunlight as a source of energy for biological photosynthesis. Light-dependence is minimized by the incorporation of anaerobic steps in the waste recycling loop. MELISSA has four successive microbiological compartments, colonized respectively by thermophilic *Clostridia* for waste liquidation-fermentation, anaerobic *Photorhodochromogens* for the removal of soluble organics, nitrobacteria for nitrification of ammonium ions, and the blue-green algae *Spirulina* for food production and carbon dioxide/oxygen recycling. The control system for MELISSA will be highly complex, featuring a high number of measurement and/or control parameters.

Biological Air Filtration is a proposed concept for air quality control. The principle of the Biological Air Filter (BAF) system is based on the combination of a support/sorbant material colonized by selected micro-organisms in a near-resting state converting the various contaminants to harmless compounds, mainly CO_2, H_2O and salts. The BAF consists of two different phases separated by membranes (see figure VI.11). The contaminants present in the gaseous phase are transferred through the membrane into the liquid phase which contains the micro-organisms and nutrients, resulting in the formation of a biolayer on the membrane surface.

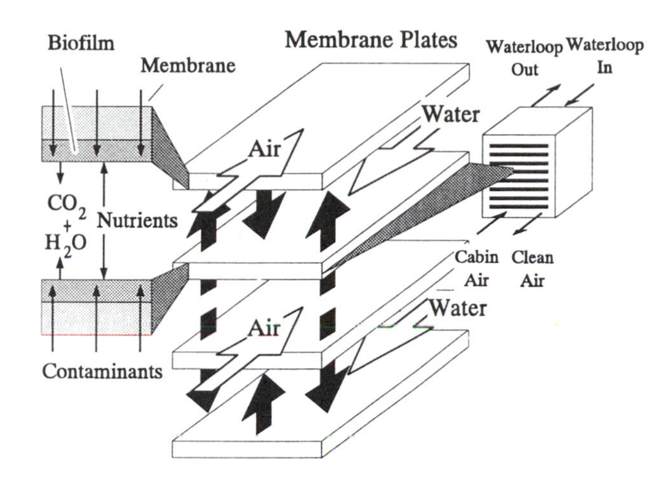

Figure VI.11: **Principle of the Biological Air Filter [77]**

By using a membrane filter and a closed liquid phase, micro-organisms are contained inside the BAF, which is of great importance for closed air circulation systems as used in space cabins.

For most of the contaminants addressed, one or more bacterial strains have been selected that have the ability to effect complete aerobic mineralisation to carbon dioxide, water and salts. More than 85% of the major atmospheric contaminants are found to be biodegradable. The efficient operation of a BAF is largely dependent on the kinetic properties of the microbial community in the biofilter. A variety of chemical compounds have to be degraded by a mixed population of micro-organisms with different degradation capacities. Complex interaction effects of mixed populations degrading a number of different contaminants have been investigated in chemostat experiments. The elimination capacity for, e.g., toluene, chlorbenzene, and dichlormethane was more than 99.4% during a test period of more than 50 days. In the experiments it has been found that contaminants can be degraded at very low concentrations, the biodegradability of a mixture of contaminants by co-cultures increased in performance compared against results obtained with pure contaminants, and the stability and performance of some microbial strains are considerably improved by culturing under BAF conditions.

An experimental BAF module was incorporated into an European space flight simulation facility with a crew of four, where experiments have been performed with a real atmosphere. In this experiment methane, acetone, toluene and (iso)-propanol amongst other compounds were removed by the BAF system. The classical physico-chemical approach involving a combination of oxidation catalysts with pre- and post-adsorption and chemisorption filters suffers from a number of important drawbacks, such as the high susceptibility to poisoning by certain compounds and the risk of creating chemical decomposition products which are possibly more difficult to remove or more toxic than the initial contaminants. In comparison with physico-chemical systems, the BAF is expected to have several advantages inherent to the living nature of its biological catalyst, such as adaptability to unexpected contaminants, filter efficiency, and ability to recover normal efficiency after accidental poisoning. [6, 63, 76]

Waste treatment, i.e., biological decomposition of organic wastes in a CELSS, can principally be accomplished by the use of either activated sludge systems or composting. These subjects are further dealt with in section VI.4.6. [63]

VI.3 ALGAL SYSTEMS

As mentioned earlier, a biological process that is likely to be central to the successful development of a CELSS is photosynthesis. Photosynthesis has the potential for simultaneously carrying out most of the major functions of a life support system, namely, the production of food, oxygen, and water, and the removal of carbon dioxide. Among those organisms capable of photosynthesis, there are two kinds of microorganisms that have several characteristics making them attractive candidates for inclusion in a bioregenerative life support system:

- Green Algae (*Chlorophyta*)
- Blue-green Algae (*Cyanophyta*)

The advantages of green and blue-green algae with respect to an application for life support in space are:

- Rapid growth
- Controllable metabolism
- High harvest index
- Gas exchange characteristics compatible to human requirements

Unfortunately, there are also a number of problems associated with the introduction of micro-algae in a CELSS, e.g., the questions about the adequacy and acceptability of algal-derived food, harvesting and processing of the algae, and the long-term stability of algal cultures. Anyway, techniques for fractionation of bacterial or algal cells, and manufacturing of nutritious food from parts of the cell material, although in their infancy, are promising solutions to some specific problems, such as supplying protein to humans, not only in space, but on Earth as well. However, if it is used directly by man, algal food will probably have to be supplemented with food from other sources. Indirect use of cheaply produced single cell food, e.g., through feeding algal food to animals which may then be used as human food, is also a promising approach. [4, 35]

Since the early 1960's, both in the U.S. and the former Soviet Union, experiments were conducted that were linking the gas exchange of animals, and later men, with algae-based systems. The probably most exiting experiment with algae was conducted in the former Soviet Union in the 1960's. In this experiment, one man lived in a hermetically-sealed, 4.5 m^2 room for 30 days, sustained by a 30 liter algae apparatus. During this time, his oxygen was completely supplied by *Chlorella,* grown in a special reactor. This system went through 15 cycles of regeneration and achieved a high degree of stability, e.g., carbon monoxide and methane accumulation stabilized after a couple of days. Moreover, water for drinking, food preparation, and hygiene was collected from condensation, filtered and reused, as was the urine, and 50 g of dried *Chlorella* were included in the diet. [44, 61]

VI.3.1 Algal Species and Attributes

Most of the work on algal physiology related to sustained culture under space-related conditions has been done on a relatively few species of green (eucaryotic) and blue-green (prokaryotic) algae. The primary species currently considered for space-related applications are:

- *Chlorella* (green) - several species
- *Scenedesmus* (green)
- *Anacystis* (blue-green)
- *Spirulina* (blue-green)

However, there is no ideal alga. For example, *Chlorella* is an extremely productive alga, capable of doubling its mass in nine hours under favorable conditions, and an efficient producer of oxygen. On the other hand, *Spirulina* is superior with respect to food value. Thus, in some cases, there may be a definite trade-off between ease of culturing and harvesting, nutritional value, etc., e.g., *Chlorella* vs. *Spirulina*. Also, the prokaryotic, N_2-fixing cyanobacteria (blue-green) species deserve attention, because of some of their special attributes, like minimal nutrient requirements, short regeneration time of some species, a wide tolerance for environmental stresses, high protein content, and a good growth in a broad range of light (between 600 and 650 nm). The effect of light is probably the most important consideration in the design of an algal culture system. Since algae are capable of adapting to different light conditions, changes in light intensity can elicit corresponding to an alga's physiology. Productivity follows some general rules, i.e., the maximum productivity occurs at

- Relatively high population densities
- A specific growth rate equal to about one-half the maximum growth rate

Under natural conditions, essentially every part of the spectrum from 300-950 nm is utilized by one or another organism. In green algae, maximal absorption is due to chlorophyll-a and chlorophyll-b, with absorption maxima at 675 and 650 nm, respectively. However, the blue-green algae contain accessory pigments (phycobiliproteins) that absorb in the range of 500-650 nm, and this energy can be used photosynthetically by transfer to chlorophyll.

The effects of CO_2 and O_2 on algal growth have not been worked out as quantitative as those of light. One finding is that some algae are able to pump HCO_3^- (bicarbonate), thus concentrating inorganic carbon in a manner similar to that found in C_4 plants. The main features of carbon dioxide utilization of algae are:

- CO_2 is a substrate, and air levels (0.03 %) are not saturating. Algal cultures generally run on 1-5 % CO_2 in air.
- The enzymology of CO_2 uptake is well understood.
- HCO_3^- by algae is recognized but not well understood.
- Mass transfer and interface problems have to be addressed.

O_2 is both a product and an inhibitor of photosynthesis. In general, deleterious O_2 effects at atmospheric levels can be circumvented by using high CO_2 tensions. The concentration relations between O_2 and CO_2 are quite complex, and depend on such factors as specific physiology, HCO_3^- pumping, and mass transfer. The responses of algae to temperature are not unusual, compared with those of other organisms. In general, algae have a wide range for growth, with given species having an optimal growth range of 2°-3° C. Thermotolerant *Chlorella* have an optimal growth range of 39° C, and some thermophilic blue-greens have been reported with optimal growth ranges greater than 50° C.

The main mineral requirement for algal growth is fixed nitrogen in a proportion of about one nitrogen atom per six carbon atoms. Urea, ammonia, and nitrate have all been used as nitrogen sources, and there have been some experiments using human wastes. Other requirements are K, Mg, S, and P. In addition, about 10 microelements seem to be required for optimal growth. Algal growth can be considered as a balanced chemical equation:

$$6.14\,CO_2 + 4.7\,H_2O + HNO_3 + light \rightarrow C_{6.14}H_{10.3}O_{2.24}N + 8.9\,O_2$$

Assimilatory quotient = 0.69

$$6.14\,CO_2 + 3.7\,H_2O + NH_3 + light \rightarrow C_{6.14}H_{10.3}O_{2.24}N + 6.9\,O_2$$

Assimilatory quotient = 0.89

As a result of developments in molecular biology and genetics, it appears that in the near future it may be possible to specifically alter the nutritional and/or other qualities of an alga. The question remains, though, how the modified organisms may behave in culture. [4, 44]

VI.3.2 Algae for Provision of Food

In an algal regenerative system, a given amount of regenerated oxygen will be associated with a given total biomass of algae (X) and a fraction of this biomass (Y) will be utilizable as food. The biomass X is a function of the species of alga selected, of growing conditions, of harvesting methods, and of other factors.

The biomass fraction Y is obviously a function of X, but it is also a function of space, of growing conditions, and of harvesting methods (see section VI.3.3). Moreover, the fraction Y is a function of the number of food processing variables, such as methods of purification, method of preparation, and palatability. [4]

Algae are quite rich in vitamins and protein, e.g., *Chlorella pyrenoidosa* contains 40-60 % protein, about 20 % carbohydrates, 10-20 % fat, and 5-10 % ash. However, data compiled to the present show that the permissible portion of algae in the food supply of human beings does not exceed 20 %. Experiments in which 50-150 g dry algae were added to human food, caused dyspeptic phenomena, i.e., belching, nausea, and loss of appetite, which disappeared after a few days. However, no objective disturbances in the gastrointestinal tract were observed. A disadvantage of unicellular algae as food is their poor assimilability caused by indigestible walls. Enzymatic destruction of algal cell walls by means of the cellulase enzyme of certain bacteria, lower fungi, and mollusks and subsequent increase in the digestibility of algae protein has been demonstrated. Since the direct consumption of single-cell biomass without purification seems not to be feasible in amounts that would be of significance to biomass recycling, it will be important in the future to:

- Establish the feasibility of extracting purified edible components from biomass
- Establish the feasibility of converting these components of food
- Scale down relevant processes to space conditions
- Analyze the relationship between the degree of biomass utilization and the weight of required equipment

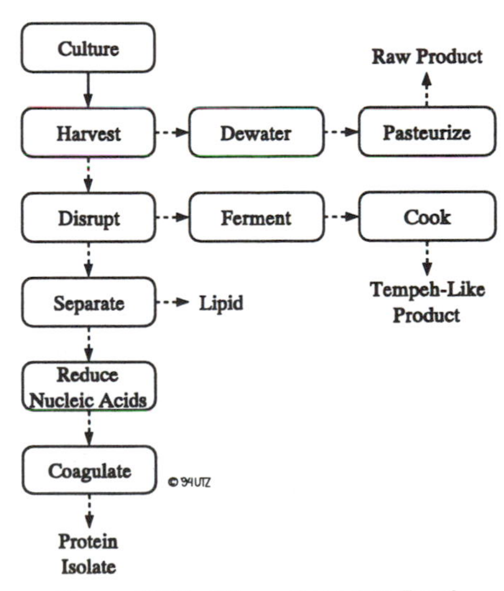

Figure VI.12: **Alternatives for Food Production from Algae [4]**

The recovery of part of the protein may be the least difficult task, followed by recovery of part of the liquid. However, recovery of carbohydrates, which in a large measure are in cell walls, will require substantial chemical processing, and the carbohydrates may not be of nutritional value. Some alternatives in processing are shown in figure VI.12.

A preliminary estimate of available Joules per kg of dry algae is given in table VI.6. The estimate is speculative but it does provide a basis for preliminary consideration of nutritional requirements. It suggests that additional carbohydrate would be required for nutritionally balanced diets.

	Approximate Weight Distribution				Nutritive Value	
Constituent	Total		Wall	Cytoplasm	Unit	Total
	[%]	[g]	[g]	[g]	[J/g]	[J]
Protein	45	450	67	384	18.84	8478
Nucleic Acid	5	50	-	50	-	-
Lipid	15	150	40	110	39.78	5967
Carbohydrate	25	250	219	30	15.49	3873
Fiber	5	50	50	-	-	-
Ash	5	50	-	50	-	-
Total	**100**	**1000**	**376**	**624**	**-**	**18318**

Table VI.6: **Estimated Mass and Energy Distribution of 1 kg of Algae [4]**

In particular, cyanobacteria seem to have many advantages concerning the nutritional value and also the processing of algae as food. Most of the major proteins of cyanobacteria should be readily accessible for fractionation. Being prokaryotic, it is in principle relatively easier than with eukaryotic algae to design methods combining enzyme treatment with mild mechanical treatment or osmotic shock to break the organisms open and free their internal contents. Cyanobacteria also have already been widely used as a source of food for animals and humans in various parts of the world. Indeed, in China, it has been used for centuries. [4]

VI.3.3 Algal Growth in Space

Because of the complex requirements for productive algal cultures coupled with weight and energy limitations on spacecraft there have been only very few operational photosynthetic reactors yet for use as prototypes of spacecraft photosynthetic systems. Ground-based algal growth reactors use processes that are dependent on gravity, such as gas-bubble sparing and mixing, and overflow harvesting. The processes have to be replaced with functionally analogous processes that can operate in the space environment. Onboard COSMOS 1887, *Chlorella vulgaris* was flown in a sealed Aquarium unit as part of an algae-bacteria-fish ecosystem. Active growth of *Chlorella* occurred during the flight, and more than two generations were produced. An algae-microorganism-fish ecosystem was also flown on COSMOS 2044 and there were also no major changes in growth of the algae compared to ground control. The main results of studies of unicellular growth in space to date are:

- μ-g strongly enhances cell proliferation in bacteria, algae, and protozoa.
- Biological periodicity is undisturbed, but cell division are at faster rate.
- Single cells in culture are sensitive to gravity, i.e., important cellular functions are affected by low-g conditions.
- There is a marked reduction in glucose consumption, indicating that the metabolism of the cells is changed in microgravity. [4, 15, 49, 59]

Although short-term growth of algae would seem to present no problems, long-term exposure to the space environment may affect culture stability or viability, since, e.g., the Space Shuttle cabin atmosphere has been reported to contain many volatile organic compounds outgassed from Space Shuttle structural materials. Continual exposure of algal cultures to such compounds may allow their accumulation until they reach toxic levels. Other problems that may only be manifested after long-term exposure of the algae to the spacecraft environment include mutagenic effects of ionizing radiation, microbial contamination of the algal cultures, and application of techniques of heat and mass transfer on liquid and particle behavior. Biological parameters that might be affected include rates of growth, photosynthesis and respiration, algal composition, and the excretion of organic and inorganic compounds. To assess properly the long-term behavior of algal cultures, spaceflight opportunities lasting for several hundred algal generations, corresponding to several months of continuous growth, will be required. [4]

VI.4 HIGHER PLANTS

As mentioned before, one of the most important aspects in biological life support is to develop and validate techniques fulfilling one important life support requirement, namely food production. It was shown in the previous chapters that algae use growing space efficiently, produce oxygen, are rich in protein, may be used as food supplements, and are also efficient in processing metabolic wastes. However, it was also found that productive algal systems are difficult to maintain for long periods of time, do not provide a balanced palatable diet and require cumbersome maintenance and harvesting procedures. Thus, the production of algae can be seen as a first attempt to tackle the question of food production, but sound and acceptable food for crew consumption can only be anticipated from higher plant biomass. Although the daily protein intake can easily be fulfilled with single-cell proteins, only plants can provide most, if not all, of the major food needs of man: calories, proteins, fats, carbohydrates, minerals, vitamins and trace elements. When introducing a higher plants compartment into a life support system, the plants may not only serve as food providers, but can also be utilized for:

- *Atmosphere revitalization*
  When the necessary light energy is given to plants they are consuming water and the atmospheric carbon dioxide, transforming it into carbohydrates and oxygen (photosynthesis).

- *Water regeneration*
  All photosynthetic reactions are producing much heat that is removed from the plant leaves by transpiration. Plant transpired water can be collected and used as hygiene or potable water.

This means that by the use of higher plants for food production on long-duration or large scale manned space missions the large storage and resupply costs associated with carry-on food and oxygen may be minimized. Nevertheless, there is in general no consensus on the breakeven point (concerning mass) in flight duration for such a biological life support system compared with food supply and resupply alternatives. It is only clear that the higher the degree of closure is, the later the breakeven point for a plant production system will be reached. [63, 72]

Parameter	Amount
CO_2	40-300 g/m²/day
Water	5-10 kg/m²/day
Minerals	10-100 mg/m²
Lighting period	8-24 h
Lighting power	13-170 W/m²

Table VI.8: **Plant Requirement Values [12]**

Parameter	Amount
O_2	30-220 g/m²/day
Transpiration water	5-10 kg/m²/day
Edible biomass	20-40 g/m²/day
Inedible biomass	4-20 g/m²/day

Table VI.9: **Plant Performance Values [12]**

The plant species that are to be cultivated in a CELSS have to provide an optimal and, if possible, complete food supply. Beyond this, a maximum of oxygen and water yield and carbon dioxide assimilation is to be obtained by the production of biomass. It is generally recognized by crop physiologists that the nutritive value of plants grown in controlled environments varies considerably and has nutritive values that are similar to that of plants grown in field environments. Thus, requirement and performance data of a plant production system can be estimated based on terrestrial plant production experience. A summary of these values is given in tables VI.8 and VI.9. It should be noted, though that these requirements are varying very strongly depending on the plant species.

Consumable	Required Area per Person
Water	3-5 m²
Oxygen	6-10 m²
Food	15-20 m²

Table VI.10: **Area Requirements of Higher Plant Growth [12]**

If the numbers of the tables above are taken into account, the areas values required to supply one human with a sufficient amount of consumables produced by higher plants can be estimated as indicated in table VI.10.

The environmental factors that have to be considered in a plant production system design are:

- Temperature
- Light intensity
- Lighting duration
- Spectral composition
- CO_2-concentration of atmosphere
- Irrigation
- Water quality
- Plant protection
- Fertilization
- Cultivation techniques

Furthermore, according to space conditions and restricted availability of volume and energy the following criteria for plant production are of importance:

- Fast growth (short growth periods and/or continuous harvesting)
- High harvest index
- Low-input genotypes' growth on minimal space, light, water, etc.
- Growth on low-quality water if necessary
- High transpiration rate for fast water turnover
- Balance of CO_2-assimilation rate and human respiration rate [7]

One of the great concerns when growing plants in a closed environment is the possible accumulation of toxins within the system, which might damage or kill plants or perhaps harm humans. Potential contaminants may be carried in the air, root medium or water. They may originate from man, the spacecraft, or the plants themselves. Plants are known to give off at least 200 discrete substances including hydrocarbons, aldehydes, alcohols, ketones, ethers, carbon monoxide, various organic acids, lactones, flavones, ethylene and teropenes. Humans also release 150 volatile substances that may concentrate in a closed system and disrupt plant growth. The significance of plant toxins was demonstrated in BIOS-3 (see section IV.5.2). It was shown that higher plants grew vigorously without the algae system, but the plants died shortly after the algae were introduced into the unit. Presumably, unknown toxins given off by the algae killed the plants. Also, care should be taken that seed stock and culture systems are reasonably free of pathogenic organisms. Plants growing vigorously in controlled environments rarely develop disease problems unless pathogens are introduced by some external source.

Precise control of environmental conditions will be necessary to obtain maximum crop yields within a regenerative life support system. Most plants have been found to have best growth with alternating light and dark periods that provide maximum yields with 16-18 hours light and 6-8 hours dark. Another aspect that is critical for maximum yield is the spacing of plants. It has been established that the optimum leaf area index, i.e., vertical density of leaves, for planting varies depending on light intensity and quality of light. Additionally, for the optimization of the production parameters the following options are available concerning the phenotype of the plants:

- *Genetic factors:* Obtaining plants with the desired characteristics by selection and cultivation
- *Production expenditures:* All controllable factors in a closed system

A limited manipulation of the genetic information (genotype) of plants, in order to optimize production parameters, is also possible by plant breeding and genetical engineering. In summary, a program concerning the integration of higher plants into life support systems should therefore have multiple goals:

1. Definition of man's nutritional needs as a function of length and type of space mission.
2. Selection of plants according to their:

    - Ability to provide sufficient edible biomass
    - Growth condition compatibility with the space conditions
    - Ability to grow on media regenerated from waste

3. Studies on plant models of the intrinsic problems of growth in closed and limited conditions (Plant contamination control and monitoring, water and carbon dioxide supply, oxygen removal, nitrogen sourcing, light intensity and spectral quality, and biological rhythms).
4. Safety problems related to the genetic variability of the chosen strains due to mutations induced by cosmic rays and particles.
5. Effects of microgravity on plant life-cycles (from seedling to fructification).
6. Optimization of the control system for managing the different subsystems of the higher plant compartments and the connections with preceding, surrounding and following compartments. [72]

VI.4.1 Plant Physiology

VI.4.1.1 Basic Plant Processes and Requirements

The diversity of organisms on Earth is determined by two factors: the genetic information inside (endogenous factor) and the environment outside (exogenous factor) the organism. Here, the environmental factors and requirements of plant growth are dealt with. These basic plant growth requirements are:

- Carbon in the form of CO_2
- Hydrogen and oxygen in the form of H_2O and O_2
- Radiation between the wavelengths 400 and 700 nm

Because of these requirements, during the growth of plants the following physiological processes occur:

1. During the daylight period, photosynthesis dominates over respiration and leads to the formation of plant matter. As a net effect, CO_2 is consumed and O_2 is produced.
2. At night, i.e., during the time period with no illumination, the respiration leads to CO_2 production and O_2 consumption.
3. During the light and dark phases, the plant evaporates water which it sucks by means of its roots from the soil, i.e., the substrate. The evaporation rate in the light phase is higher than in the dark phase.

Many of the factors, summarized in table VI.11, that affect plant growth on Earth will also affect growth in a space environment. They include growth area, light photoperiod, the plant root environment, CO_2 concentration, temperature, humidity, radiation, and reduced gravity. In space, control of these parameters will not be provided by the forces of nature, but must be engineered into the system. [32, 42, 65, 67]

Consumption	Amount	Production	Amount
Carbon Dioxide	40-300 g/m²/day	Oxygen	30-220 g/m²/day
Water	5-10 kg/m²/day	Transpiration water	5-10 kg/m²/day
Mineral nutrients	10-100 mg/m²	Edible biomass	20-40 g/m²/day
Illumination duration	8-24 h	Inedible biomass	4-20 g/m²/day
Energy	13-170 W/m²		

Table VI.11: **Requirements and Productivity of Higher Plants [16]**

VI.4.1.1.1 Plant Physiological Parameters

The major goal of plant growth, whether on Earth or in space, is usually to obtain maximum crop yield. Although it may also be possible to improve it by manipulation of the genotype of a plant, e.g., by genetical engineering, the yield is mainly dependent on the growth conditions. Thus, a list of the most important environmental factors and parameters of plant growth is given below:

- Incident photosynthetic photon flux (PPF) in [μmol/m²/s]
- Absorption of the incident PPF by photosynthetic tissue
- Photosynthetic efficiency, i.e., moles of CO_2 fixed per mole of photons absorbed
- Respiratory carbon use efficiency (CUE), i.e., net carbon fixed in biomass per unit carbon fixed in photosynthesis
- Harvest index, i.e., ratio of edible biomass / total biomass
- Carbon dioxide assimilation rate in [$gCO_2/m^2/d$]
- Transpiration rate in [$gH_2O/m^2/d$]
- Leaf area index
- Dry mass production rate in [g/m²/d]
- Respiration rate in [$gO_2/m^2/d$]
- Vegetation duration in [d]
- Nutritional requirements
- Cultivation procedures
- Seed requirements
- Light requirements (photoperiod, intensity)
- Area requirement
- Temperature requirements

Harvest Index	Plants
0.1 - 0.2	Sunflower
0.2 - 0.3	Sweet lupine
0.3 - 0.4	Broad bean, ground-nut, common bean, linseed, rape, soybean
0.4 - 0.5	Barley, pea, oat, maize, rye, mustard, topinamur, wheat
0.5 - 0.6	Cucumber, melon, celery, tomato, zucchini
0.6 - 0.7	Carrot, potato, red beet, white cabbage, sugar beet
0.7 - 0.8	Aubergine, rutabaga, pepper, parsnip, sweet potato, yams, onion
0.8 - 0.9	Broccoli, kohlrabi
0.9 - 1.0	Mushrooms, Chinese cabbage, endive, sweet chard, peppermint, leek, lettuce, spinach

Table VI.12: **Harvest Indices of Different Higher Plants [16]**

Crop Adaptability Inventory					
Photosynthesis characteristics	Crop group				
	I	II	III	IV	V
Photosynthesis pathway	C_3	C_3	C_4	C_4	CAM
Radiation intensity at max. photosyn. [J / (cm²·min)]	0.8 - 2.5	1.3 - 3.4	4.2 - 5.9	4.2 - 5.9	2.5 - 5.9
Max. net rate of CO_2 exchange at light saturat. [mg / (dm²·h)]	20 - 30	40 - 50	70 - 100	70 - 100	20 - 50
Temperature response of photosynthesis					
optimum temp.	15 - 20 °C	25 - 30 °C	30 - 35 °C	20 - 30 °C	25 - 35 °C
operative temp.	5 - 30 °C	10 - 35 °C	15 - 45 °C	10 - 35 °C	10 - 45 °C
Max. crop growth rate [g / (m²·day)]	20 - 30	30 - 40	30 - 60	40 - 60	20 - 30
Water use efficiency g / g	400 - 800	300 - 700	150 - 300	150 - 350	50 - 200
Crop species	Field mustard, potato, oat, tomato, rye, grape, pyrethrum, sugarbeet, bread wheat, chickpea, french bean, arab. coffee, sunflower, olive, barley, cabbage, lentil, linseed	Groundnut,fig, french bean, rice, soybean, cowpea, ses-ame, tomato, hyacinth bean, raselle, tobac-co, sunflower, grape, castor bean, sweet potato, sweet orange, bana-nas, lemon, avocado,pear, coconut, cot-ton, mango, robusta cof- fee, olive, oil palm, cocoa	Japanese barnyard millet, foxtail millet, common millet, finger millet, pearl millet, hungry rice, sorghum, maiz, sugarcane	Japanese barnyard millet, foxtail millet, common millet, sorghum, maiz	Sisal, pineapple

Table VI.13: **Physiological Data of Higher Plants [16]**

	Vegetative (Leaf, lettuce)	**Root** (Onion, Radish, Carrot)	**Fruit** (Tomato, Pepper, Cucumber)	**Sprout** (Alfalfa, Bean, Radish)
Light (μmol $m^2 g^1$) Photoperiod (hrs)	250 - 275 18 - 24	275 - 400 18	300 - 400 18	0 - Ambient Ambient
Temperature (°C)	22 - 28	15 - 25	20 - 28	20 - 28
Humidity (% Rh)	50 - 85	50 - 70	50 - 75	High - 90
Gas Composition (ppm)	CO_2 = 300-1500	TBD	CO_2 = 300-1500	Good aerobic germination
Biocompatibility	No vinyl plastics or copper	No aluminium	No vinyl or ammonia	Standard
Hoagland's Nutrient Solution	Half normal	Half normal	Half normal with 100% iron+calcium	Water
Air Flow (m/s)	0.1 - 1.0	0.1 - 1.0	0.1 - 1.0	Relative to oxygen humidity
Substrate, Nutrient Delivery	No substrate: aeroponics and hydro-ponics OK	No substrate	No special substrate	None / wetting
Plant Volume Root Zone	15x15x15cm 2.5cm deep	5x5x15cm 47cm³ for bulb 25cm top for onion	30x30x30cm	10x10x5cm Food Pack
Contamination Control	Ethylene and gaseous am-monia control	TBD	Ethylene control	Standard
Access	No special access, ex-cept for varying plant spacing	Substrate contain-ment during harvest	Pollination methos requires special access	Standard

Table VI.14: **Plant Growth Requirements for Several Vegetables [16]**

These factors may be varied in order to optimize the crop yield and the integration of the plant growth unit into the overall system. As examples, some individual growth requirements for vegetative, root, fruiting vegetable, and sprout crops are specified in tables VI.13 and VI.14, and the harvest indices of several higher plants are given in table VI.12. [10, 54, 74]

VI.4.1.1.2 Photosynthesis

In the photosynthetic process, the carbon, hydrogen and oxygen are combined to form glucose, and excess oxygen is evolved. This process takes place in the leaves of plants:

$$6\,CO_2 + 6\,H_2O \xrightarrow{Light} C_6H_{12}O_6 + 6\,O_2$$

The leaves export most of this glucose to other parts of the plant, where it is converted into new plant biomass, including new leaves. The composition of this new biomass varies, but a typical example might be:

$$1g\ C_6H_{12}O_6 + 0.1g\ NO_3^- + 0.003g\ SO_4^{2-} + 0.12g\ O_2 + 0.034g\ \text{Other minerals}$$
$$\rightarrow 0.65g\ \text{New biomass} + 0.34g\ CO_2 + 0.27g\ H_2O$$

This equation shows that plants recycle some of their own carbon, hydrogen, and oxygen, and that many other chemical elements are required. The photosynthetic process takes place within the chloroplasts of green plants. Here, the energy from photosynthetically active radiation (PAR) of wavelengths between 400 and 700 nm is utilized for photolysis of water to hydrogen and hydrozyl ions. The reducing potential of the hydrogen ions is utilized to reduce carbon dioxide to the level of simple sugars, while the oxidizing potential of the hydroxyl ions is utilized to evolve oxygen. These biochemical processes of photosynthesis are summarized in the following reactions:

$$4H_2O \xrightarrow{PAR} 4H^+ + 4OH^-$$

$$4H^+ + CO_2 \rightarrow CH_2O + H_2O$$

$$4OH^- \rightarrow H_2O + O_2$$

Thus, the main products of the photosynthetic process are the basic carbon units of carbohydrates and the regenerated oxygen required for respiration. From the carbon units, the amino acids, vitamins, carbohydrates, and other organic components essential for human nutrition are synthesized. Leaf photosynthetic response of C_3 and C_4 plants to PPF is shown in figure VI.13 and may be represented by three parameters: dark respiration R_d, the slope α at the origin that represents the apparent quantum yield in [mol CO_2/mol photons], and the maximum assimilation at high PPF A_{max}. In most C_3 plants α is about 0.05-0.07, while A_{max} varies from 2 to 60 μmol/m²/s depending on the plant species (mostly 15-45 μmol/m²/s), and R_d is normally about 8-10 % A_{max}.

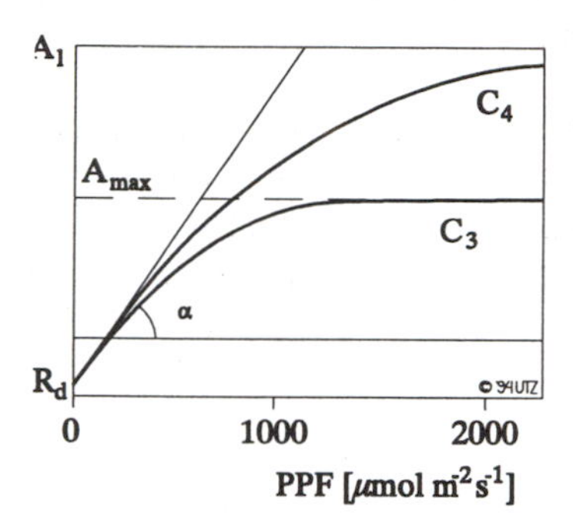

Figure VI.13: **Leaf Assimilation as a Function of Light (PPF) [50]**

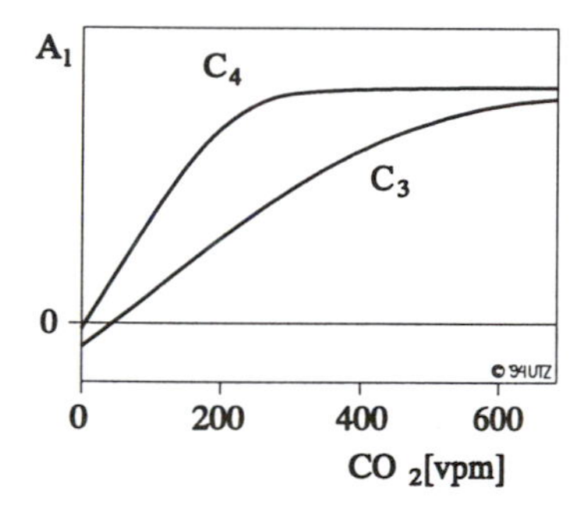

Figure VI.14: **Leaf Assimilation as a Function of CO_2 Level [50]**

Variation of A_{max} with external CO_2 concentration is presented in figure VI.14. In C_4 plants A_{max} rises quickly to a plateau that is reached at about the normal atmospheric CO_2 concentration, i.e., 350 ppm CO_2. In C_3 plants due to photorespiration A_{max} is negative at zero CO_2 concentration and increases more slowly with CO_2 up to a plateau that is reached only at CO_2 concentrations of about 1000 ppm.

In order to predict photosynthesis rates, i.e., plant growth, as a function of environmental conditions, like humidity, temperature, and radiation, models have been developed. These may help in the development of closed artificial environments. Such a simple model of biomass production may be written as:

$$P_n = \Delta B/\Delta t = e\, I_a = e\, I_0\, (1\text{-}a)\, (1\text{-}e^{-kL})$$

Net production of total biomass: .... P_n
Leaf area index (from vegetation top to soil surface): . L
Total plant biomass: .... B
Albedo of a closed canopy for visible radiation (0.05): a
Time: .... t
PPF absorbed by canopy: .... I_a
PPF received by canopy: .... I_0
Conversion efficiency of absorbed PPF into biomass [mol dry matter / mol photon]: .... e
(e_{max} = 0.05)
Extinction coefficient for PPF: .... k
(0.4 (erect leaves) < k < 0.9 (horizontal leaves))

Normally, 25 to 57 g CO_2 may be fixed by plants per square meter growing area per day. Since photosynthesis rates decline near maturity, it will be necessary to harvest promptly to maintain the atmosphere cycling capacity of the vegetation. [20, 42, 51, 55]

VI.4.1.1.3 Gravitropism and Phototropism

In the process of evolution, many living organisms have developed mechanisms, responsible for perceiving the force of gravity. The properties of plants for reacting to its effect are called geotropism. As geotropic organisms, plants grow their roots down and their shoots up in response to gravity. Furthermore, there are gravity-related mechanisms that control branch angles, leaf orientation, and other behaviors. Also, the orientation and special shape of the plant plus polar movement of certain plant hormones contribute to differentiation processes. These effects are due to the fact that of all environmental factors affecting plants, gravity is most constant in magnitude and direction. It exerts its influence most dramatically on plant cell growth, elongation, and development.

When plants sense gravity, the first step of their response is the perception of gravity. This is the conversion from the physical stimulus to a biological signal. In other words, gravity induces mechanical responses in the cells of plants indirectly by inducing vectorial and hydrostatic pressure in the vicinity of single cells, and by direct effects in which primary receptors for gravity, i.e., statoliths in higher plants, are located inside the cell. The indirect effects are responsible for most of the adaptations of Earth plants, such as the strength of stems and trunks to stand against gravity. Statoliths are specialized plastids which contain dense starch grains. They function in gravitropism by settling to the lower surfaces of cells in the root cap, and in parenchyma cells adjacent to vascular bundles in primary shoots. This biological signal has to propagate to that site of the shoot, which is responsible for the curvature response, i.e., to the outer cell layers of the growth region of the shoot. Besides, the entire process of hormonal regulation is also related to gravity through polar transport systems. [30, 42, 76]

A very similar response is phototropism. In the case of unilateral irradiation, the plant reacts on the light gradient within its tissue and performs a curvature towards the light or away from it, depending on light intensity. It is obvious that phototropism and gravitropism have distinct receptors due to different physical types of stimulus. But because of the similar growth response which leads to the curvatures, it is very probable that there are common pathways within the signal transduction chain. Light-dependent plant physiology is mainly performed by irradiation of seedlings with monochromatic light. The plants have to be grown up under standardized conditions, concerning substrate, humidity and temperature. Gravitropism related experiments have the problem that gravity cannot be turned off. One possibility is to use a clinostat, which simulates zero gravity by a low speed rotation. The plant recognizes the gravity vector permanently from another direction, so that the integral is equal to zero. Comparison with experiments in space in a non-gravitational environment have shown that the clinostat can simulate many effects qualitatively but not quantitatively. To elucidate the signal transduction chain of gravitropism and phototropism it is very helpful to have the possibility of switching on and off light as well as gravity. Presently, this is possible only in orbit under microgravity conditions. Phototropic reactions would not become independent from gravitropic compensation, but it would be possible to perform experiments for testing the reversibility of photoinductions by an additional gravistimulus and vice versa. For this purpose different concepts, like the Botany Facility or the Gravitational Physiology Plant Facility, have been proposed (see section VI.4.1.2.3). [7]

VI.4.1.2 Effects of the Extraterrestrial Environment

The most important physical environmental factors which interfere with biological processes are radiation and gravitational forces. Both factors are changed if organisms live outside Earth, where radiation increases in quantity and quality and where gravitational forces decrease. Very little is known about the growth patterns of living organisms in reduced gravity environments. In gravity perceiving systems of plants, animals and man sedimentation processes are probably the first step triggering the mechanism of signal transduction in a stimulus-response-chain, which ends with the orientation of the organisms in space. For example, cress roots growing under 1-g conditions (ground control experiment) and under microgravity has shown that roots from ground control experiments show oriented growth which is more or less parallel to the g-vector. Under microgravity roots are nearly straight and form angles with the seed plate that are more or less alike. This demonstrates that germination and root growth in a almost stimulus free environment are under the control of endogenous factors (genetical determination).

Gravity strengths in the inner solar system range from almost zero-g on orbiting space stations and interplanetary spacecraft to about 1/3 g on Mars. During its life cycle a CELSS plant may experience transitions between several gravity levels, as would occur during transportation between Earth and Mars. There are no known methods for simulating these reduced gravity fields on Earth, other than using aircraft or sounding rockets and fall towers to create microgravity environments lasting only several minutes or seconds, respectively. Meaningful experimentation with plant growth in reduced gravity will require much longer exposure periods of days to months. Most of the current knowledge on the long-term effects of reduced gravity comes from microgravity experimentation performed on the space stations and satellites of the former Soviet Union. Galactic cosmic radiation is another problem where the effects on plants are not fully understood. Among the potential problems are reduced yields and genetic effects on subsequent generations. On Earth, most harmful GCR is blocked by the atmosphere. This will not be the case in space, where artificial methods are required to protect living organisms from radiation poisoning. Radiation dose limits for plants are generally higher than for humans, but exceeding these limits will have the same lethal results. [15, 35, 65, 67]

The absence of gross morphological deformities in higher plants grown in space from seeds developed on Earth might be thought to indicate that no major disturbances occur in metabolic pathways, growth hormone distribution, transport phenomena, etc., at least over a period of 1-2 weeks. However, more detailed examination of the plant material reveals that disturbances in cell division, nuclear and chromosomal behavior, metabolism, reproductive development, and viability occur under microgravity conditions. [68]

Nevertheless, it can be assumed that basic metabolic processes such as photosynthesis and respiration function in space. Whether these processes function at rates comparable to Earth-based plants is not known. Thus, research is needed to utilize higher plants effectively in space-farming systems. Research needs fall into two categories: physical and biological parameters. The early research has to focus on the physical parameters, because control of these physical factors would be required for successful conduct of most biological-parameter experiments. Physical parameter research needs include:

- *Water and nutrient delivery through solid media*
  Because the gravity-dependent movement of water cannot occur under weightlessness, other mechanisms of water movement must be employed. Solid media may be used in early experimentation phases of CELSS for ease of maintaining satisfactory growing conditions (see section VI.4.3.2.1).

- *Liquid transport to and from roots in hydroponic and/or aeroponic systems*
  Hydroponic approaches using thin films of water or misting systems most likely will be required in an operational CELSS to reduce the system's overall weight and to simplify regulation and control in the recycling of elements (see sections VI.4.3.2.2 and 3).

- *Oxygen and carbon dioxide solubility and diffusion in liquid and solid media*
  Physical properties of gas movement in microgravity should be known before adequate biological assessments can be made. This should include rates of oxygen and carbon dioxide diffusion through liquid and solid growing media and exchange between these media and the atmosphere. Once solubility and diffusion rates are determined, then solution, water-film, and aeroponic systems should be evaluated to determine how to provide adequate gas exchange at the root surface with each system.

Biological parameter concerns include:

- *Seedling establishment and seed coat shedding*
  Problems in seedling establishment that should be studied in CELSS experimentation include the difficulties for radicles from germinating seeds to penetrate a solid medium and the inability of cotyledons to shed their enclosing seed coats. Both problems have been encountered in Earth-based studies when seeds were germinated on media surfaces with no physical restriction around them.

- *Orientation of roots, stem, and leaves to maintain plant productivity*
  The orientation of shoots and roots needs to be studied carefully during spaceflight to examine what effects this might have on plant productivity and cultivation techniques. Shoot growth, which is usually negatively gravitropic, must be studied to determine if light can be used as an orientation stimulus in the absence of gravity. The growth of each candidate species should be investigated under irradiance levels utilized for food production to establish whether effective orientation can be obtained. Similarly, root growth of each of the selected species should be examined.

- *Flower initiation, pollen transfer, and fertilization*
  Many of the selected crop species require flower initiation, flower development, pollination, and fertilization, both for production of edible biomass and reproduction. Transfer of pollen from anthers to stigmas, growing the pollen tube to effect fertilization, and embryo development should be studied in each of the appropriate species under weightlessness. Although studies of this sort would normally require intact flowering plants, it is conceivable that these processes be examined using excised shoots or buds in short-duration testing. Ambient ethylene concentration should be carefully monitored in studies because of its known effects on flower and fruit aborting floral sex expression, and fruit maturation.

- *Accumulation of edible biomass*
  Efficient carbohydrate and protein storage in edible plant organs and the normal development of these organs are crucial for the effective use of higher plants in space-farming. Yet there is no evidence that vegetative organs, such as seeds of wheat or soybeans, develop and accumulate carbohydrates and proteins normally under microgravity. The specific growth and development of edible organs of each of the proposed crops should be studied in detail.

- *Apical dominance*
  Effects of microgravity on apical dominance in the proposed crop species should be studied. Maintenance of healthy main shoot development may be crucial to high production of edible biomass in species where seeds are utilized for food.

- *Plant production and exchange rates*
  A very important phase of space experimentation for CELSS will be to determine whether plant productivity in a space-farming system is comparable, or even better, than that found in Earth-based systems. All aspects of productivity should be evaluated in these experiments, including:

- The amount of edible biomass produced
- Proportion of inedible to edible biomass
- Oxygen evolution
- Carbon dioxide consumption
- Water regeneration

However, these studies cannot be attempted until sufficiently large plant growth units are available to sustain plants to maturity. These units must be capable of controlling of major environmental variables, e.g., light, temperature, humidity, and CO_2, for the complete growth cycle of each plant species. If productivity is significantly less than that in comparable ground-based systems, then application of artificial gravity through centrifugation should be explored as a means of enhancing production. [69]

VI.4.1.2.1 Gravity

Some of the fundamental questions concerning the growth of plants in space are how cells "feel" gravity and how their development and function is influenced and changed in the space environment. Once these question have been answered, this may also lead to a general understanding of how the development and the processes of plants, organs, and cells are influenced by gravity. Growth and development of plants are determined by several factors, e.g., enzymes, proteins, and hormone-like acting substances. The transport and behavior of these substances is also influenced by gravity. According to the current knowledge, the biophysical effects of microgravity can be summarized as follows:

- *Decreased pressure / strain in cells*
  An obvious effect of reduced gravitational force is that the physical stresses and loads in cells change. The reduction in pressure and strain are of utmost importance in many areas of plant physiology.

- *Sedimentation of particles and buoyancy of bubbles*
  Another obvious, but fundamental effect of reduced gravitational force is that sedimentation of particles in fluids is diminished or even totally absent. Correspondingly, gas-filled volumes and vesicles will not move as effectively, because of reduced buoyancy, or will simply remain in place under microgravity conditions.

- *Stirring / convection*
  When density gradients are present in a liquid or gas under normal 1 g conditions, stirring and mixing takes place. Because of this movement, the gradients ultimately disappear. A thermal gradient induces convection and mixing in the same way. In microgravity, such mixing effects simply do not occur.

- *Changes of surfaces*
  Under normal 1 g conditions, gravity often inhibits increases in the size of bubbles and drops. Surface forces and phenomena depending on flow are easier to study under microgravity conditions.

The main results of plant cell biology studies to date are as follows:

- Seeds appear to germinate and grow into plants that show essentially the same gross morphological features as these grown on Earth.
- In higher plants, the most striking gross difference is loss of the normal relative orientation of the shoot and root, because the gravity-sensing guidance system (the gravitropic response) has failed to operate.

Experiments, conducted by the former Soviet Union, which observed the growth and development of plants through a complete life cycle and the formation of viable seed in microgravity, indicated that a viable seed can be produced, but there was also a large number of abnormal seeds. There is doubt, however, that the viable seeds subsequently produce normal seedlings. Thus, more detailed study on the long-term developmental effects in plant grown in space is needed. In summary, the microgravity environment provides not only the opportunity to assess the feasibility of using plants and microorganisms to provide life support, but also for conducting basic plant and cell research, i.e., to elucidate the fundamental mechanisms regulating normal plant and cell growth and development on Earth. [28, 49]

VI.4.1.2.2 Radiation

As already explained in section III.1, the intensity of ionizing radiation increases in space. In order to investigate the effects of cosmic radiation on plants, in the former Soviet Union, e.g., the VOSTOK and SOYUZ spacecraft were used in conducting experiments with lettuce, leaf cabbage, onion, pea, barley and other crops. In the U.S., e.g., two million seeds of 120 different varieties representing 106 species, 97 genera, and 55 plant families were flown onboard the Long Duration Exposure Facility (LDEF). [1, 38]

The LDEF was launched in April 1984 and deployed one day after launch for what was expected a one year mission. Mainly because of the Challenger accident in January 1986, it was finally retrieved in January 1990. With three exceptions all of the plants grown from seeds from LDEF, developed normally and produced seed. Thus, it could be concluded that the mutation rate of the seed that flew in the experiment was very low, occurring less than one in a thousand. This impression was also strengthened by the Space Exposed Experiment Developed for Students (SEEDS), flying about 12.5 million Rutger's tomato seeds on LDEF. The seeds were packed into five aluminum canisters which were sealed at 101 kPa pressure and 20 % relative humidity. A passive maximum temperature thermometer was placed in each canister and thermoluminiscent passive dosimeters were placed between the layers of seeds. The canisters were fastened inside a tray which was loaded onto the LDEF. An equal number of seeds was placed in a controlled environment of 21° C, 101 kPa pressure, and 20 % relative humidity on Earth.

The SEEDS was the first experiment removed from the LDEF. Radiation data indicated that the seeds had received 3.5 to 7.25 Gy. Afterwards, SEEDS kits were distributed to high schools and colleges. Each kit contained one packet with 50 Earth-based seeds and 50-100 space-exposed seeds. The students compared germination and growth characteristics of the seeds and returned data to NASA for analysis. A summary of the results is given in table VI.15. It is obvious that there are no big differences between the values for space-exposed and Earth-based seeds. Similar results could be obtained from the seeds of lettuce, radish, and garden cress that were stored onboard the MIR station for 240 days.

	Space-exposed Seeds	**Earth-based Seeds**
Germination rate (seeds germinated 14 days after planting)	66.3 %	64.6 %
Average time required for germination	8.4 days	8.5 days
Average height at 56 days	21.2 cm	20.9 cm
Average width at 56 days	12.0 cm	11.9 cm
Flowering rate	73.4 %	72.3 %
Average time to first flower	46.7 days	46.9 days
Plants producing fruit	74.6 %	76.1 %
Average time from planting until fruit formed	94.3 days	94.4 days

Table VI.15: **Summary of the Results of the SEEDS Experiment [15]**

Anyway, other results of experiments of both the U.S. and the former Soviet Union show that under space conditions damage to the genetic system of the plants is well expressed and the number of mutations increases. Also, changes in the physiological functions of the plants, the number and the shapes of the blossoms and leaves were observed. During brief flights small doses of ionizing radiation caused stimulation of the development of plants. They grew more intensively, enzymatic activity in them were stimulated, etc. However, in experiments with increasing doses of ionizing radiation to the levels which should be expected during long-term flights, the phenomenon of radiostimulation of plants was yielding to the harmful effects of cosmic rays. According to these results, in preparations for long-term flights the problem of protecting plants against cosmic radiation seems to be urgent. [15, 19]

In general, it is not easy to eliminate the radiation danger because radiation is capable of penetrating to different depths. The application of lead barriers in space has to be excluded due to weight and cost constraints. According to research in the former Soviet Union, in plant production it is also scarcely possible to use known chemical substances used for the purpose of decreasing the radiosensitivity of warm-blooded animals, because in plant cells these protective substances are not only poorly assimilated, but are also rapidly destroyed. Instead, it is necessary to use methods which favor the biosynthesis of protective substances in the plant organisms itself. Experiments with lettuce, leaf cabbage, pea, and other model plants demonstrated that the harmful effect of ionizing radiation could be decreased if the plants were exposed to some physiologically active substances. These substances served not only as radioprotectors, i.e., the plants were protected against ionizing radiation by processing them with protective substances prior to irradiation, but also manifested protective properties even when they were used after the plants had been subjected to irradiation. The fact that the tolerance of plants to ionizing radiation could be substantially increased by changing the percentage of mineral substances in the nutrient solution was also of great importance. Nevertheless, the Soviet researchers concluded that plant organisms, in comparison with animal organisms, are more resistant to ionizing radiation. [40]

VI.4.1.2.3 Biological Experiments in Space

The environmental conditions in free space, like the state of microgravity and the solar and galactic cosmic radiation, cannot satisfactorily be simulated in ground tests. For example, long exposures to accelerated particles required for low dose rates would be impractical to carry out at accelerator facilities. Moreover, they are based on single sources of monoenergetic radiation species applied in unidirectional fashion. Therefore, biosatellites are designed to study the effects of microgravity, radiation, fundamental biochemical processes in cells, the development and growth of plants and animals in space, and the removal from Earth's 24-hour periodicity on numerous organisms. Many satellites were used for performing biological experiments in space: starting with several SPUTNIK satellites (1957-61, USSR), over DISCOVERER-29 and MERCURY-5 (1961, USA), BIOS-1 to -3 and OFO-1 (1966-70, USA), to the numerous satellites of the COSMOS program (since 1962, USSR). The COSMOS satellites have been flown for periods up to 21 days and have been successfully recovered since 1975. Since the early 1970's of course many experiments were also conducted aboard the space stations SALYUT, SKYLAB, and MIR, and aboard the Space Shuttle and Spacelab.

The first experiments on the effect of space environmental factors on plants were onboard the satellite SPUTNIK 2. Then, e.g., *Chlorella*, and seeds of various varieties of onions, peas, wheat, and corn completed their own flight, and were the first to return successfully to Earth. *Chlorella* cultures also flew into space on manned VOSTOK capsules. After this, plant organisms traveled into space on many spacecraft, capsules, and the COSMOS biological satellites of the former Soviet Union. In wheat and peas it was not possible to obtain flowers, let alone seeds. The plants simply died in the stage of their formation. It was found later that this was caused by the special cultivation technology that was created for these unusual conditions. In 1978, onions were grown onboard the space station SALYUT using two methods: a scientific method and the "peasant way". It was found that if the plants are not trimmed at the top, they begin to rot, while if they are trimmed, they grow well, and do not rot. Nevertheless, it was not possible to bring these plants to flowering.

For a mission to SALYUT-6, tulips were prepared such that they only had to open up in space, but they also did not bloom. Yet, it is not understood why. Next, epiphytic tropical orchids were chosen for tests in space. This time, parts of the orchids were dispatched already in bloom. The flowers fell off almost at once, but the plants themselves provided growth, with not only leaves, but also aerial roots forming on them. The exotic plants stayed for sixth months onboard SALYUT-6 without ever blooming in space. But they needed only to be returned to a hothouse on Earth and they were immediately covered with flowers again. It was unclear why the plants did not bloom in space. In another experiment, plantules of some plants were grown in a small centrifuge. It created a constant acceleration of up to 1 g onboard the space station. It turned out that, in the physiological sense, the centrifugal forces are equal to the force of gravity. In the centrifuge, the plantules were clearly oriented along the vector of the centrifugal force. In a stationary block, conversely, total disorientation of the seedlings was observed. In a magnetogravistat unit the effects of another factor, a heterogeneous magnetic field, were studied. Its effects on plantules of crepis, flax, and pine also compensated the absence of a gravitational field. On another mission onboard SALYUT-6, *Arabidopsis*, a self-pollinator which grows perfectly in artificial soils, bloomed in a plant growth chamber. This experiment proved that it is principally possible to grow plants in weightlessness conditions, through all stages of development - from seed to seed. [11, 28, 37]

A series of plant growth experiments were conducted onboard the MIR station in a device called Svetoblok. For example, in 1991 an attempt was made to achieve a complete life cycle of wheat in Svetoblok. During this 107-day experiment, growth slowed after the 40th day and ceased, though new growth occurred at the 100th day. Plants returned to Earth and grown under higher levels of illumination produced wheat spikes and seeds. A larger plant growth apparatus, first brought to MIR in 1990, is the Svet Greenhouse. In the initial experiments, radishes and Chinese cabbage were grown in Svet. Although germination was lower and the flight plants grew considerably less, it was achieved for the first time to produce a radish root crop in microgravity (with 31 % fresh weight and 61 % dry weight of the ground control plants). [15]

In 1967, the U.S. satellite BIOS-2 was recovered after 45 hours in space. The greatest effects of weightlessness were seen in young and actively growing cells and tissues. Rapidly dividing cells with high metabolic activity were more affected than mature cells which divided more slowly. The radiation effects observed depended on the nature of the organism. In some cases the effects of radiation combined with weightlessness were greater than those observed in Earth-based controls. In other cases, the effects were less. Most interesting, with respect to food production, were probably the results of the experiment with wheat seedlings. The wheat seedlings reacted with increased shoot growth, increased activity of enzymes for sugar synthesis, and disorientation of roots. In other plants disorientation of leaves and stems were observed. The basical results of BIOS-2 were that plants are obviously sensitive to changes in gravity.

Experiments to be flown onboard the Space Shuttle or Spacelab that have been developed in the U.S. include the Plant Growth Unit (PGU) and the Gravitational Plant Physiology Facility (GPPF). The PGU is equipped with three 15-W plant growth fluorescent lamps, a timer to provide day-night cycling, temperature sensors, electronically controlled fans, heater strips for temperature modification, a data acquisition system, and internal batteries. The GPPF was flown on the Space Shuttle IML-1 mission. The equipment design allows for growth (in darkness) of wheat and oat coleoptiles initially exposed to an artificial 1-g force by rotation on the culture rotors to provide straight coleoptiles for use in subsequent procedures to investigate aspects of the seedling gravi- and phototropic response. Mainly, the experiment investigates how plants respond to phototropic stimuli in the absence of the complicating gravitropic counter-responses in plants grown at 1 g and is also designed to explore the phototropic dose-response relationship at 0 g. [5, 28]

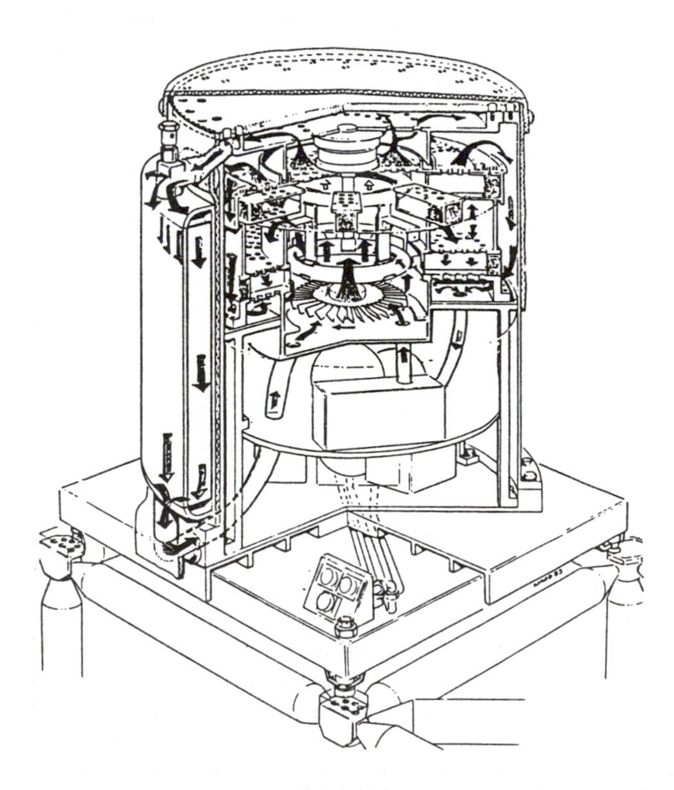

Figure VI.15: **The EURECA Botany Facility [32]**

The latest series of NASA plant growth flight tests are the ASTROCULTURE™ flight experiments. The ASTROCULTURE™-1 (ASC-1) flight experiment was flown on STS-50 as part of the U.S. Microgravity Laboratory-1 mission to validate the performance of a matrix-based Porous Tube Nutrient Delivery System (PTNDS). This system was developed to provide nutrients and plants while maintaning good aeration at the roots and preventing water from escaping in reduced gravity. The PTNDS, basically consisting of porous tubes, circulation pumps, and pressure controllers, successfully worked under microgravity conditions. ASTROCULTURE™-2 was flown on STS-57 as part of the SPACEHAB-01 mission. For the ASTROCULTURE™-3 (ASC-3) experiment (STS-60, SPACEHAB-02) a zeoponic substrate (see section VI.4.3.2.4) was used. The water movement into and out of the substrate performed well and the substrate was releasing plant growth nutrients into the solution as planned. While ASC-3 was another engineering test, i.e., no plants were grown in the substrate, plants will be grown in zeoponic solutions on future SPACEHAB missions, e.g., during STS-63 and 73. [39, 43]

European contributions to plant growth research in space are the BIORACK, which was flown on the IML-2 mission in 1993 and the EURECA Botany Facility as shown in figure VI.15. The BIORACK equipment consists of incubators, a freezer, a cooler, stowage equipment, observation facilities, and a glovebox. During the IML-2 mission experiments concerning the growth of yeast cells, roots growing, and radiation measurements have been conducted in the BIORACK. The EURECA Botany Facility is an "orbital greenhouse" to be flown on the EURECA platform. The features of this facility are summarized in table VI.16. [32,70]

Purpose	Growth of higher plants and fungi from seed to seed and spore to spore
Carrier	EURECA
Mission Duration	6 months
Features	18 μ-g cuvettes (5 x 5 x 15 cm³) 6 1-g control cuvettes (5 x 5 x 15 cm³) Illumination Video observation Data acquisition Pollination device Fixation device
Life Support Subsystem Functions	Contamination control Air humidity control Control of soil water content Control of atmospheric pressure Control of atmosphere composition

Table VI.16: **Features of the Botany Facility [32]**

VI.4.2 The Plant Selection Process

Plant cultivation in space has a good number of peculiarities, e.g., the area of a space greenhouse is limited and therefore the productivity of plants is of enormous importance. To support long-duration manned spaceflight, mixed crop systems will be needed. These crops must be highly efficient in light utilization and provide a balanced diet. Productivity effects on parametrics examines the following three crop characteristics:

- Edible biomass produced per unit volume
- Growth period from planting to harvesting
- Biologically recoverable calories per gram of dry edible biomass

To begin with the plant selection process, the nutritional needs for a human crew member on a long duration space mission have to be clearly defined because they are the cornerstone on which all of the selection work is build. Once complete nutritional requirements are established, one has to define which part of them has to be fulfilled through regenerative means according to the different possible space mission scenarios and the different sources of food available. At this point, the biological compatibility of the selected species also has to be taken into account. For a trade-off of food sources, microorganisms, algae, and higher plants have to be considered. The trade-off criteria between those may be divided into four classes:

- Operating criteria
- Safety and reliability
- Food characteristics
- Impact on life support subsystems

Under space mission conditions the use of higher plants as source of food suffers major drawbacks, mainly in operating criteria, while their food characteristics are very good. On the contrary, e.g., microalgae are very good candidates as far as food production is concerned, but they may be only considered as partial contributors to the daily diet (25-30 % dry weight of the diet - see section VI.3). [48, 64]

VI.4.2.1 Selection Criteria

In the selection process of plant species for use in bioregenerative life support systems, many aspects have to be taken into account that may be divided into the following categories:

- Biomass production and nutritional aspects
- Plant requirements and growth conditions
- Atmosphere regeneration capability
- Water regeneration capability
- Waste regeneration capability
- Behavior in extraterrestrial environment
- Interaction with other systems

Biomass Production and Nutritional Aspects

Plants grown in a CELSS must be able to provide a nutritionally and psychologically satisfactory diet for the human inhabitants. The selection of plants has to yield a harmonized diet with the exact balance of proteins, lipids and carbohydrates and all necessary minerals and vitamins. Also, the biomass should be palatable and stable during storage. Thus, the selected plant species must be evaluated in terms of a number of criteria, including:

- Proportion of edible biomass
- Yield of edible plant biomass
- Energy concentration
- Nutritional composition
- Palatability
- Acceptable serving size and frequency
- Flexibility of usage
- Storage stability
- Toxicity
- Degree of human nutritional experience

In this respect, the interfaces between plant production system and man may also be adapted via the plants, by selection of the appropriate selection of species and processes. A minor adaptation via man is only possible through the respiratory quotient (RQ). The RQ depends on the lipid and carbohydrate contents of the food. The higher the lipids contents, the more O_2 per CO_2 is converted (see section IV.3.1.2). Of course, low-input genotype plant species, i.e., plants that require little space, little light, little water and nutrients and yield maximum edible biomass, are preferable. Also, fast growing plant species with a short vegetation period and continuous harvesting possibility will be advantageous. The harvest index, i.e., the ratio edible biomass / total biomass (ideal 1.0), of any selected plant species should be as high as possible to minimize waste and loss in mass transfer.

Plant Requirements and Growth Conditions

Those species and cultivars should be selected that produce the maximum quantity of edible biomass and the minimum quantity of inedible biomass at minimum processing requirements. Thus, the following parameters concerning the plant growth unit have to be taken into account:

- Volume of space required
- Labor requirements
- Weight of the plant-growing system
- Electrical energy utilized
- Purchase and maintenance cost of plant-growing system

The canopy structure of selected plants should intercept maximum incident light and efficiently use the available space. All chosen plants should share a common environmental tolerance (temperature, pH, light intensity, quality and periodicity, pCO_2, humidity) to decrease the need for different climatic growth conditions. In any case, the selected plants should have maximum flexibility and tolerance to uncontrolled environmental extremes. Concerning pollination and propagation, these plants should be autogamous or at least anemogamous. If needed, the plants should be suitable for hydroponic or aeroponic culture conditions. Further criteria include:

- Environmental tolerance
- Photoperiodic and temperature requirements
- Pollination and propagation
- Vegetation period (in relation with mission design)
- Harvesting methods
- Continuous harvesting
- Possibility of using low quality water
- Compatibility with other plant species
- Ecological and genetic stability and resistance against infections
- Degree of loop closure of the whole system
- Water consumption and transpiration rate
- O_2 production and CO_2 assimilation

Atmosphere Regeneration Capability

As actively growing plants photosynthesize, carbon dioxide is fixed into organic carbon and oxygen is released. This oxygen can be used for human respiration if the atmosphere of the plant growing area is effectively integrated with the human habitation areas of the space habitat. Plants can also help in maintenance of carbon dioxide levels in a regenerative life support system, i.e., carbon dioxide exhaled by humans in the spacecraft can be cycled to the plant growing area. Only exchange associated with digestible food obtained from each plant is useful. The oxygen that must be consumed during plant decomposition of this non-digestible fraction is equal in quantity to the oxygen released in the photosynthetic production. Also, not until non-digestible plant parts have been decomposed (oxidized) to CO_2 and water will be truly regenerative. This is another reason why plant species should preferably be selected to provide maximum edible and minimum inedible biomass. The minimum surface area required to provide adequate oxygen for one person varies according to the plant species being grown and the extent of surface area that can be continuously covered with photosynthesis tissue.

Water Regeneration Capability

Higher plants transpire predictable quantities of water , at rates that are considerably greater during light than during dark. The transpired water can be used to maintain humidity in the plant growing areas or directed to other compartments of the spacecraft for humidity maintenance. Some of the water may be condensed for use as potable water (see section VI.4.7).

Waste Regeneration Capability

The possible use of higher plants for recycling human wastes within a regenerative life support system appears feasible. If certain nutritive elements are unavailable, proper mineral supplements of the activated sludge may be necessary (see section VI.4.7).

Behavior under Extraterrestrial Conditions

This point is largely unknown and can only be determined under space conditions. Seed to seed flight experiments have to be conducted. Extraterrestrial radiations are known to be mutagenic and therefore either means have to be found to protect the compartment or plants that are resistant against these radiations (see section VI.4.1.2).

Interaction with other Systems

The higher plant compartment should adequately fit with the rest of the life support system and with the rest of the station or base. [21, 54, 62, 72]

VI.4.2.2 Attributes and Nutritional Values

It is possible for an adult person to obtain sufficient energy on a strict vegetarian diet. The amount of energy required is related to age, weight, sex, degree of physical activity, including physical and psychological stress, and environmental conditions. Protein is of major concern in a vegetarian diet both for supply of energy and needed amino acids for the body. Plant proteins occur generally at low concentration and their quality is mostly poor. Also, it was found that most plant proteins are not fully utilized in the intestinal tract. The digestibility coefficient varies from 20 % to 89 %, and is mostly close to 75 %. Lipids form another important part of the diet as a source of energy and for culinary reasons. Sufficient amounts of most vitamins and minerals can be supplied from plant sources. However, it is difficult to get sufficient riboflavin and vitamin B_{12} since it does not occur in plants. Among the minerals and trace elements, iodine may occur at suboptimal concentrations, and the low bioavailability of calcium, iron, zinc, and copper may render these deficient although they may nominally be present in sufficient amounts. Supplementation may be considered for these minor nutrients (see section IV.3.1.2). [72]

Several diets have been proposed to satisfy human nutritional requirements and to provide an interesting variety of fresh food. All of these diets include several stable crops, such as rice, wheat, corn, beans, and potatoes. Other proposed kinds of food include broccoli, spinach, lettuce, peas, peanuts, turnip greens, tomatoes, onions, carrots, and beets. Fruits may also be included, such as strawberries, raspberries, and pineapples. Further research is required to determine which species and combinations of species are best. Studies may in fact prove that a vegetarian diet alone is not enough to maintain health for extended periods. Nutrients from animals, fish, or other living organisms might also be required. [67]

By order of NASA, a group of crop specialists selected some crop species that are of major interest as human food sources in bioregenerative life support systems. The selected plant species were divided into two groups:

- *Group 1:* Species that are commonly used food plants and can provide the major nutritional needs of man.
- *Group 2:* Species with lower nutritional values but high psychological value.

The plant species of group 1 that have been selected and their respective growth and nutritional characteristics are:

- *Wheat:* has a high caloric density and is the basis for many different types of food. The edible portion of the biomass is high. Also, wheat has a high starch content and contains a reasonable amount of protein (up to 14%), phosphorus, iron, thiamin, and niacin. The photosynthesis of wheat ceases during seed maturation.

- *Rice:* has a high caloric density and contains 8 % of nutritionally balanced protein, phosphorus, iron, thiamin, and niacin.

- *White Potatoes:* are high calorie food and require minimum processing. White potatoes have a high carbohydrate concentration and about the same protein concentration as rice. Also, they are a good source of vitamin C and potassium. Tuber production of potatoes will require unique mechanical systems for support.

- *Sweet Potatoes:* like white potatoes, they are a high calorie food, can be eaten with little processing and have about the same protein concentration. Moreover, they are adapted to warm environments and contain about 30 % more carbohydrate than white potatoes. Sweet potatoes have potassium, vitamin A, and vitamin C, and their leaves and young shoots are edible. Unique mechanical support systems for the enlarged roots will be required and the vine-type growth of their stems may be a disadvantage.

- *Soybeans:* are a major source of dietary protein. Their seeds have the highest protein concentration of all common food plants (45-50 %). This protein is highly digestible. Also, soybeans contain well balanced amino acids, oil (20 % - must be processed), phosphorus, iron, potassium, and thiamin.

- *Peanuts:* are a major source of protein (25%). Their essential amino acids are not that well balanced, but they contain a lot of oil (45 % - without processing), phosphorus, iron, potassium, thiamin, and niacin. A problem is the excessively complicated growing and harvesting procedures.

- *Lettuce:* contains vitamin A and vitamin C.

- *Sugar beets:* provide sugar, but there is a weight penalty for equipment to extract the sugar. They can be eaten raw and their tops are also edible.

	Spirulina	Lettuce	Spinach	Tomato	Rice	Wheat meal	Potatoes	Soybeans	Sunflower seeds
Energy [kJ/100g]	255	69	122	101	1488	1484	326	660	2611
Protein [g/100g]	10.8	1	2.9	0.9	8.4	12.7	2.1	13	19.3
Carbohydrate [g/100g]	1.95	2.1	3.6	4.3	77.7	70	17.1	11	24.1
Lipid [g/100g]	1.1	0.2	0.3	0.7	1.1	2.5	0.1	6.8	49.8
Water [g/100g]	85	95.8	91.5	93	12	13	79.8	67.5	5.3
Harvesting Ratio [%]	100	85	70	45	45	40	80	50	33
Essential Amino Acids [mg]									
Isoleucine	670	75	145	21	370	435	92	570	1140
Leucine	950	70	225	33	730	670	105	925	1660
Lysine	505	75	175	33	320	275	113	775	935
Methionine	275	28	90	20	325	375	46	275	945
Phenylalanine	480	78	240	38	710	865	58	1050	1835
Threonine	575	53	120	22	295	290	83	515	1350
Tryptophan	170	8	39	7	90	125	22	155	350
Valine	720	62	160	23	510	470	113	575	1350
Hydrosoluble Vitamins [mg]									
Ascorbic Acid C	1,5	3,9	28	18	0	0	9	29	0
Biotine H	0.006		0.007	0.0015	0.003	0.009	0.0001		
Choline									
Cyanocobalamin B12	0.03	0	0	0	0	0	0	0	0
Folic Acid	0.008	0.055	0.195	0.01	0.029	0.052	0.025		0.235
Nicotinic Acid	1.75	0.185	0.7	0.6	1.5	4.4	1.5	1.65	4.5
Pantothenic Acid B5	0.17	0.046	0.07	0.25	0.55	1.1	0.2	0	0.7
Pyridoxine B6	0.05	0.04	0.2	0.05	0.17	0.39	0.39		0.8
Riboflavine B2	0.67	0.03	0.19	0.05	0.04	0.12	0.04	0.175	0.25
Thiamine B1	0.83	0.046	0.08	0.06	0.13	0.66	0.1	0.435	0.11
Liposoluble Vitamins [mg]									
Retinol A	26				0	0	0		0
Tocopherols E	2.9	0.52	2.35	0.44	0.46	1.16	0.013		41.6
Vitamine K			0.00035	0.008					
Macrominerals [mg]									
Calcium	14.5	19	100	7	32	37	7	197	116
Magnesium		9	80	11	28	113	20		354
Phosphorus	180	20	50	23	127	386	53	194	705
Potassium		160	560	205	85	435	407		690
Sodium		9	79	8	8	3	3		3
Microminerals [mg]									
Copper		0.028	0.13	0.077	0.2	0.93	0.311		0.173
Fluorine			0.01	0.024	0.19	0.053	0.045		
Iodine			0.012	0.003	0.0018	0.0041	0.004		
Iron	8	0.5	2.7	0.5	0.1	4.3	0.6	3.55	6.77
Molybdenum			0.026		0.015	0.036	0.003		
Selenium			0.0017	0.001	0.02	0.063	0.01		
Zinc		0.22	0.53	0.11	1.3	3.4	0.58		5.06

Table VI.17: **Nutrient Composition of Microbial and Higher Plant Raw Products [64]**

The plant species of group 2 and their specific characteristics are:

- *Taro:* Is a tropical root crop.
- *Winged beans:* Are a good protein source. Their tops and roots can be eaten and they are adapted to warm temperatures.
- *Broccoli:* Contains vitamins A, B1, B2, B7, and C.
- *Strawberries:* Contain vitamins B2, B7, and C, and are of very high psychological value.
- *Onions:* Have no significant nutritional value and very special growth habits.
- *Peas:* Are a rich protein source and also contain a high amount of methonine, little fat, and large quantities of minerals, e.g., potassium, copper, iron, sulfur, and phosphorus. Peas have no strong stem and, thus, require a special culture system.

The detailed nutritional values of several representative higher plants and, for comparison, the microorganism *Spirulina* are given in table VI.17. [72]

VI.4.3 Plant Growth in a CELSS - Design Considerations

The major rating criteria concerning conceptual designs for plant growth in a bioregenerative life support system or CELSS are to:

- Satisfy caloric requirements for the crew members
- Satisfy biological requirements for plants
- Minimize human involvement

Thus, once certain crops have been selected to be included into such a system, the optimum environment conditions for plant growth have to be determined. The growth environment parameters which have been shown to have most significant effects on productivity are:

- *Light intensity, quality (wavelength), and duration of illumination*
  The lighting system of the plant compartment will be determined by the optimal light conditions for proper plant growth, the power requirements and the heat load. The indirect use of sunlight through mirror collectors and fiber optic cables may be considered.

- *Air Speed, Humidity, and CO_2 concentration*
  A constant atmosphere circulation has to provide adequate CO_2 and has to remove the produced O_2, the plant transpiration water and toxic organic gases generated by plants. CO_2 control levels should be between 350 and 2000 ppm. Relative humidity levels of 70 % have been encouraged for general use at 15° to 30° C during light and dark periods.

- *Nutrient delivery*
  The nutrients will come from the compartments dealing with the regeneration of waste and water. The methods of delivery will depend both on the plant growth medium and on the presence or absence of gravity. Sensors and actuators have to be provided to exactly deliver them.

- *Crop density and planting interval*
  Optimum plant spacing must be determined from seeding to harvesting for each species. Also, optimum intervals between plantings to maintain a continuous food and oxygen supply have to be determined for each species (length of harvesting period, length of storage period, amount of photosynthesizing crop canopy needed at particular times to supply human O_2 requirements).

- *Temperature*
  Temperature optima at different stages of development for each species must be established. Day night differentials should be studied to determine these optima. Leaf temperatures should be measured to provide an accurate basis for air temperature control. Media or root temperature should be monitored to optimize the growth of plants.

In order to control these parameters within the desired levels, a higher plants compartment will necessitate complete and effective sealing. The crops that have been investigated concerning these parameters to date include wheat, potatoes, soybeans, sweet potatoes, lettuce, rice, and tomatoes. In order to give an idea about the influence of the different factors described above, some sample numbers from Salisbury and Bugbee are given below. In table VI.18 the parameter influence on the components in wheat is given, while table VI.19 outlines some typical growing conditions for different plant species.

Conditions	Plants/ m²	Spikes/ m²	Total seeds per spike	Mass per seed [mg]	Total yield [g/m²]
Cool temp. (20°C day, 15°C night), 14hr photoperiod	1150	2007	21	29	1154
Cool temp. (17°C), 24hr photoperiod	1076	2387	16	30	1054
Warm temp. (27°C), 14hr photoperiod	1030	3234	5	27	279
Warm temp. (27°C), 24hr photoperiod	830	2128	10	29	872
Conditions	**Days to harvest**	**Edible yield** [g/(m²·day)]	**Harvest index**	**Edible yield**	**Yield efficiency** [mg/mol]
Cool temp. (20°C day, 15°C night), 14hr photoperiod	77	15.1	46	16.3	300
Cool temp. (17°C), 24hr photoperiod	66	16.3	35	17.5	189
Warm temp. (27°C), 14hr photoperiod	66	4.3	11	4.3	85
Warm temp. (27°C), 24hr photoperiod	61	14.2	29	14.4	164

Table VI.18: **Influences of Temperature and Photoperiod on Yield Components in Wheat [53]**

	Wheat	Potatoes		Lettuce	Soybeans	
		Canopy	Caged plants		Expt.1	Expt.2
Irradiance [mol/(s·m²)]	1000	475	400	430 (0-11 d) 900 (11-19 d)	700	550
Lamp source	(HPS)	(CWF)	some CWF side light	(INC)	(CWF + INC)	
Photoperiod	24	24	24	20	9	12
Total PAR [µmol/(m²·d)]	86.4	41	34.6	30.9 (0-11 d) 64.8 (11-19 d)	22.68	2376
Temperature [°C]	17.5-22.4 (+1°C / week)	16	16	25	30/26	26/20
CO_2 [µmol/mol]	1000	365 and 1000	360	360 (0-11 d) 1500 (11-19 d)	400	675
Oxygen	Ambient	Ambient	Ambient	Ambient	Ambient	Ambient
Nitrogen	Ambient	Ambient	Ambient	Ambient	Ambient	Ambient
Relative humidity [%]	60-80	70	70	75 (days) 85 (dark)	75-95	75-95
Plant spacing [m²/plant]	0.00083	0.21	0.2	0.0125	0.0924	0.0924
Plant density [plants/m²]	1200	5	5	80	10.8	10.8
Air velocity [m/s]	~ 1.0 (0.1-4.0)				0.33	0.33
Cultivar	Yecora Rojo	Norland Early Red		Waldmann's Green	Ransom	

Table VI.19: **Typical Growing Conditions for different Plant Species [53]**

The integration of such a plant production system into a bioregenerative life support system brings about problems on a system level in the following categories:

1. Harmonization of energy- and mass-flow rates:
    - Plant physiological parameters
    - Adaptability and variability of plants
    - Plant off-gassing
    - Plant growth under reduced atmosphere pressure
    - Continuous cultivating/harvesting
    - Recycling of nutrient solution
    - Recycling of transpiration water
    - Biomass composition
    - Partial pressure of O_2 and CO_2
    - Use of human waste by crop plants
    - Techniques for monitoring and control, limiting values
2. Miniaturization
3. System stability:
    - Accumulation of substances and exceeding of limiting values
    - Formation of deficiency symptoms by element fixation
    - Infection by microorganisms
4. Man in an artificial ecosystem:
    - Basic body functions
    - Hygienical parameters
    - Composition of the diet
    - CO_2 and O_2 concentration limiting values
    - Harmful off-gassing from plants and materials

The quality and quantity of these many problems to be solved make it obvious that the development of a higher plants compartment for a controlled ecological life support system will require interdisciplinary cooperation. Among the disciplines to be involved into the design process are:

- Plant cultivation technology
- Plant production technology
- Phytopathology

- Trophology
- Material science
- Monitoring and control technology
- Process technology
- Medicine
- Epidemiology
- Air conditioning technology

Another important aspect that has to be considered once the above mentioned problems have been solved is that the plants used in a CELSS should be free of pests and diseases. While it should be relatively easy to exclude bacterial and fungal disease, viruses pose a more difficult problem because some of them may be transmitted through plant seeds. [13, 36, 48, 53, 62, 72, 74]

VI.4.3.1 Growth Area

A CELSS farm must be large enough to produce the quantity of food required by the crew, yet small enough to fit within the space constraints dictated by the mission. The size of a CELSS farm can be reduced by minimizing the plant growth area required to support each crewmember. Plant growth experiments suggest the minimum growth area required to support one crewmember will range between 13 and 50 m^2/person. Minimizing growth area requires maximizing plant productivity, measured in yield per unit area. Yield per unit area can be increased by optimizing plant density. Planting many seeds in a small area is ideal for increasing yields as long as plant relationships are synergistic. Density reaches a maximum when synergistic behavior starts to be replaced by destructive behavior, where plants fight over available resources. The productivity of some cereal plants, especially wheat, has been investigated in the past in order to determine the required cultivation area. For wheat production some data is listed in table VI.20.

	Absolute Seed Yields [g/m^2]	**Life Cycle** [days]	**Total Biomass** [g/m^2]	**Harvest Index** [%]
Field	300-700	90-130	700-1800	45
World Record	1450	120	3200	45
Salisbury	1200	56	2667	45
Gitelson	1000	56	2860	35

Table VI.20: **Wheat Production Data [45]**

Assuming a diet that is only composed of wheat, the required daily food quantity would be about 618-855 g, and the absolute seed yield would be 1200 g/m², the required growth area can be calculated as follows:

$$A = \frac{\text{Required Food} \cdot \text{Life Cycle Days}}{\text{Absolute Seed Yields}} = \frac{618 \text{ to } 855 \text{ g/day} \cdot 56 \text{ days}}{1200 \text{ g/m}^2} = 28.8 \text{ to } 39.9 \text{ m}^2$$

Assuming that some other plants are required in order to obtain a nutritional balance, it seems reasonable to assume 40 m² of required plant cultivation area per person. [45, 67]

VI.4.3.2 Growth Media

One of the significant problems of higher plant growth in a controlled environment system is the selection of an appropriate rooting medium that will afford optimum nutrition and yet be practical in terms of total amount of salt and water required. The nutrient medium for plant growth is critical because it can influence yield under any conditions, and nutrient supply to roots faces special problems in microgravity. Thus, the selected plant growth method will depend on the plant species and the spaceflight requirements. Basically, two kinds of artificial plant cultivation methods, with and without substrate, have been studied, including:

- *With Substrate:*
  - Earth-like soils
  - Artificial soils
- *Without Substrate:*
  - Hydroponics
  - Aeroponics

The distinction between hydroponics and aeroponics lies in the method by which the nutrient solution is supplied to the plants. A big problem concerning the non-substrate cultivation of plants is that it cannot be assumed to serve for tens of generations of self-reproducing populations. Even in excellent hydroponic cultures, plant flowering stops after several life cycles. To obtain plant growth for long periods it may be necessary to use artificial substrates. A comparison of the main features of these methods is given in table VI.21. [27, 54, 63, 74]

Considering what is mentioned above, the criteria for the selection of a certain growth method are the following: [72]

- Required mass of nutrient solution
- Turnover rate of nutrient solution
- Required mass of substrate (if needed)
- Recycling rate of substrate

- Required controlling equipment (temperature, humidity-pumps, ventilation, compressors, phase separator at μ-g, light, tanks)
- Required equipment for the plant growth procedure (vessels, conveyor belts, plant fastening, spray nozzles)
- Interfaces for maintenance (easy access for harvesting and maintenance, microbial control, adaptation of different atmospheres)
- Energy consumption (thermal, light, electrical, automation, continuous cultivation)
- Harvesting methods (separation of edible and inedible biomass, treatment of the edible portion, treatment of the inedible portion)

Factors	Hydroponics	Aeroponics	Soil
Nutrient System (General description)	Liquid solution (bathes plant roots)	Mist (thin film on plant roots)	Soil solution (thin film on media particles and roots)
Amount required per plant	High volume of solution	Low volume of solution	Moderate portion stored in rood media
Weight per plant	High	Low	High
Respiration oxygen needs	Requires aeration system	Misting results in aeration	Present as result of porous root media
Nutrient concentration	High (e.g. nitrogen 310-620 ppm)	High (e.g. nitrogen 310-620 ppm)	Low (e.g. nitrogen 101-150 ppm)
Nutrient sources	Restricted to nutrient solution	Restricted to nutrient solution	Replenished from root media
Supply and maintenance of nutrients	Needs frequent maintenance and adjustments. Amenable to adjustments.	Needs almost daily replenishment. Amenable to adjustments	Adequate amounts of nutrients can be stored on media matrix. Not readily adjusted - additionally nutrients can be added.
pH	Susceptible to variation of remaining nutrient ratios	Susceptible to variation of remaining nutrient ratios	Buffered by ion exchange capacity of media

Table VI. 21: **Comparison of Different Growth Media [47]**

VI.4.3.2.1 Soil

In a space environment mineral-enriched soil may not be attractive as a natural resource in which to grow plants due to transportation costs, potential problems with microbiological contamination, and the handling difficulties introduced during planting and harvesting. Also, since in microgravity drainage does not occur, a solid medium could become water-logged and hence anaerobic. Aeration of solutions is difficult because of the absence of convection, and growth in artificial media also can result in the uptake of undesirable chemicals into plants. Furthermore, solutions might leak out around the base of plant stems, with droplets of solution floating loose in the area where the plant shoots are. Various techniques have been developed to solve these problems such as porous tubes containing nutrient solution under slight negative pressure, with roots in direct contact with the tubes. Low gravity like on Moon or Mars may be sufficient to overcome these problems. Nevertheless, a plant can grow without solid rooting media as long as nutrients are supplied in some manner. In most cases structural support is also required to assist plant growth. The two methods in this respect proposed for growing plants in space are hydroponics and aeroponics (see below). [54, 67, 74]

VI.4.3.2.2 Hydroponics

In hydroponic systems, plants are grown with their roots in an aerated, circulating liquid nutrient solution that provides nutrients, oxygen, and water. The plant roots are located in any or no substrate for plant support. Figure VI.16 shows a diagram of a hydroponic device using a substrate.

Pure hydroponics, i.e., circulating nutrient solutions with no solid medium, offer a precise control of nutrient availability, composition, and pH. Such systems allow adequate root-zone oxygen and adequate water, which is to say that low watering is never a limiting factor. In hydroponics it is easier to provide ample root-zone oxygen than in any other approach except aeroponics. But in aeroponics it is not as easy to provide ample nutrient and water. Furthermore, hydroponics minimize problems such as clogged irrigation nozzles and cleaning of culture media between crops and allows for more precise control of the root zone environment. An advantage of liquid nutrient solutions over solid substrates is that much of the solution can be recovered as water and reused. Plants have the capability to respond to the root environment by producing more roots when nutrient or water are limited. Typically, plants grown with hydroponics in a good facility do have small root systems and large leaves compared to those grown in solid media. In good hydroponic systems, the root system may make up only 3 to 4 % of the total dry weight of the plants compared with 30 to 40 % for plants grown in soil. Nevertheless, hydroponics will require probably more equipment than a system based on a solid medium, and more equipment means more

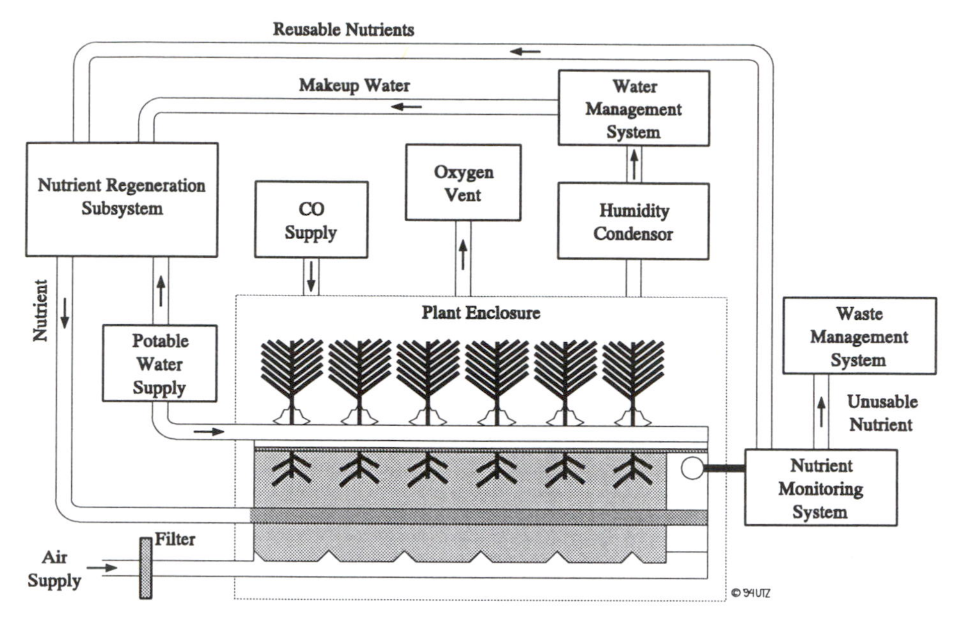

Figure VI.16: **Hydroponics Growth [42]**

possible breakdowns and the need for more careful monitoring. Also, the nutrient solution must remain in contact with plant root tissue to prevent dehydration and nutrient starvation, which is a difficult task in microgravity. Thus, hydroponics are better suited for a low gravity environment, such as on Moon or Mars, where natural liquid/gas separation helps guarantee that roots are immersed in liquid. [52, 54, 58, 67, 71]

VI.4.3.2.3 Aeroponics

In the aeroponics method, the nutrient solution is supplied to substrate-free plant roots by sprayers in the form of a fog. A diagram of an aeroponics device is given in figure VI.17. An advantage of aeroponic culture is the small amount of nutritive solution used per unit of planting area or biomass of the cultivated plants, which allows a significant reduction in weight of the cultivation devices. However, aeroponics cannot be recommended as a method for prolonged culture until clogging in the sprayers has been eliminated. The decrease in volume of the nutritive solution needed in aeroponic devices has the disadvantage of more rapid accumulation of toxic products than when plants are grown on an unchanging medium.

Although the use of aeroponics can minimize significantly the total weight of the plant-culture system, this system requires frequent and regular care to maintain desirable salt balance and solution pH. The solution pH should be maintained at

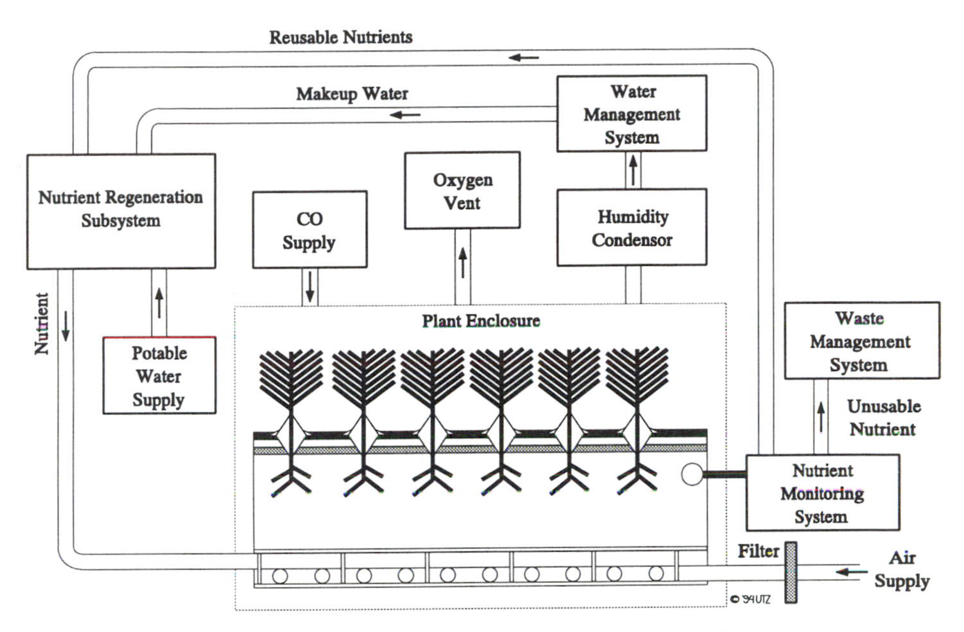

Figure VI.17: **Aeroponics Plant Growth Tray [47]**

4 - 6 and has to be adjusted at least once daily, and the use of automatic pH controllers is recommended. Moreover, an aeroponic system in a microgravity environment must be designed to provide total enclosure of the root system to prevent nutrient solution from escaping into the atmosphere. The need for confinement is twofold. First, the nutrient solution is moderately corrosive and will corrode any unprotected electronic/structural component it contacts. Also, the nutrient solution may contain bacterial, fungal, and viral agents that may contaminate other plants, vehicle hardware, or even the crew. Aspiration may remove exhausted nutrient solution from root zone. The aspirated nutrient solution may then be separated from the air, analyzed for correct composition, renewed with necessary nutrient salts, and returned to the supply tank. A scheme of an aeroponic system is given in figure VI.18.
[47, 63, 67, 72, 74]

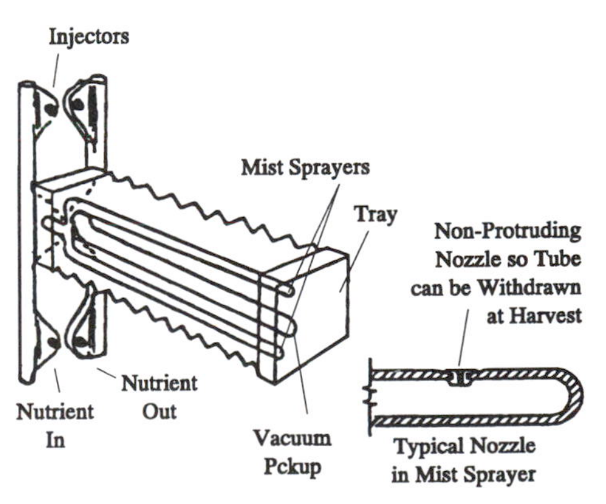

Figure VI.18: **Aeroponic System [47]**

VI.4.3.2.4 Zeoponics

Hydroponic and aeroponic nutrient delivery systems are complex and require pumps and sophisticated control and monitoring equipment. Therefore, it would be desirable to have a substrate that provides the plant essential elements in a static watering system which eliminates this equipment. To address these needs, a highly-productive, synthetic soil (or substrate) is being developed at NASA's Johnson Space Center in Houston, TX. These synthetic soils have been termed zeoponic plant growth substrates. Thus, a zeoponic plant growth system is defined as the cultivation of plants in zeolite mineral substrates that contain essential plant growth nutrients.

Zeolites are crystalline, hydrated minerals that contain loosely-bound ions, e.g., Ca^{2+}, K^{+}, and Mg^{2+}, within their crystalline structures. Zeolites have the capability to freely exchange some of their constituent ions with ions in solution without change to their structural framework. In addition to the zeolites, the zeoponic plant growth substrates consist of a synthetic calcium phosphat mineral called synthetic apatite. The apatite has several essential plant growth nutrients incorporated into its structure, e.g., Ca, Mg, S, P, Fe, Mn, Zn, Cu, B, Mo, and Cl. The nutrient content of the synthetic apatite may be altered to best suite the plant requirements or the conditions of the substrate in which they are added.

The substrate, consisting of zeolite and synthetic apatite, is designed to slowly release the plant growth nutrients into the "soil" solution where they become available for plant uptake. Moreover, it has the capability to buffer the solution's ionic composition and pH without sophisticated monitoring and control systems. A zeoponics substrate system is illustrated in figure VI.19.

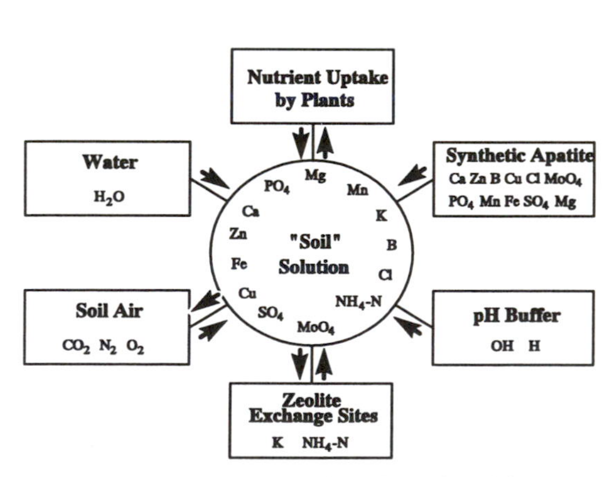

Figure VI.19: **Schematic of the Functional Components of a Zeoponic System [39]**

Nevertheless, zeoponic systems are still in the development phase. Wheat has been grown in several generations of zeoponic substrates in terrestrial plant growth chambers and nutrient imbalances have been detected in wheat tissue tests. In early 1995, the first plant growth test in space using a zeoponic system has been conducted in the ASTROCULTURE™ flight experiment (see section VI.4.1.2.3) onboard STS-63. The overall objective of current research is to develop zeoponic substrates wherein all plant growth nutrients are supplied by the plant growth medium for many growth seasons with only the addition of water. [40]

VI.4.3.3 Atmosphere Requirements

Like in the habitat area of any space station or planetary base in the higher plants compartment of a bioregenerative life support system, the following parameters have to be controlled:

- Atmosphere composition
- Temperature
- Humidity
- Ventilation
- Toxicant levels

Another serious problem concerning the atmosphere of a higher plants compartment are the vacuum of space and the low atmospheric pressure on Mars, respectively. There will always be leaks, so it will be necessary to carry stored gases and to regenerate the necessary atmospheric gases with high efficiency. To ease the engineering problems of building structures in space that must contain an artificial atmosphere, pressure within the structures can be reduced, particularly by reducing N_2 partial pressures (not too much, to avoid fire hazards). Partial pressures of O_2 and CO_2 must be maintained at levels suitable for humans and plants, respectively. [49]

VI.4.3.3.1 Atmosphere Composition

On Earth, where gas ratios are relatively constant, changing these ratios is both difficult and impractical. In a controlled space environment, where all atmospheric gases are regenerated and added, gas ratios can be optimized to produce maximum yields. As described in earlier chapters, a sufficient atmospheric O_2 partial pressure is critical for humans. This applies much less to plants, which may even photosynthesize better when the O_2 level approaches zero. Instead, the carbon dioxide level is critical for plants. Crop yields are greatly increased by higher CO_2 concentration around the plant leaves from 350 ppm to several times that ambient value (see section VI.4.1.1.2). The maximum concentration for optimum growth will depend on the plant, and may change for each stage of a plants' life cycle. Upper limits for most plants have yet to be determined, but there is a threshold beyond which plant growth may be reduced. Although CO_2 is an essential raw material for photosynthesis, it even becomes toxic for plants at levels above about 2000 ppm (humans: 20000 ppm CO_2). The exact toxicity levels have to be carefully determined for various plant species. Reducing O_2 pressure is another way to increase the efficiency of photosynthesis. Further research into the manipulation of these and other gases may lead to new methods for maximizing food production. Of course, gases that are toxic to either plants or humans must be maintained at suitably low levels within a CELSS. [54, 67]

VI.4.3.3.2 Temperature and Humidity

As the atmospheric component level, temperature and humidity in a higher plants compartment can also be controlled to maximize plant yields. Temperature is extremely important for both living and dead components of a bioregenerative life support system, and the thermal control subsystem must maintain module interior temperatures within a relatively narrow range (12°-28° C) for efficient plant growth. Anyway, optimizing temperature and humidity depends not only on the plant species, but also on other control parameters, such as CO_2 concentration, irradiance level, and photoperiod. For example, high temperatures (25° C) in conjunction with long photoperiods have been shown to promote rapid growth rates and short life cycles in wheat, but also lead to poor pollination and reduced yields. On the other hand, cooler temperatures (20° C) tend to increase yields, but slow down growth rates and lengthen the plant life cycle. Thus, like many other factors, temperature should be optimized for each phase of the life cycle to obtain the best yields per unit time.

Suitable humidity levels are important for the human perception of well-being, and they also influence plant transpiration. Thus, relative humidity in a higher plants compartment must be optimized to maintain efficient transpiration rates. High humidity levels reduce transpiration rates, and this could be advantageous in a CELSS because the plants transpire much more water than is needed for the other occupants of the system at lower humidity levels. On the other hand, if humidity is high and water transpiration rates decrease, leaf cooling capacity is reduced. A low cooling capacity can result in unacceptably high leaf temperatures when irradiance levels are high. Low transpiration rates have also been shown to hamper mineral absorption from nutrient solutions, particularly if CO_2 concentration in the plant growth chamber is high. Also, if humidity is too high this could lead to corrosion and growth of various microorganisms.

Air movement is important for convective heat transfer (cooling) and transpiration. In microgravity, the lack of convection makes it essential to produce a forced air flow with fans in order to remove heat from plants that are irradiated with high levels. Thus, forced air circulation and the natural transpiration of moisture from plants are used to maintain growth temperature range. Of course, cooling air velocities have to be kept low to avoid inhibiting plant growth. In any case, temperature and humidity will have to be controlled by physical components, such as heat exchangers and fans. One of the largest heat loads will be generated by the plant irradiance source, particularly if the source is high irradiance lamps. The temperature and humidity control subsystem must be properly sized to prevent irradiance and other heat loads from hindering plant growth or burning plants. Temperature control will be extremely challenging on Moon or Mars. There, unlike in space, excess heat cannot easily be dumped with radiators and, thus, active refrigeration will be required. On the other hand, solar radiation may have to be used as a source of heat, since heat generation will be necessary during darkness. [48, 54, 67]

VI.4.3.4 Lighting Analysis

VI.4.3.4.1 Lighting Requirements

One of the most significant drivers in the design of a CELSS is the means by which light is supplied to photosynthetic organisms. For a plant lighting system basically three options can be considered:

- Solar collector based light systems
- Artificial-based light systems
- Hybrid light systems

Sunlight near noon on a clear day in temperate latitudes provides a photosynthetic photon flux (PPF) of about 2000 μmol/m²/s or 450 W/m² of photosynthetically active radiation (PAR). While humans can see quite well in relatively low light (10-25 μmol of photons/m²/s), plants require much more light to carry out photosynthesis. Plant growth rate increases with increasing PPF, e.g., plant growth is barely acceptable at 200 μmol/m²/s, but highly productive at 2000 μmol/m²/s. These values indicate that plant productivity is a strong function of irradiance level, although efficiency, i.e., energy fixed into chemical-bond energy of the biomass divided by light energy absorbed, drops off somewhat with increasing light. While, in principle, maximum irradiance is desirable, irradiance levels will be limited by heat and power constraints. Incidentally, algae might be somewhat more efficient in photosynthesis, but the gain in power savings would probably be lost in food preparation. In any case, it is clear that the power requirement for photosynthetic light is a major problem for the design and construction for a CELSS that depends on plants and algae.

Light has to be provided in the wavelengths most usable by the plants, i.e., the visible wavelengths. To understand how different types of light distributions effect plants, a brief discussion of the light absorption spectra of plants is necessary. In general, the wavelength of the light greatly influences the photosynthetic rate of the plant. Figure VI.20 shows light absorbance vs. light wavelength for green plants. It can be seen that plants will not effectively absorb light at a wavelength of 550 nm, which is the green wavelength. The plants reflect this wavelength, which is why the plant appears green. For efficient use of

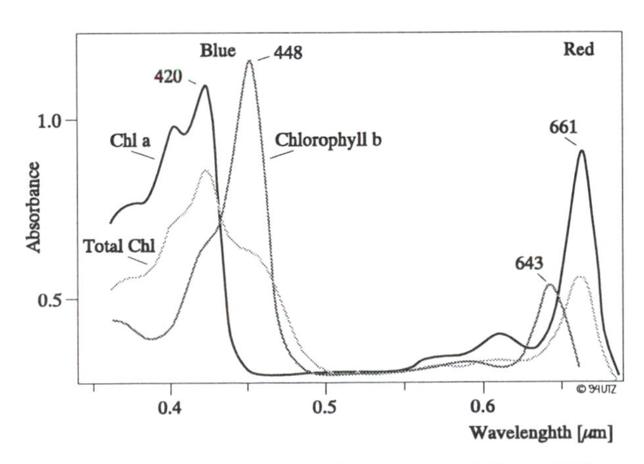

Figure VI.20: **Light Absorption Spectra of Plants [75]**

light, the plants require wave-lengths of 430 and 670 nm, i.e., red and blue. Also, the photoperiod, i.e., the amount of time a plant is irradiated each day, must be optimized to achieve maximum yields. Continuous exposure to light promotes fast growth in most cases, but does not always lead to the highest yields. In addition to optimizing the photoperiod for individual species, the photoperiod should be modified for each phase of plant development. In general, it is so that long photoperiods, i.e., 18 h or more, permit lower irradiation levels. This indicates very clearly that also the dependence between photoperiod and irradiance level has to be determined for every plant species. On the other hand, the variable irradiance level and photoperiod requirements for different crops can also be accommodated by accepting somewhat lower yields for specific varieties grown under non-optimal conditions.

In terrestrial greenhouses, where sufficient volume is available to prevent hot lamps from burning plants, high efficiency, high intensity discharge electric lamps work well. Nevertheless, in the small volume of a space-based CELSS, high intensity lamps are thermally impractical. Fluorescent lamps, e.g, generate lower thermal loads, but do not provide the light intensity required to optimize growth for many species being considered for a CELSS. Power consumption can be reduced by employing solar energy, which can be distributed to plants using windows, mirrors, fiber optics, or a combination of these devices. The main problem with using sunlight is that some missions will face long periods of darkness, such as the 14-day night experienced by a non-polar lunar base. Such missions may require a combination of sunlight and artificial lamp light to maintain photosynthesis while minimizing power consumption.
[48, 54, 58, 67, 75]

VI.4.3.4.2 Solar Light

In space, near Earth, the radiation energy density, i.e., the solar light constant is 1353 W/m². The part of the solar spectrum that is interesting for plant growth, i.e., the 400-700 nm wavelength band is about 516 W/m², or 2375 µmol/m²/s. For comparison, the 400-700 nm band on Earth's surface is about 435 W/m², or about 2000 µmol/m²/s, at sea level at midday on a cloudless summer day. Thus, each square meter of collection surface exposed to solar radiation in space or on the lunar surface could theoretically provide about 8 m² of plant growth area with a PAR of 300 µmol/m²/s. Nevertheless, depending on the location, it might be necessary to provide artificial light for plant growth during dark-side orbits or during nights. In principle, there are three methods of using solar radiation directly for illumination of plants:

- Light collection and distribution systems
- Transparent structures
- Light reflectors

A collector system consists of four major parts: the light concentrating and filtering device, the tracking system, the light transport device, and the structure of the system. For light collection, basically, two kinds of systems may be considered:

- Flat Plate Collector (Hamiwari design)
- Geodesic Parabola Concept

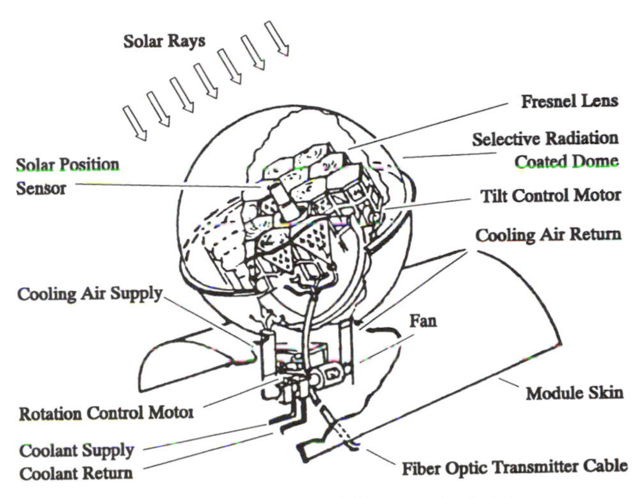

Figure VI.21: **Hamiwari Fiber Optic Light Collection System [47]**

The flat plate collector utilizes a fiber optic system called the Hamiwari, as shown in figure VI.21. In this concept, solar light is collected by a number of Fresnel lenses focusing solar rays onto the end of glass fibers. A tracking mechanism maintains exact alignment of the solar ray collector with the Sun. These lenses are specifically constructed to focus each wavelength at a specified point. The optic fibers are positioned to collect only those frequencies useful to the plants. Harmful frequencies like IR and UV are not collected for transmission. Also, glass additives may be used to filter undesirable frequencies. The collected light energy is transmitted through fiber optic cables formed from numerous smaller cables fused into a trunk line. The trunk line pipes the light to distribution buses from which it is directed to plant growth units. In case that only solar light, but no artificial light provides the plant illumination, only power to keep the collector Sun-oriented is needed. In table VI.22, sample physical data on a flat plate, fiber optic collection system is presented. [48, 58, 75]

Lens Quantity	Collector Area [m²]	Mass [kg]	Tracking Motor Power [W]
37	2.59	1012	221
61	4.26	1579	278
127	8.87	3129	433
900	62.9	5503	373

Table VI.22: **Physical Data for Fiber Optic Solar Radiation Collectors [48]**

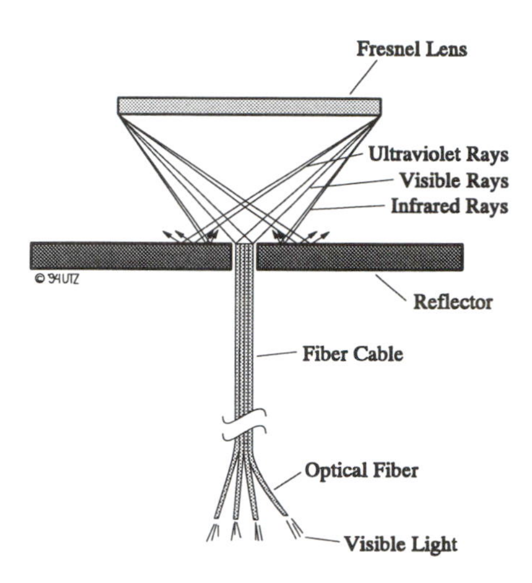

Figure VI.22: **Principle of a Fresnel Lens [58]**

The basic principle of a Fresnel lens is outlined in figure VI.22. The Solar rays are perpendicular to the lens, which is very important since for a Fresnel lens to be most effective, the Sun's rays need to be less than 2° from perpendicular. As light passes through the lens, the angle of refraction is dependent on the wavelength. IR has a smaller angle of refraction than UV radiation. If a polished fiber cable end is placed at the focal point of the green wavelength, the cable collects the light in the range of 350-850 nm, i.e., the light plants need for photosynthesis.

The geodesic parabola concept is based on a design that has been tested on Earth with collector efficiencies of about 50 %. It is designed to concentrate the incoming light onto a lens. This lens filters out the harmful radiation and focuses the visible light onto the end of a bundle of fiber optic cables. The advantages of this design are its light weight and the ability to filter out harmful radiation. The major disadvantages are the large amounts of heat that the geodesic parabola generates, and the reduction in efficiency. Three possible designs for the lens system may be considered: a Fresnel lens design, a converging lens, which focuses on a light tube or oversized fiber cables, and a reflecting mirror which uses a converging lens to focus the light onto a tube or cable.

The second principal method for supplying natural sunlight to plants that may especially be considered for Moon or Mars and is also the lowest mass alternative, is a transparent-walled greenhouse structure on the lunar surface. It would also have to include artificial lamps to provide PAR during the lunar night. The major problems with this alternative are the heating/cooling that the structure would experience on the lunar surface, the selection of a transparent wall material which would be low in mass, yet tolerant of the solar UV radiation load and exposure to cosmic radiation. Two solutions to these problems can be envisioned. One is to use lunar glass, fabricated in situ. This option is attractive for a number of reasons, but requires an analysis of the machinery required to manufacture the glass, an analysis of the capability of the glass to withstand the temperature and humidity conditions it would be exposed to, and an analysis of the mechanisms that could be used to mount glass panes with minimum leakage.

The second solution utilizes light plastic films, coated to prevent or retard degradation by UV radiation. These materials have to be analyzed for both resistance to lunar surface environmental conditions and for structural and mechanical characteristics which would typify the wall of a greenhouse structure.

As a third alternative, the use of light, inflatable reflectors and light guides as sunlight collection mechanisms may also be considered. These devices hold a great deal of potential for enabling direct use of sunlight at a very low mass, without transparent-walled structures. This concept would provide full or near full Earth-surface PAR values. Although this concept would not eliminate the need for artificial lighting during dark periods, it would provide a means of protecting the plants from radiation by covering the plant growth unit with regolith, while still using natural sunlight. [48, 58]

VI.4.3.4.3 Artificial Light

Collection of solar light for plant growth is, in general, an appropriate technology for use on orbit or during interplanetary transits, and may decrease the initial launch mass, but it may not significantly improve the case for the use of extensive bioregenerative life support systems on planetary surfaces. For example, lunar bases will be subject to nights that are 14 Earth days long, and substantial electrical power resources will be required for base operations during such lengthy periods without sunlight. Thus, fully regenerative technology may require nuclear power generation as part of a planetary base, or some method of collecting and storing power generated from sunlight during dark periods. [36]

Providing effective plant growth illumination levels and frequencies with artificial light is possible using available electrical lamps, e.g., fluorescent tubes and high-pressure sodium (HPS) lamps. HPS lamps are also called high intensity discharge (HID) lamps. Table VI.23 summarizes the power allocations of different types of electrical lamps. Although the highest efficiency (27 %) for conversion of electrical power to Photosynthetically Active Radiation (PAR) in the 400-700 nm range is provided by low pressure sodium lamps (LPS), these lamps provide an essentially monochromatic light which may not be suitable for all varieties of higher plants. A number of other lamp types have conversion efficiencies in the 20-25 % range and provide emission spectra which are more acceptable to a diversity of higher plants. On the basis of the data in table VI.23, only four lamps may be considered efficient enough for CELSS use. These are HPS, LPS, MHB, and CWF. Table VI.24 shows the installed lamp wattage per m^2 of growing area to produce 300 $\mu mol/m^2/s$ PAR. Also, mass estimates, derived from the required lamp wattage values, are given. As these values indicate, the most efficient lighting systems from a perspective of mass are the 1000 W HPS and MHB systems.

Lamp Type	Input Power [W]	Visible Radiation (400-700 nm) [%]	Nonvisible Radiation [%]	Conduction, Convection [%]	Ballast Loss [%]
Incandescent 100A	100	7	83	10	0
Incandescent 200A	200	8	83	9	0
Cool White Fluorescent (CWF)	225	20	37	39	4
Warm White Flu-orescent (WWF)	46	20	32	35	13
Fluorescent Plant Growth B (PGB)	46	15	35	37	13
Clear Mercury (HG)	440	12	63	16	9
Metal Halide B (MHB)	460	22	52	13	13
High Pressure Sodium (HPS)	470	25	47	13	15
Low Pressure Sodium (LPS)	230	27	25	26	22

Table VI.23: **Power Allocation of Light Sources [58]**

Lamp Type	Lamp Wattage Required for 300 µmol/m²/s PAR [W]	Lamp Wattage [W]	Mass [kg/m²]
HPS	255	250	8.4
MHB	319	250	8.7
CWF	319	215	20.9
LPS	237	180	23.5

Table VI.24: **Lamp Wattage Required for 300 µmol/m²/s PAR and Mass Estimates [58]**

Not only the power requirements for artificial illuminations are very high, but also the warm-up time for lamps requires energy expenditure without appreciable light generation. Thus, short warm-up times are preferred in lamp design. HPS lamps require relatively long (several minutes) warm-up times and generate intense, localized heat loads. Fluorescent tubes require no appreciable warm-up time, and do not generate intense, localized heat loads. HPS lamps are three to four times more efficient than fluorescents in terms of lumens produced per watt. The intense heat generation makes them unsuitable for direct plant illumination at the required very close plant-to-lamp spacing. An alternative approach using fiber optic light pipes solves the heat problem. However, the efficiency losses encountered in focusing and transmitting HPS light reduce the overall efficiency by 30-60 %. Compared to fluorescent light in fixtures directly over the plants, the power requirements are nearly identical for each lumen at plant canopy.

The fluorescent lamp frequency spectrum is adequate for plant growth. Modifications in phosphor coatings can increase red frequency output, which may enhance plant growth. The fluorescent lamp mercury content is a major health hazard. Specially designed sealed lamps and/or fixture may be required to prevent mercury escape. High-intensity light-emitting diodes (LEDs) may also be considered for use as artificial light sources. Like the LPS lamp, these devices are essentially monochromatic light sources. Two particular advantages of LED technology are that it does not present a problem of mercury contamination if the device is broken, unlike conventional lamps. Furthermore, LED lifetimes are significantly longer than that of conventional lamps, providing as much as 100000 h of illumination with only a 20 % decrease in output. Most conventional lamps have lifetime figures of 10000-20000 h, and some lamp types can lose as much as 40-50 % of their initial output value over their lifetime. Based on current technology estimates, an LED-based illumination system would be about 65 % the mass of 1000 W HPS and MHB systems and would only use 50 % more power. [48]

VI.4.3.4.4 Light Transportation and Distribution

For light transportation, three alternatives can be considered. The first consideration is to pipe the light by continuing fiber optic bundles into the plant growth area and taking them to a light diffuser directly over the plants. Secondly, the solid dielectric pipe to transport the light to the diffuser may be considered, or finally, the use of a hollow dielectric fiber.

As described earlier, fiber optic light cables from each lens may be bundled together into trunk lines. These bundles may be broken out into individual cables. Then the cables have to be routed to the plant growth unit where they have to be connected to light diffusers. Different cable arrangements to each diffuser may vary the light intensity, but also internal illuminator controls may permit reducing light intensity without moving cables. As cable, pure quartz fiber seems to provide the best flexibility, radiation resistance, wavelength characteristics, and allowable incidence angles. The losses for a pure quartz cable are 4 % for 10 m of cable. A potential problem with quartz fibers is that the efficiency for light transmission drops considerably at temperatures below -20° C. Also, it may not be forgotten that breaks in the cables will be necessary for handling, maintenance, and system upgrading causing between 2 and 35 % light loss depending on the technique used.

The hollow fiber has the lowest weight and the highest efficiency. Here, light enters the greenhouse via fiber optic bundles extending from a light collector. The light emanating from the end of the fiber bundle is channeled into a hollow dielectric pipe. This pipe, precisely machined into an optical waveguide, is made of tempered, low expansion brosilicate. The light travels through the waveguide and is sent into a light diffusing pipe that is placed over the plant growth bed. This light diffuser has to diffuse the light from the pipe to the plants in a uniform and efficient manner. To accomplish this, two alternatives may be considered. Both consist of an optical diffuser. The first shines the light down onto the plants while the second shines the light up to a mirror that is above the pipe and then back down onto the plant bed. The advantage of having the mirror is that it could direct the light over the entire width of the plant bed. However, the diffuser that reflects downwards weighs much less and is more efficient because there are no reflection losses. [48, 75]

VI.4.3.4.5 Hybrid Concepts

Combining solar lighting systems during light-side orbit or light periods, with artificial light during dark-side orbits or dark periods, provides continuous light and reduces power demands. The hybrid system provides any length of day required by plants up to a continuous 24 hours of illumination. Nevertheless, reduced illumination intensities during dark-side operations may be necessary to conserve power. Also, illumination intensity equivalent to collected solar light is precluded by high electrical power requirements for artificial light. One concept may be to install fluorescent tubes in the same luminarities with fiber optic beam splitters. One light diffusor may then serve both types of light emitters. In that case, heat loads are comparable for both systems and common cooling devices may handle luminary heat loads for all lighting conditions. The principle of such a system is shown in figure VI.23. [48]

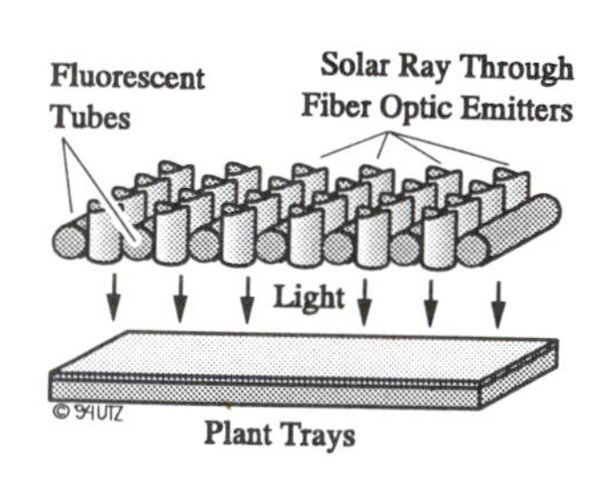

Figure VI.23: **Integrated Direct Solar and Lamp Illuminators [47]**

VI.4.4 Microorganisms

Microorganisms, i.e., bacteria, yeast, fungi, protozoa, and viruses, are ubiquitous in and on the human body. From birth, people live in a microbial biosphere composed of innumerable microorganisms representing types, variants, strains, species, genera, etc. The composition of this microbial environment is dynamic. Numerous additions and deletions, both qualitative and quantitative, constantly take place. Therefore, microorganisms will be present in spacecraft designed for transporting and housing humans. The presence of microorganisms and their effects on the bioregenerative life support system must be factored early into the design process.

Microorganisms may be divided into two groups according to their activities: the harmful ones and the useful ones. The first kind, the pathogenic microorganisms, cause diseases merely because they are able to grow on or in the bodies of humans, animals or plants and damage the host in greatly varying degree. Of more than 1700 known kinds of bacteria, about 70 are known to cause disease in human beings, and of these only about a dozen, as a rule, are considered dangerous. The second kind live in the outside world and are normally harmless to humans. In nature, these types of microorganisms (saprophytic or saprozoic) are important in decomposing plants and animals so that the components can be used to produce new biomass. In this role, microorganisms may be used to recycle waste matter, in conjunction with a plant growth facility for growing food.

Uncontrolled activity of microorganisms can be extremely damaging to the operation of a closed bioregenerative life support system. Thus, the prevention of excessive microorganism growth is required for maintaining system health, although it is both undesirable and impossible to create a completely sterile environment. As in humans, also in plants some microbial interaction is beneficial, and in some cases even necessary. Nevertheless, plant life is particularly susceptible to diseases caused by microorganisms. Thus, a method for monitoring and evaluating microbial population in the plant growth system must be included in a CELSS to help avoid widespread death of the plants that perform key life support functions. [1]

High plant density and high humidity provide an ideal environment for rapid spread of diseases. To keep pathogens out from a bioregenerative life support system, pathogen-free seeds and resistant cultivars may be used and also the system may be decontaminated between crops. In order to block off the spread of plant diseases, to provide specific environments for specific crops and to prevent the spread of leaks, compartmentalization may be helpful. In the duct work UV and ionizing radiation that are lethal to spores of pathogens may be used. Furthermore, special microbial communities with proper balances of organisms may be produced that ward off the pathogenic ones.

Formation of biofilm, an adhesion of microorganisms to moist surface, should also be considered as a concern for longer duration spaceflights. Bacteria tend to attach to surfaces in nature, where it can be protected from the environment by secreting a tough film of adhesive polymer called a glycocalyx. Biofilms represent a major area of interest both from the standpoint of the potential contaminant generation and the possibility of their deliberate controlled use in bioregenerative life support systems. Biofilm can be of benefit when the microorganisms attached are able to remove substances diluted in the water that can be harmful to humans. On the other hand, they can clog filters and tubing and induce corrosion of materials. The glycocalyx can also provide protection against biocides providing a safe haven to pathogenic organisms. [2]

Bacteria and fungi, either as airborne contaminants or as components of closed waste and water recycling systems, may pose additional complications. For example they may be a source of chemical corrosion of containment vessel surfaces which are usually extremely thin, and a secondary source for the generation of toxic chemical species, which, depending on the starting contaminant concentration and the microbial load, may not be trivial. Detection and identification of microorganisms is generally a time consuming and tedious process, not suited to application on spacecraft, where information may be needed within hours, or even "real-time". Current microbial monitoring techniques do not meet all the requirements needed to be installed as part of spacecraft hardware. Some of these requirements include high sensitivity, small or no crew intervention, small amount of expendables, low detection time, small volume, low weight, and small power requirement. [62]

VI.4.5 Atmosphere Regeneration

Carbon dioxide removal and oxygen production are both natural processes of photosynthesis. One problem introduced when plants are used for these two purposes is that some O_2 is lost, because the rates of production and intake of O_2 and CO_2 differ between plants and humans. Humans produce about 0.85 moles of CO_2 per mole of O_2 consumed, while plants consume about 0.95 moles of CO_2 to produce one mole of O_2. This translates into 0.1 moles of O_2 lost between the time a human beeing inhales and a plant "exhales". Since maximum oxygen production and carbon dioxide utilization generally occur during periods of maximum plant growth, frequent planting and harvesting will be required to assure desired gas exchange. The minimum surface area required to provide adequate oxygen for one person varies according to the plant species being grown and the extent of surface area that can be continuously covered with photosynthesis tissue. Thus, oxygen production and CO_2 incorporation by plants in the system could be balanced against human activity patterns in various ways such as by:

- Providing multiple plant growth chambers so that a portion of the plant population would be photosynthesizing, while other portions are in darkness.
- Altering radiation or temperature levels to moderate photosynthetic rate as needed.
- Providing a system of gas storage

Another problem of CELSS atmosphere regenerators concerns the removal of trace gases. While beneficial trace gas removal is a natural process for most plants, many of the same plants produce trace gases that will require removal by some other devices. [67]

VI.4.6 Water Recovery

A bioregenerative life support system can contribute significantly to water recovery because plants act as natural water purification devices. Plant roots take in water, some of which is then transpired as water vapor from the leaves. Transpiration water can be condensed and collected for drinking water. In a hydroponic plant growth system the quality of the nutrient solution water may be low, yet the transpiration water recovered is potable in many cases. The rate of water release from plants can be controlled in several different ways to satisfy moisture and water demands within the space habitat. Controls include varying the humidity level of the plant growth area, varying the intensity and length of the light period and amount of plant surface maintained. It has been suggested that plants may give off harmful substances with the transpired water, and thus water condensed from the air may need further purification before being used for drinking or in supporting the growth of other plants. [62, 67]

VI.4.7 Waste Processing

Some of the liquid and solid wastes produced by humans and/or animals in a bioregenerative live support system cannot be directly used by plants without some form of processing. Plants will also produce waste in the form of inedible biomass. To help complete CELSS closure, a waste processor will be required to break down human and plant waste to recover useful constituents, such as CO_2, nitrogenous compounds, and nutrient salts. In the terrestrial ecosystem, waste processing is performed by microbes that ingest organic material and produce water and carbon dioxide. However, if it appears that biological waste treatment systems will be too large for use in space, physico-chemical oxidation processes, such as SCWO (see section V.3.2.1), may have to be applied. An overview of the advantages and disadvantages of a biological waste treatment system are given below:

Advantages of biological waste treatment:

- Liquid and gas effluents would be compatible with plant growth and would contain fewer chemical toxins than effluents from physico-chemical systems.
- Biological systems generate CO_2 at fairly constant rates, whereas mechanical systems produce periodic pulses.
- The biological waste management system could serve as a better place to buffer or store carbon stocks.
- The formation of a partially oxidized feedstuff from waste can better be accomplished by a biological process.
- In general, biological systems favor the production of nitrates over ammonia and cause less denitrification.
- A potentially large knowledge of activated sludge systems exists because of the frequent terrestrial application.
- Biological systems can be quite easily repaired in case of a breakdown by initiating a new compost or sludge with a new innoculum.

Disadvantages of biological waste treatment:

- Relatively slow turnover time
- Potential for epidemic populations
- Slow start-up times
- Difficulty in control
- Production of more humic substances that may require further treatment, e.g., incineration

The requirements of a biological waste treatment system depend on the microorganisms that are used in the process. In general, the microorganisms are composed of several different populations of organisms. The relation of the different organisms is in dynamic equilibrium depending on flux rate, temperature, pH level, waste composition and other factors. Thus, by controlling these parameters, the whole waste treatment reactor may be controlled. In principle, biological decomposition of organic wastes in CELSS could be accomplished through the use of either aerobe or anaerobe methods. In an anaerobic bioreactor, e.g., a methane fermentation reactor, the organic substances are decomposed by bacteria. This process yields low-molecular compounds, e.g., CO_2, H_2O, NH_3, NO_2, NO_3, and bacterial sludge. The sludge has to be filtered and yields a nutrient solution for the plants. Aerobe methods include activated sludge systems or composting. Both of these alternatives are described in detail below. Before, in table VI.25, the characteristics of aerobic and anaerobic bioreactors are compared.

	Aerobic Bioreactor	Anaerobic Bioreactor
Use	- Liquid wastes - Maybe composting of solid wastes	- Solid wastes / Sludge - Possibly treatment of liquid wastes
Wastes	Biologic. decomposable fractions of liquid and solid wastes	Biologically decomposable fractions of liquid and solid wastes
Output	CO_2, H_2O, NH_3, (NO_2, NO_3), biomass / sludge	CH_4, CO_2, (H_2), biomass / sludge
Advantages	- Transition from an insoluble to a soluble state may limit the production rate - Basic system for the treatment of liquid wastes	- No gas consumption - No leakage problems - Methane may be used as fuel - Small amounts of well stabilizes sludge
Disadvantages	- O_2-consumption - CO_2-production - Microbial activity may be influenced by toxic components	- Methane production - Danger of inflammation of Methane - Slow decomposition - Very good controlling required - Microbial activity may be influenced by toxic components

Table VI.25: **Comparison of Aerobic and Anaerobic Waste Treatment [74]**

Activated Sludge

Activated sludge systems use oxygen to convert a wide range of organics to CO_2, H_2O, and cellular residues. Some organic molecules are resistant to biological breakdown and are degraded very slowly. The catalysts for these reactions (generally operating at 15° to 50° C) are bacteria, yeasts, protozoa, and some fungi - organisms normally present in the input streams. The ratio of various types of organisms to one another is in dynamic balance and can be altered by changing flow rates, temperature, pH, etc. A typical aqueous activated sludge unit consists of an aeration basin, settling tank, and a recycle pump. The influent sewage stream to be treated is normally dilute (300-30000 ppm COD). Typically, the ratio of the liquid volume in the aeration to influent flow rate results in a liquid residence time of 2-10 hours. The effluent from the aeration basin is sent to a gravity settling basin or clarifier. A clarified effluent, nearly free of suspended material and cells, exists at the top, and a concentrated sludge occupies the bottom. The sludge (0.5-10 % solids) contains cells and residues of incompletely consumed particulates. Most of these solids are returned to the aeration basin. Generally, aqueous systems are operated with mean cell residence times of 3 to 15 days. Mean cell residency time cannot be increased to infinity because the portion of material not susceptible to reasonably rapid biological breakdown in the tank would go to infinitely high levels. Consequently, a portion of the sludge from the clarifier is typically wasted (usually 20-25 %, under special conditions maybe 5-10 %). It could be treated further by physico-chemical or biological techniques, e.g., composting.

Composting

In contrast to the aqueous biotic oxidation system, semisolid biotic oxidation follows terrestrial models of composting, solid substrate fermentations, or natural decomposition. The low water content encourages the growth of many fungi and discourages the growth of many bacteria. In addition, the low water content would reduce the weight of the system. Fungi tend to grow more slowly but are generally more efficient at the biological oxidation of lignin or cellulosic materials. In the absence of a well-mixed aqueous phase, mass- and heat-transfer gradients are established, resulting in a heterogeneous reaction system. Often the physical process of mass and heat transfer will control reaction rates. The rate of decomposition in such a system can often be accelerated by reducing material volume, then placing it in a rotating drum reactor. The residue from such decomposition serves as an ideal terrestrial substrate for plant growth. In terms of a CELSS, semisolid biotic decomposition might be most usefully employed on the portion of the waste high in lignin and cellulose residues, e.g., plant residues and the waste sludge from an aqueous biotic oxidation unit. [32, 67, 74]

VI.4.8 Nutrient Supply

Two major components compose the nutrient supply system. The first is a nutrient reservoir where the nutrient salts are combined to create fresh base solution. The second is a nutrient regeneration subsystem. This regeneration subsystem has to constantly adjust the nutrient solution to maintain optimum solutions for plants. The continual reuse of the nutrient solution for cropping is an important means of conserving water in a controlled bioregenerative life support system, since the treatment of spent nutrient solution is costly in terms of equipment, energy and space. The principle of such a system is given in figure VI.24.

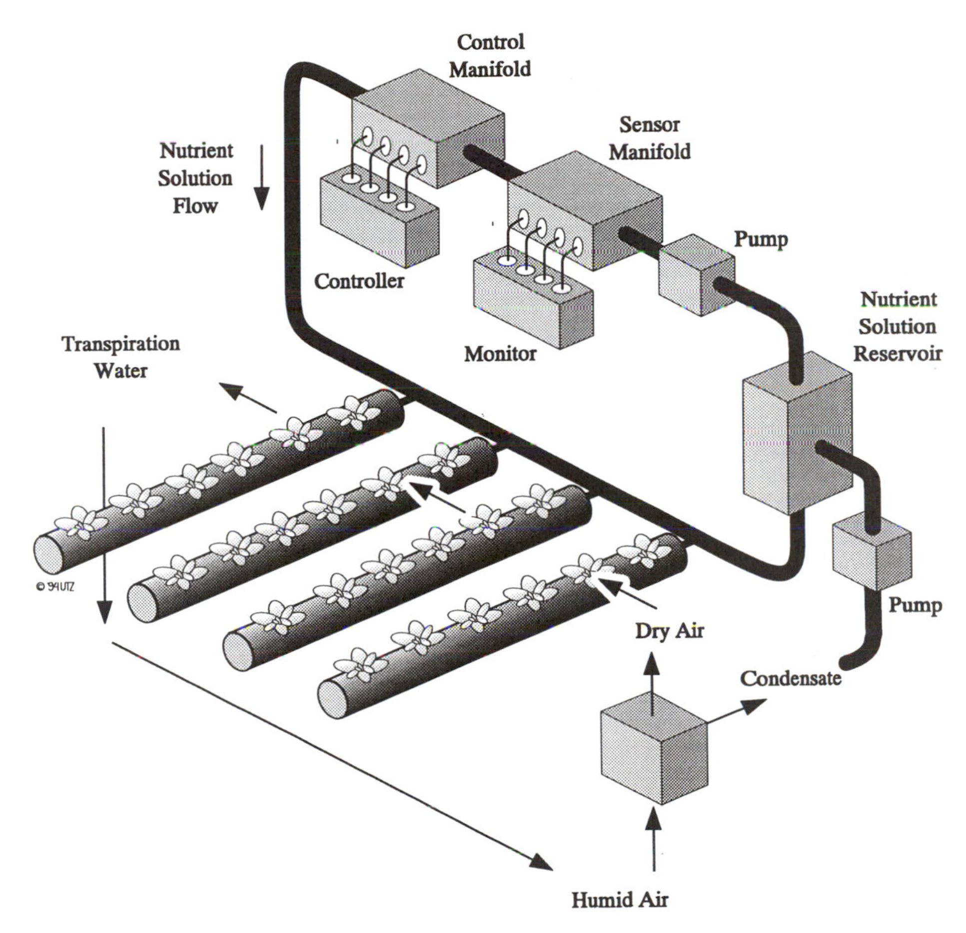

Figure VI.24: **Sample Nutrient Supply System [48]**

The nutrient supply system has to keep plant roots moist and has to provide the required nutrients. Nutrient solutions have to be specifically formulated to promote optimum growth for each plant species. Thus, the solutions may have to be separately monitored and adjusted for each plant species grown. A constant aerated nutrient supply has to be pumped to the roots. The exhausted nutrient solution may be removed from the root zone by aspiration. Aspirated air, with entrained nutrient solution, then may be passed through a water separator.

As nutrient solutions are recovered from the plant growth units, they have to be monitored for composition and pH. When analysis indicates low nutrient levels in solutions, make-up constituents have to be added as required. pH is constantly adjusted and buffer added. Solution concentration is monitored through conductivity measurements that trigger water injections to make-up for water loss to transpiration. Fungicides, bactericides, and other pre-control substances may be added to control specific problems. These photogenic organisms are identified by central microbial analyzers that periodically receive nutrient samples from each nutrient regeneration unit. Particulate contamination may be filtered out as the nutrient solution is returned to the supply tanks. When dissolved contamination reaches limit levels, the solution may have to be dumped to the waste regeneration system and fresh solution is added from the main nutrient solution supply reservoir. From there, the solution may be recycled to the roots.

Element	Hoagland #1 [mmol]	Peter's Hydrosol [mmol]	Salisbury/Bugbee [mmol]
N	15.0	3.6	15.0
P	1.0	1.5	0.2
K	6.0	5.4	3.0
Ca	5.0	Added as $Ca(NO_3)_2$	12.0
Mg	2.0	1.3	4.0
S	2.0	1.2	2.0
	[µmol]	[µmol]	[µmol]
Fe	50.0	50.0	100.0
Si	-	-	300.0
B	46.0	46.0	80.0
Mn	9.0	9.0	8.0
Zn	0.8	2.3	0.8
Cu	0.3	2.4	0.3
Mo	0.5	1.0	0.1

Table VI.26: **Comparison of Nutrient Solutions [59]**

There are many different kinds of nutrient solutions and modified versions thereof, which have been published in the past 50 years. The University of California (Berkeley) agricultural experimental research station by Hoagland and Arnon (1938) is recognized as the basis for many formulations currently being used by investigators and commercial Companies. Depending on the species and environmental conditions, adjustments to the recipe were frequently made for improved growth. Fertilizer mixers, such as Peter's Hydrosol (1987), are commonly used in industry because they are ready-made. The slightly lower salt concentrations in Hydrosol also allow for nutrient manipulation as needed for the culture of a wide variety of plants. However, copper and zinc concentrations are well above the average found in many other nutrient solutions. Bugbee and Salisbury (1985) modified the Hoagland/Arnon formula to grow wheat hydroponically under high irradiance and elevated CO_2 environments. The relatively low mobility of boron, calcium, and magnesium lead to an increase of their concentrations in the nutrient solution. Lowering the phosphorous concentration allowed for an increase in iron uptake by the plants. In addition, Bugbee and Salisbury added silica to their solution to minimize micronutrient toxicity symptoms. An overview of the nutrient solutions mentioned above is given in table VI.26. Table VI.27 summarizes the nutrient content ranges for nutrient solutions in hydroponics as indicated in several references.

Component	**Amount [g/l]**
Potassium nitrate	0 - 1.1
Calcium nitrate	0 -1.29
Ammonium nitrate	0 - 0.1
Calcium biphosphate	0 - 0.31
Potassium sulfate	0 - 0.63
Calcium sulfate	0 - 0.76
Magnesium sulfate	0.17 - 0.54
Ammonium sulfate	0 - 0.14
Total Salts	**0.95 - 3.17**

Table VI.27: **Composition of Nutrient Solutions for Higher Plants [59]**

Another interesting aspect concerning nutrient solutions is the potential recovery of inorganic nutrients from inedible biomass. Recovery of water soluble, inorganic nutrients from the inedible portion of wheat, i.e., straw and roots, was found to be an effective means of recycling nutrients within hydroponic systems. Through aqueous extraction (leaching), 60 % of the total inorganic nutrient weight, and about 20 % of the total organic carbon could be removed from the biomass. Thus, leaching holds considerable promise as a method for nutrient recycling in a CELSS. [17, 48, 56, 59, 71]

VI.4.9 Crew Time Requirements

It is obvious that it has to be attempted to minimize the crew time requirements for the servicing, maintenance, etc., of any subsystem of a bioregenerative life support system. The probably best way to reduce crew time requirements is the automation of certain processes. To make specific estimates, these processes

Activity	Time Requirement
Higher Plants	**[man-hours / day / m^2]**
Planting	0.0199
Harvesting	0.0199
Wheat Grinding	0.135 man-hours / day / 100 g
Observation	0.0158
Preventative Maintenance	0.0475
Nutrient Solution Maintenance	0.03
Algal Reactor	**[man-hours / day / reactor]**
Sampling and Analysis	0.76
Harvest (Centrifuge and Dry)	0.0508 man-hours / day / 100 g
Nutrient Solution Preparation	0.22
Monitoring Operation	0.278
Preventative Maintenance	0.833
Domestic Activities	**[man-hours / day / crew-member]**
Food Preparation, Eating, Clean-up	1.7
Water Preparation	0.14
Personal Hygiene	0.39
Monitoring Operation	0.278 man-hours / day / reactor
Living Compartment Hygiene	0.27

Table VI.28: **Crew Time Requirements for Various Activities (Adapted from BIOS-3) [58]**

will first have to be defined. Due to the state of development of bioregenerative life support subsystems, of course, hardly any of these definitions exist. Thus, although the amount of time required to service and maintain a CELSS is such an important issue, little experimental information is available to serve as a guide for time estimates. Probably the only experimental data were obtained during the BIOS-3 experiments. Thus, some data derived from BIOS-3 outlined in table VI.28.

Nevertheless, it has to be taken into account that this data comes from an experiment that has been conducted on Earth, i.e., under 1-g gravity conditions. Depending on the processes and technologies that may be applied, these values may be completely different concerning time distributions and requirements under microgravity or low-gravity conditions. [58]

VI.4.10 Higher Plant Growth Facilities

Higher plant growth facilities have not yet been flown in space. For a future application of these systems, a stepwise approach may be recommended. On the long run, a complete supply of a crew with a diet that consists exclusively of crops that are grown onboard a space station or space habitat, may be envisioned. Before, smaller devices will have to be flown in order to test the functioning of, first, higher plant growth, and later, a CELSS in space. Thus, a test program may principally consist of the following three steps and devices that are under development and discussion:

- *Step 1:* Salad Machine
- *Step 2:* CELSS Test Facility (CTF)
- *Step 3:* Plant Growth Units (PGU)

The growth of salad vegetables onboard of space stations and during other long duration missions can provide psychological and dietary benefits to crew members. For this reason, a unit called the Salad Machine, capable of producing 600 g of edible salad vegetables per week, enough for one salad three times a week for a crew of four, is being planned at NASA Ames Research Center. With this unit it will also be possible to recover the water which is transpired by plants during the growth cycle. Candidate crops for the Salad Machine include leaf and/or head lettuce, carrot, green onion, cherry tomato, cucumber, dwarf bell pepper, and three varieties of sprout: alfalfa, bean, and radish. [24]

The Salad Machine is shown in figure VI.25. It is planned to be contained in one standard space station double rack, which provides a user volume of 1.3 m^3. Approximately 0.8 m^3 will be used for the plant growth chamber, and the remainder will be filled by support hardware and a small germination/sprout growth chamber.

The total plant growth area will be 2.8 m², consisting of four rectangular shelves (76 x 94 cm), each with an area of 0.7 m². To achieve the desired biomass output, inputs of approximately 1.5 kW continuous peak power (0.5 kW off-peak), and 0.6 kg water per week will be required.

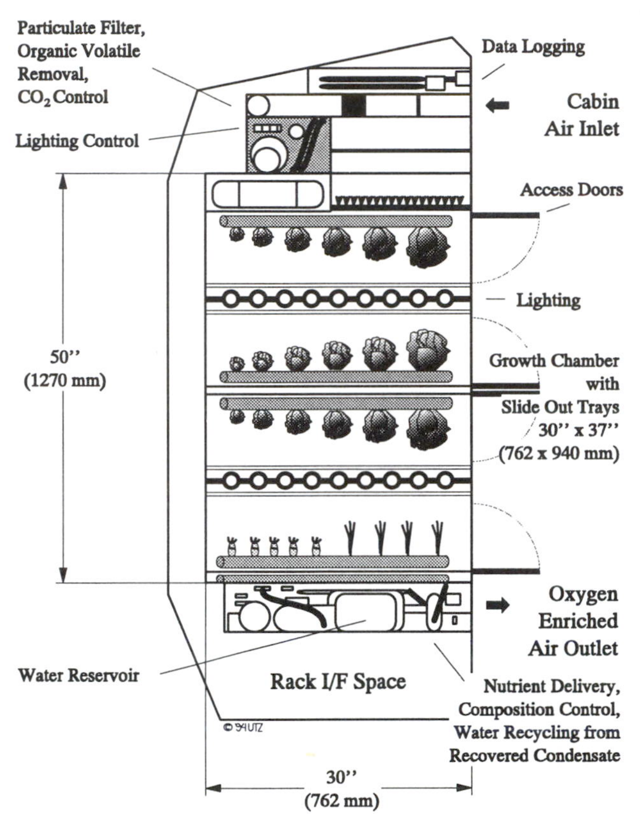

Figure VI.25: **Salad Machine Design Concept [36]**

The next step will be the CELSS Test Facility (CTF). This unit will be a space-based crop research chamber which is also to be flown onboard the space station. It is estimated that a total growing area of about 1 m² will be available, with a growing height of about 85 cm. This chamber will have tightly controllable internal environments that will essentially be closed to the entrance or exit of materials to allow refined monitoring of crop responses to environmental factors. The system consists of several subsystems, each of which must work as part of the integrated total. In such a configuration the functioning of any one subsystem will affect the operational parameters for one or more of the other subsystems. All of the subsystems must be controlled, both independently and as part of the system. This means that a multiplicity of feedback loops will be inherent in the subsystem designs. Because of this, each subsystem will be required to be part of a rather sophisticated hierarchical control system, capable of integrating sensed information and control functions. Like the Salad Machine, the CTF will occupy two standard space station racks. The initial estimate for power usage for the CTF is in the order of 2 kW peak, accountable primarily to the lighting system. The mass of it is estimated to be in the order of 1145 kg. A design concept of the CTF is illustrated in figure VI.26. [36]

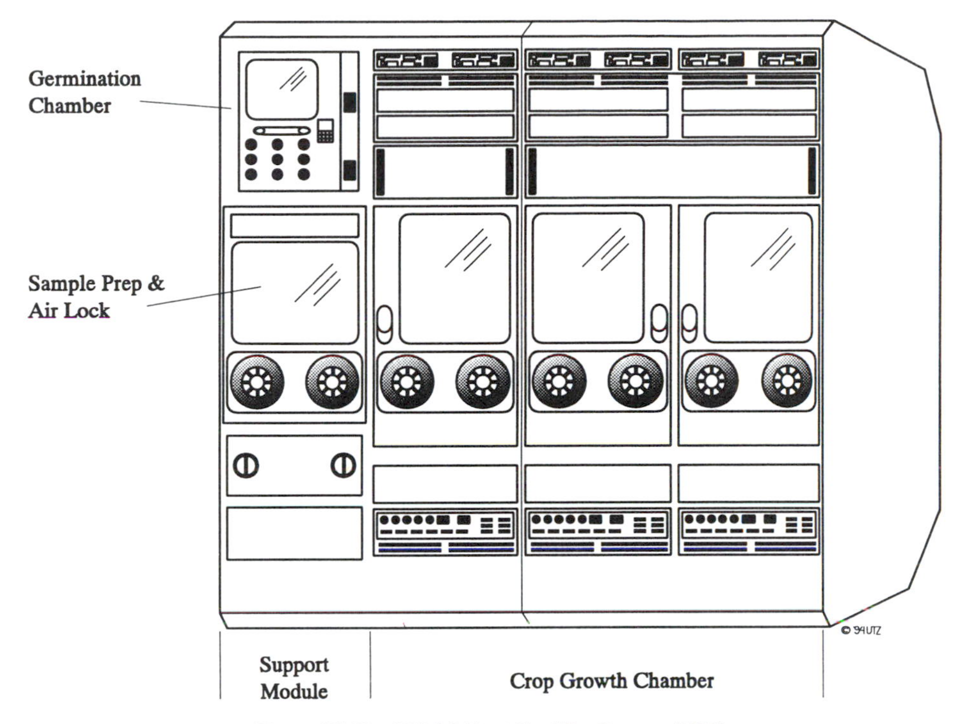

Figure VI.26: **CELSS Test Facility Concept [36]**

Next, a plant growth unit (PGU), providing place for growing plants and a means of holding them in position during a growth cycle, may be tested at the space station. A PGU will comprise the largest single unit in a CELSS module design. The method of retaining and holding the plants has to permit exposure to light and nutrients. Also, the plant growth unit design must confine nutrients to the root zone and provide a means of starting the plants from seed. Mature plants may be automatically collected for food processing. PGU designs may permit sequencing of crop planting and harvesting in order to have daily food collection. For plant growth in a CELSS several different kinds of plant growth units have been proposed. Brief descriptions of the most promising options are given below:

Warehouse Tray Stack

The warehouse tray stack concept has trays on vertical racks serviced by a robot that moves along the center. As shown in figure VI.27, trays fit into different sized slots, which places lights as close as possible to the plant canopy while allowing area for growth. As plants grow, a robot moves the trays into progressively larger slots that accommodate growth. At maturity, the trays may be removed entirely from the stack and transferred to the harvesting equipment. After harvest and tray recondition, the trays may be reseeded and placed in the smallest size slots. Nutrient will be supplied by pressure-fed injectors. A vacuum system removes excess and spent nutrients. When a tray is removed, nutrient and vacuum systems are disconnected. Valve arrangement prevents leakage. Lighting panels are located above openings in the stacks. This distributes light on a tray-by-tray basis and allows regulation of light intensity for different growth phases. The warehouse tray concept is conventional in construction. It can be built of simple-angle shapes attached by mechanical fasteners. No mechanisms are required within the structure for tray manipulation - the robot performs these functions. However, this concept does not optimize volume usage, light distribution is not optimal, and seeds have to be spaced for mature plant spacing requirements.

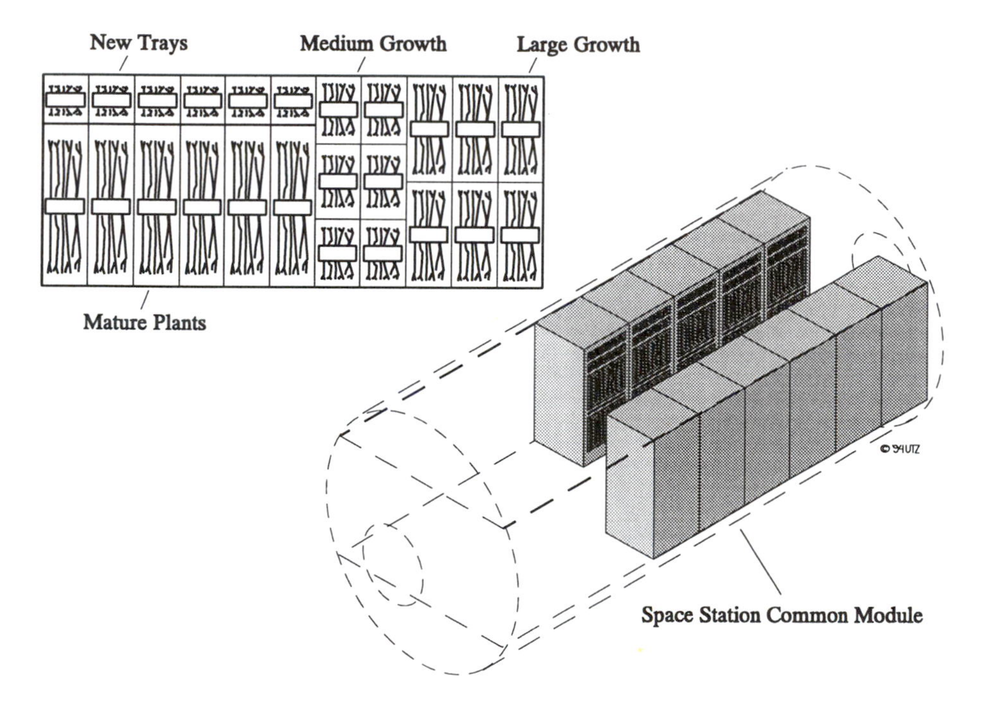

Figure VI.27: **The Warehouse Tray Stack Concept [48]**

- *Accordion Tray Stack*

  This concept, shown in figure VI.28, is centered around the use of trays that are accordion-pleated so they expand longitudinally. Tray ends are removable for harvesting operations. This allows robot access to the tray interiors to push out root masses. The trays may be arranged so that

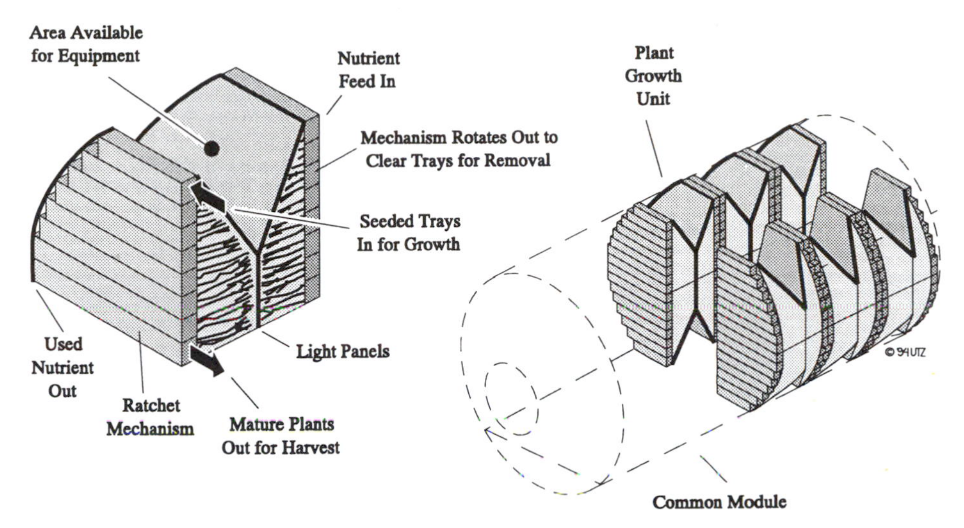

Figure VI.28: **The Accordion Tray Stack Concept [48]**

plants grow out on one or both sides, and are stacked vertically, one tray abutting the other. They are pushed down incrementally and expanded by a mechanism. The bottom trays contain the most mature plants. These trays are removed from the stack and transferred to the harvesting area, making room for the remaining trays to move down into the next growth position. After reconditioning and reseeding, the tray is compressed and inserted at the top of the stack to repeat growth procedure.

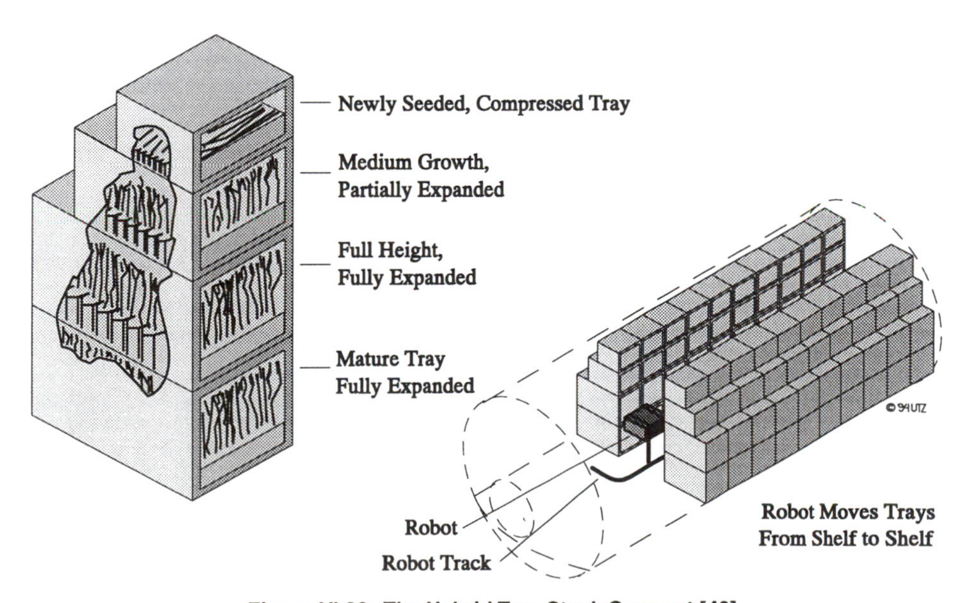

Figure VI.29: **The Hybrid Tray Stack Concept [48]**

Each time trays are moved down, the nutrient delivery system disconnects and moves clear of the travel envelope. When trays are repositioned, nutrient injectors are returned to engage tray nutrient delivery orifices. A vacuum pickup system collects spent and excess nutrients for recycling. This unit requires minimal volume by expansion to follow module hull contours as plants grow, and reliable mechanisms, while supporting maximum light efficiency, since smaller lighting systems can be used in the earlier stage of growth.

- *Hybrid Tray Stack*
  This concept has trays on vertical racks, accessible from an aisle. As shown in figure VI.29, the racks extend from aisle to module inner hull surface. This creates progressively deeper slots with the deepest slot at module center line. Trays are built with accordion folds so they may be collapsed to fit the shallow top slot. The trays are moved to deeper slots as plants mature. This allows the tray to be expanded, thereby providing more plant growth area per tray. Trays are moved from slot to slot and finally to the harvester by a robot.

Further concepts that have been proposed are:

- *Conveyor Belt PGU*
  Uses two conveyors facing a common light source. Newly seeded trays are inserted at the shallow side where lights are close to the conveyor. Plants grow as the conveyor belt slowly moves. The deep side, where lights are far away from the conveyor, has mature plants. They are removed from the conveyor for harvesting.

- *Honeycomb Tray Concept*
  Has six-sided trays facing toward six-sided light sources. Plants grow on two or three sides, depending on tray location in the pattern. Trays plug in longitudinally to growth unit.

- *Parallel-to-Hull Concept*
  Grows plants in a false wall between module interior and hull. These false walls permit module interior use for other purposes. Light sources are mounted on the inside surface of the hull. Plants grow on the outside surface of the walls, growing out towards the lights. Nutrient delivery to plant roots is located in the walls. A robot may travel against the hull, carrying a harvester with it.

- *Cone-Shaped Growth Chamber*
  Has a continuous tray moving through a cone with light source facing inward from the cone surfaces. The growth surface is a collapsible continuous tube. A slit-in tube allows the injection of nutrients and the removal of roots at harvest.

- *Radial Tray Concept*
  Places trays facing outward with a robot at the module center. Circumferential arrangement of the trays uses a large available surface area for plant growth.

- *Baloney Slice Concept*
  Has vertical panels moving laterally to allow growth and adjust lighting distance from the plant canopy. Panels are removed from the system for harvest.

- *Clamshell Growth Concept*
  Grows plants on a core facing toward the inside of a sphere that has a light source. Plants grow on most of the core excluding only the tube that supports the core and provides nutrient plumbing.

- *Rotating Drum Concept*
  Has a slowly rotating drum (one revolution per growth cycle of 60 to 115 days). Seeding and harvesting are performed continuously as the drum rotates. [48]

VI.5 Fungi and Conversion of Inedible Plant Material for Food Provision

Higher plants produce large quantities of inedible biomass in the form of roots, stems, and leaves. In wheat, for example, up to 70 % of the plant dry matter produced is inedible material consisting of approximately 25-40 % cellulose, 25-40 % hemicellulose, 4-10 % lignin, and 4-12 % crude protein. This biomass must be processed in a CELSS to prevent its accumulation. The processing can be based upon total oxidation of the plant matter or by bioregenerative recovery of the plant nutrients.

In order to make use of the inedible biomass left over from crop harvest, carbohydrates, proteins, and lipids may be extracted from these leftovers. A brief description of the extraction methods is given below:

- *Carbohydrate Extraction / Fungal Growth*
  A large amount of inedible plant material composed primarily of the carbohydrate materials cellulose, hemicellulose, and lignin is generated as a result of plant growth in a CELSS. The enzymatic hydrolysis of this inedible plant material into simple sugars that can be incorporated into the human diet lessens the disposal load and decreases the area needed to produce plant foods. In a CELSS environment, enzymatic hydrolysis is particularly attractive since toxic materials are not needed, degradation products are not a problem, and, if needed, more enzymes can be produced by fermentation.

Cellulose is a linear homopolymer of glucose, which when properly processed will yield glucose, a sugar which can directly be added to the human diet. Commercially available food-grade cellulases can be employed for cellulose conversion. Experiments have shown that approximately 20 % of the cellulose in inedible plant material can be converted to glucose within a 24 h period. Primary impediments to the conversion of cellulose to glucose are both the physical presence of lignin and hemicellulose, and the form of the cellulose, with crystalline cellulose being more recalcitrant than the amorphous form. Hemicellulose is physically associated with the cellulose, while lignin forms a "seal" around the cellulose, preventing cellulase action. Pretreatment of the plant material is necessary to reach cellulose conversion rates of over 50 %. Most of these pretreatments require chemicals, which would present problems for a CELSS in the containment, use, recovery, and recycling of the treating materials. Therefore, hot water is thought to be an ideal treatment agent. The primary effects of hot water treatment are the removal / solubilization of hemicellulose and partial solubilization of lignin at temperatures of 180° C, while the crystalline cellulose is unaffected by the treatment. It is assumed that a hot water pretreatment process followed by enzymatic hydrolysis would generate 0.4 g edible glucose per 1 g plant material.

Hemicellulose is a heteropolymer of hexoses and pentoses that can be treated to give a sugar mixture that is potentially a valuable fermentable carbon source. Such fermentations yield desirable supplements to the edible products from, e.g., soybean, cowpea, and rice.

Lignin is a complicated, three-dimensionally branched aromatic polymer. Food-grade, commercially available enzyme systems for lignin degradation do not exist, but white-rot fungal cultures posses ligninolytic enzymes and are capable of lignin degradation. *Pluerotus ostreatus*, as an example, also produces the edible oyster mushroom. Thus, fungal growth on inedible plant material could reduce the amount of lignin present, and in doing so possibly serve as a pretreatment for cellulose hydrolysis.

Another example for the conversion of carbohydrates by fungi is temphe (*fungai mycelium*). Temphe is grown on soybean cakes and wheat cakes and it is used in Southeast Asia as a high-protein food source. In general, fungi species are known that operate under many temperature and humidity combinations. They can also break down complex carbohydrates not attacked by bacteria. Mushrooms are useful to supply vitamin B2, food fiber, and niacin and can be a flavorful addition to the diet. According to terrestrial experiments, a growing area of about 3 m^2 is required for growing mushrooms for a crew of eight, providing approximately 100 g of mushrooms per crew member per day. These results will have to be compared with the growth of mushrooms under microgravity conditions in future studies. [25, 26, 66]

- *Protein Extraction*
  The method of extraction from leafy plants and vegetables has been fairly well researched and developed for experimental uses of a product which is composed of up to 60-70 % true protein, 20-30 % lipids and 5-10 %

carbohydrates. The first step is extraction of juice containing protein by bruising and pressing the plants through modified screw expellers. Coagulation of leaf protein from the juice is performed by acidification or heating at 70-90° C. Filtration separates the protein coagulum, the remaining "whey"-type juice is discarded as fertilizer. The suspension of the coagulate in acidic water is followed by filtration and the material is pressed into moist protein cakes. Another method of extraction from plant proteins suspended in the water that has been used to wash or cook plants, e.g., in recovery from potato starch mill effluents the solution is first coagulated by heating, then centrifuged, and finally dried. A similar method of protein recovery from animal carcasses utilizes rendering the material as a first step, with the last two steps being the same as for plants. Three methods for recovery of animal proteins already suspended in water include bulk protein extraction, ion-exchange, and ultrafiltration.

- *Lipid Extraction*
  Two basic categories of lipid recovery from secondary plant products are from either non-photosynthetic plant tissues or from photosynthetic plant tissues, animal tissues or microorganisms. The simplified extraction method for the first category involves blending cut tissue with chloroform and then suction filtering the homogenate. The filter is blended with methanol-chloroform and water, and the homogenate is filtered again and washed with methanol-chloroform. Water and chloroform are added and phase separation is performed. Finally, the chloroform is withdrawn and the solution is diluted with benzene. Subsequent dissolving of residual lipids by chloroform-methanol is only used for laboratory analysis. [18, 48, 58]

VI.6 ANIMALS AS HUMAN FOOD

Considerations associated with the use of animals as human food in a CELSS include the efficiency of converting feed to human food, the harvest index, the energy, mass, and volume requirements, the animal growth and reproductive rate, the palatability to humans, and the crew time required for preparation. Table VI.29 provides nominal values for production efficiency based on feed conversion efficiency and harvest index for several common domestic animals. The data show that some animal species are more efficient than previously recognized in CELSS design activities. The most efficient animal products are fish, milk, and chicken. On the basis of its area and volume requirements, milk production may be eliminated as an efficient means of producing animal food (see table VI.30). Because of the potential odor and trace contaminant control problems that poultry culture might engender in a CELSS, aquaculture seems to be the most promising. One potential problem of the inclusion of animals in a CELSS is the accidental death of any individual. This would disturb the equilibrium of the according to every animal's contribution to the total mass. Thus, it is advisable to select animals with the smallest mass and shortest life cycle to increase the system's reliability and ease of restoring equilibrium. The high metabolic rate of smaller organisms is also of value. [58, 59]

Animal	Feed Conversion Efficiency [kg Feed / kg Gain]	Harvest Index [%]	Product Efficiency [kg Feed / kg Edible Mass]
Beef	5.9 ± 0.5	49	10.2
Swine	2.5 ± 0.5	45	5.6
Lamb	4.0 ± 0.5	23	17.4
Rabbit	3.0 ± 0.5	47	6.4
Broiler Chicken	2.0 ± 0.2	59	3.1
Eggs	2.8 ± 0.2	90	3.1
Milk	3.0 (Dry wt. basis)	100	3.0

Table VI.29: **Efficiency Characteristics of Various Animal Species [58]**

Animal	Area [m^2]	Volume [m^3]	Water [l/day]	Feed [kg/day]
Beef Cattle				
Calf:	1.3	2.43	23 - 27	1.5 - 1.75 kg
1 year:	2.0	4.0	28 - 42	per 100 kg
Adult:	2.7	5.4	50	live weight
Dairy Cattle	3 - 3.5	6 - 7	< 136	10 -12
Swine (40-100 kg)	0.7 - 1.0	0.7 - 1.0	< 4.5	2.3 - 3.4
Sheep (30-40 kg)	1 - 1.5	2.3	2.6 - 2.8	1.3 - 1.4
Rabbit	0.23	0.105	< 1	6 % live weight
Chicken (Broiler)	0.1	0.05	0.5	0.06 - 0.07
Chicken (Egg/Breed)	0.05	0.025	0.25 - 0.3	0.09 - 0.11

Table VI.30: **Resource Requirements per Animal for Intensive Animal Production [58]**

VI.7 AQUACULTURE SYSTEMS

There are a number of freshwater fish species which grow to maturity rapidly (6-12 months), and therefore seem to be appropriate for a fish based aquaculture system in a bioregenerative life support system. Candidates include carp, trout, and *Tilapia*. All could be fed with vegetable materials produced in a CELSS, although a high-protein dietary supplement might be required to achieve optimum productivity. Another aquaculture system for potential CELSS application is one using crustaceans and mollusks. Freshwater crawfish are generalist omnivorous, and thus seem to be excellent candidates. Unfortunately, their harvest index is only about 15 %, and they tend to be extremely cannibalistic. Saltwater organisms have some potential, but generally take 2-3 years to reach edible size. Also, breeding these organisms is difficult, as many are adapted to deep-water spawning. The characteristics of some candidate species for an aquaculture system are summarized in table VI.31.

Animal	**Feed Conversion Efficiency** [kg Feed / kg Gain]	**Harvest Index** [%]	**Product Efficiency** [kg Feed / kg Edible Mass]
Shrimp	2.5 ± 0.5	56	4.5
Prawns	2.0 ± 0.2	45	4.4
Catfish	1.5 ± 0.2	60	2.5
Grass Carp	1.5 ± 0.2	60	2.5
Tilapia	1.5 ± 0.2	60	2.5

Table VI.31: **Efficiency Characteristics of Various Aquaculture Animal Species [58]**

The primary problem with implementing an aquaculture system is the large mass of water required to support an adequate human food supplement. A second, less significant problem concerns the 12 to 18 months required to bring an aquaculture system into steady-state production. However, the inclusion of a small amount of meat in the crew's diet may pay off psychologically as well as nutritionally. On Moon or Mars the mass requirement for an aquaculture system may be lowered significantly, if water or oxygen, which can then be reacted with hydrogen brought from Earth, can be extracted from regolith. Using a simple combination of ion-removal and submicronic filters, an aquaculture system could also provide a large water reserve for emergency needs. The basic requirements for some of the promising species are given in table VI.32.

Animal	Area [m²]	Volume [m³]	Feed [g/day]
Shrimp (Penaid)	0.005 - 0.006	0.003 - 0.004	0.35 - 0.4
Prawns	0.02	0.02	0.2
Catfish	-	0.001	4 - 4.5
Grass Carp	-	0.001	4 - 4.5
Tilapia	-	0.001	3.3 - 3.4

Table VI.32: **Resource Requirements per Animal for Intensive Aquatic Animal Production [58]**

An aquaculture system will require several different tanks. Small breeding tanks will be used to contain mixed adult males and females, in addition to the fingerlings they produce. Since fingerlings require some form of higher quality protein, the feed for this tank would include palletized, high-protein fish food in addition to the plant material. Upon reaching a certain size, fingerlings would be transferred to another small tank for sex-reversal hormone treatment, and then transferred to the main production tank for growth. The main production tanks are the largest of the aquaculture system. To minimize the total volume of the production tanks, a movable partition system in a single tank may be envisioned to separate the various sizes of fish. The fingerlings are introduced at one end, where a transverse partition keeps them separate from the rest of the population, thus preventing the larger fish from hoarding the food supply. As the fingerlings grow, the partition is moved down the tank, increasing the volume available to this set of fish. When the next group of fingerlings is ready, a new partition is placed at the end of the tank, and fingerlings are added. As the partitioned segments of the tank are moved, the spacing between partitions is increased to keep the mass of the fish per unit volume constant. Food for the fish could be obtained from the plant material remaining after the production of human food. Alternatively, biomass, forage crops, or algae could be grown specifically to provide food for the aquaculture system. [58]

A terrestrial research aquaculture project is the AQUARACK. It consists of three subunits. The first is a holding tank for the experimental animals. It is illuminated and equipped with observation windows and an automated lock for sample fixing. The animals are fed via an injection device and the tank is controlled by video cameras. The second subunit is a water recycling system which involves different components. The first is a semi-biological coarse filter. It is filled with a suitable substrate which holds back coarse particles resulting from excess feeding or naturally died animals. This device is followed by a bacteria filter. In a perlon wool or styrofoam bed bacteria of the Nitrosomonas/Nitrobacter group convert ammonia ions and ammonia excreted by the experimental animals to nitrite and nitrate ions. The latter are then eliminated by a zeolite molecule trap. The balance of respiratory gases is guaranteed by a unit consisting of a commercial blood gas exchanger with polypropylene or silicone capillaries. The water is disinfected by UV-lamps in a special reservoir and then passes a heating/cooling unit before it is driven back into the tank by pumps. The stability of the system is maintained by a computerized control unit which is equipped with probes for temperature, pressure, conductivity, oxygen, redox potential, nitrate, etc., and compensating devices.

The weak point of this initial AQUARACK concept was the ion absorber system which was exhausted after a short time and therefore had to be replaced in short intervals. Thus, the idea came up to replace the Zeolithe absorber by a highly sophisticated botanical component based on an algal reactor with a continued culture of photoheterotrophic unicellular algae. This lead to a new system called CEBAS-AQUARACK, where CEBAS stands for "Closed Equilibrated Biological Aquatic System". The basic concept of the CEBAS-AQUARACK follows a three component philosophy in which a zoological component, i.e., the tank for aquatic animals including the water recycling system, and a botanical component, i.e., a microalgal bioreactor or higher plant module, are combined to a self-maintaining closed artificial ecosystem which is stabilized and equilibrated by a process control system.

Two common experimental animals, the teleost fish *Xiphophorus helleri* as a "vertebrate model" and the pulmonate water snail *Biomphalaria glabrata* as an "invertebrate model", were selected for the CEBAS-AQUARACK project. It is disposed to be used in long-term multi-generation experiments onboard the space station and shall contribute to the development of combined zoological-botanical bioregenerative life support systems. The general question is, though, whether plant food has to be converted at all into animal protein and vice versa. In the CEBAS experiment there is primarily a production of algal biomass which has to be removed frequently from the system without delivering food for the animals. On the other hand there is the possibility to try the utilization of plant-eating fish species to convert possibly bad tasting algal biomass into animal protein. In this case it is still doubtful whether a successful reproduction of free spawning fish which release their eggs and sperm into the water is at all possible under μ-g conditions. Another aspect is the capacity of the system. Large amounts of fish in a limited water volume result in extreme conditions which have to be stabilized by additional oxygen supply, extremely effective and manpower-consuming filter systems and large amounts of food. In this case the animal system could change to a "simulated producer" because excess food and large amounts of feces could deliver nutrients for algae or higher plants. Thus a tremendous amount of work will have to be done to develop an aquatic CELSS, as the production of fish could be a suitable way for human food supply in space stations or planetary bases if autonomous production is required. [7, 8]

VI.8 FOOD MANAGEMENT AND PROCESSING

Presently, life support systems do not produce food, requiring that it be carried aboard the vehicle. However, since water constitutes a large portion of the weight of food it has been found advantageous to use dehydrated food and add water to it for reconstitution. This water is later reprocessed and reused. Approximately 0.64 kg/man-day of ashless, dry basis food with a caloric value of 10500 to 11700 kJ is required. For the future, life support systems which produce algae

System Component	Options
Food Types (Earth Supply)	- Fresh - Shelf-stable (e.g., canned or dehydrated) - Frozen
Food Types (Produced by CELSS)	- Bacteria/Algae - Plants - Aquaculture - Animals
Menu Planning	- Advance vs. Spontaneous - Individual vs. Group - Conventional vs. Unconventional Food
Diet Familiarization	- Pre-mission
Preparation	- Self vs. Staff Preparation - Scratch Cooking vs. Prepared Food - Inventory Control and Supply Management
Eating	- Congregate vs. Individual
Intake Monitoring	- Self vs. Staff Implemented
Waste Handling	- Stored vs. Recycling

Table VI.33: **Food Service Systems Options [18]**

or other plants are being investigated, because it is very likely that from a certain mission duration producing food onboard will require lower mass than resupplying it. Once food is grown onboard a space vehicle or habitat, food choices will be affected by the composition of the crew, the food service, the manner of food preparation, the social setting, and the source of the food. To use the performance capabilities of a CELSS, the food service system must be designed for a specific mission and must:

- Fulfill dietary goals by delivery of the appropriate nutrients.
- Deliver acceptable foods with appropriate sensory attributes.
- Maintain health and safety standards.
- Provide for unique needs of the space environment and activity levels of the crew.
- Provide drinking water of potable quality.

There are a number of options for supplying food, depending on the duration of the mission, specifications of acceptable foods, the menu cycle, and the food service system. An overview of food service options is given in table VI.33.

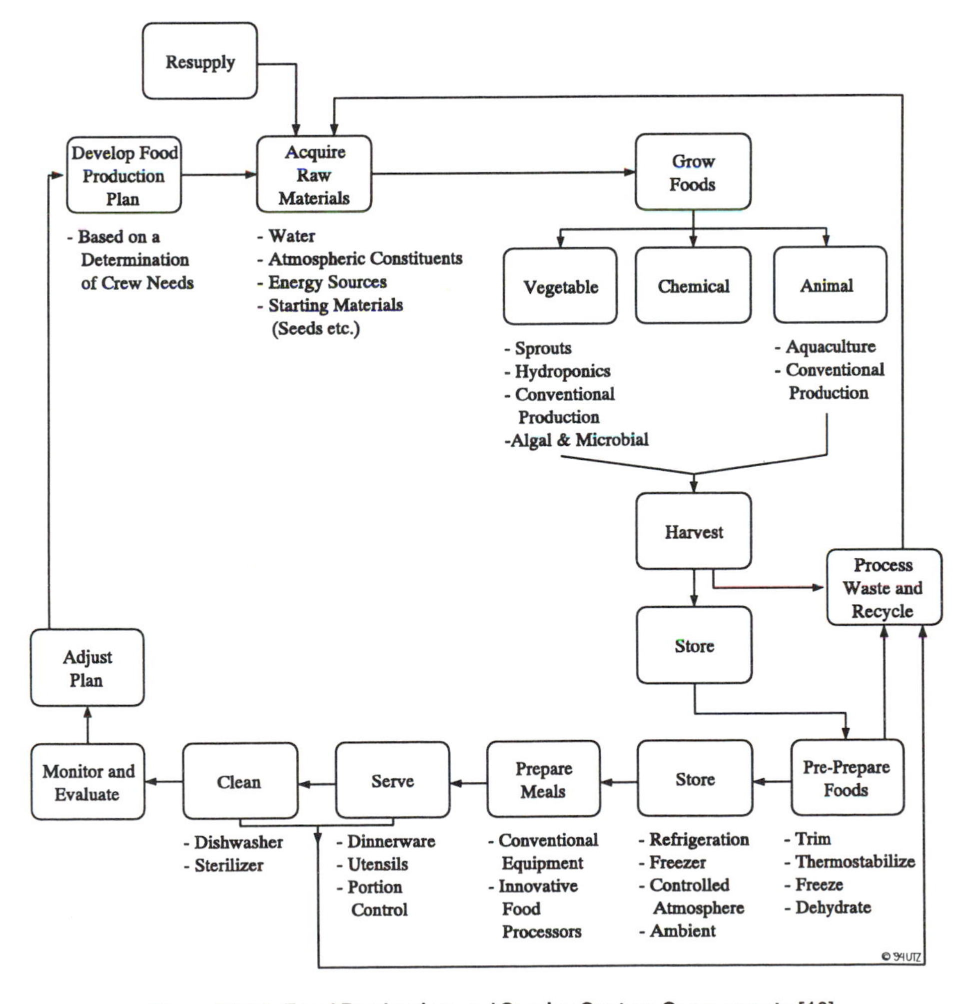

Figure VI.30: Food Production and Service System Components [18]

To provide the nutritional requirements, a combination of conventionaly, i.e., terrestrially produced, and unconventional foods, i.e., produced and recycled in a CELSS, may be selected. Based on the experience gained to date in the space programs of both the U.S. and the former Soviet Union, modifications of conventional, freeze-dried, thermally stabilized, and frozen convenience foods could provide the basis for a food service system.

The food preparation system starts with the development of a production plan, acquisition of the appropriate raw materials, either from recycling or from resupply, growing of food, harvesting, storage, pre-preparation, service, cleaning, evaluation and monitoring of the production system. This is shown in figure VI.30. [18, 66]

VI.8.1 Food Production

Food production in space will be experimental at first and will expand slowly until it reaches a significant portion in very long-term missions in the future. It will result in less resupply and fewer waste storage problems because part of the waste products will be reused. Apart from the production of conventional foods, another possible approach may be the production of engineered foods. These foods may be analogs of conventional foods, e.g., cheeses made of soy instead of cows milk, or they may be entirely new foods compounded from available ingredients and made conform to the requirements with respect to nutrition, acceptability, and stability. Here, a summary is given on the characteristics of the several potential food products and production methods in bioregenerative life support systems that have already been described in more detail in the previous chapters.

- *Agricultural Products*
  The breakeven time for growth of different foods will vary according to mass, area, growth rate, edible portion and several other factors. An agriculture could be large and complex, or simple and experimental, depending on the needs of the CELSS. It is unlikely that soil-based production of vegetables would be feasible in a CELSS, although at some future date extraterrestrial, soil-based production of grains and vegetables may be feasible. Because grains and vegetables lack essential amino acids, they would have to be supplemented by specific amino acids to balance the protein content of the food product. Anyway, there is considerable experience with the terrestrial, hydroponic growth of lettuce, tomatoes, and cucumbers.

- *Microbiological Products*
  A number of microbiological systems show promise for future food production. Photosynthetic systems such as algae may be useful in atmospheric regeneration in near-term applications and will become more attractive for CELSS if useful nutrients can be obtained from the biomass produced. At present, the quality and acceptability of foods derived directly from algae pose problems for human ingestion. But there is the possibility of using algal products as nutrients for plants or animals in a more complicated food chain, especially in aquaculture (see sections VI.2 and 3).

- *Chemical Synthesis Products*
  Another system which could be used for food production in a CELSS is the direct chemical synthesis of some nutrients, e.g., carbohydrates. Such a system, which may become feasible in the near future, might also be used to provide inputs to food production, including production of fertilizers or feeds for use in agricultural systems.

- *Animal Products*
  Animals should be the last element of the food chain to be considered for production in a CELSS. The energy required per calorie of food produced is greater than that for grain or vegetable food sources. Furthermore, the volume, environmental control, and waste disposal requirements are greater. Fish, shrimp, and small marine animals provide the most efficient biological recycling mechanism for the production of high-quality protein from aquatic plants, algal or bacterial products, or agricultural and human waste. In aquaculture, the mass of the system is the limiting factor (see sections VI.6 and 7).

- *Waste Recycling Products*
  The goal of recycling processes that are as complete as possible would require the investigation and development of simple food chains which could use bacteria, algae and solar energy to produce food from waste materials. The food production cycle would have to include harvesting, storage, preparation, service, and clean-up. [18, 22]

VI.8.2 Food Storage

The requirements of cold storage of food and other items can be met by several methods, including direct coolant loops to a space radiator, vapor-compression, thermoacoustic, and thermoelectric devices. On Skylab, the refrigerators and other areas where cooling was required were part of a coolant loop (Coolanol 15) to a space radiator. For the Space Shuttle and Spacelab, versions of the commonly used vapor-compression refrigeration units are used. In these units, heat is transferred to the atmosphere which is cooled by the temperature and humidity control system, which then transfers the heat to a liquid coolant loop. The major distinction between this cooling method and the refrigeration units under development for the Space Station is the method of transporting heat from the compressor. Onboard the Space Station the heat will be transferred directly to a coolant loop. For the low temperature freezers onboard the Space Station a Stirling cycle heat pump with helium as a working fluid is planned. This approach has fewer moving parts, low vibration, uses non-toxic working fluids such as helium, and can cool to cryogenic temperatures due to the use of helium. A major problem is that the lifetime of the Stirling cycle cryocoolers that have been developed, yet, is only 500 to 2500 hours.

In addition to mechanical heat pumps, thermoelectric heat pumps may be suitable for some spacecraft applications. These devices are thermocouples commonly made of bismuth telluride and use the Peltier effect to cool as low as to -50° C. Potential advantages include solid state operation with inherent reliability since there are no moving parts, precision temperature control, localized cooling, compact packaging, and fast response times. Devices of this type can also generate electricity and have been used on spacecraft, including Apollo lunar experiments.

Another method is thermoacoustic refrigeration which uses resonant high amplitude sound in inert gases to pump heat. This approach for pumping heat has been first developed in the early 1980's, and unlike traditional reversible heat pumps, the thermoacoustic approach uses the irreversibilities to produce the proper phasing to achieve refrigeration. The sound generator is the only moving part, resulting in a device which is inherently reliable. Testing of the concept as a cryogenic cooler has been performed onboard the Space Shuttle, and a refrigerator version is being developed. [67]

VI.8.3 Food Processing

Food products are categorized as requiring little or no processing, primary processing and secondary processing or extraction. The first category consists of food that is edible in its natural form, such as fresh fruits and some vegetables. Minor processing might consist of washing, peeling, or cutting, but little support hardware should be required. The second category includes food products that require support hardware such as juice/oil presses, grain mills, cooking/ baking utensils, etc., to make them edible. Such hardware will require adaptation to the stringent limitations of power, mass, and volume in a space environment. The third category consists of biological products which were not edible in raw or primary processed forms, but which contain potentially digestible and nutritious food for human consumption. The importance of this category lies in the need to reduce resupply requirements and increase self-sufficiency. In any case, food processing technologies must be evaluated in terms of total productivity within a CELSS. An overall view of the functional flow in the food processing subsystem is illustrated in figure VI.31. [58]

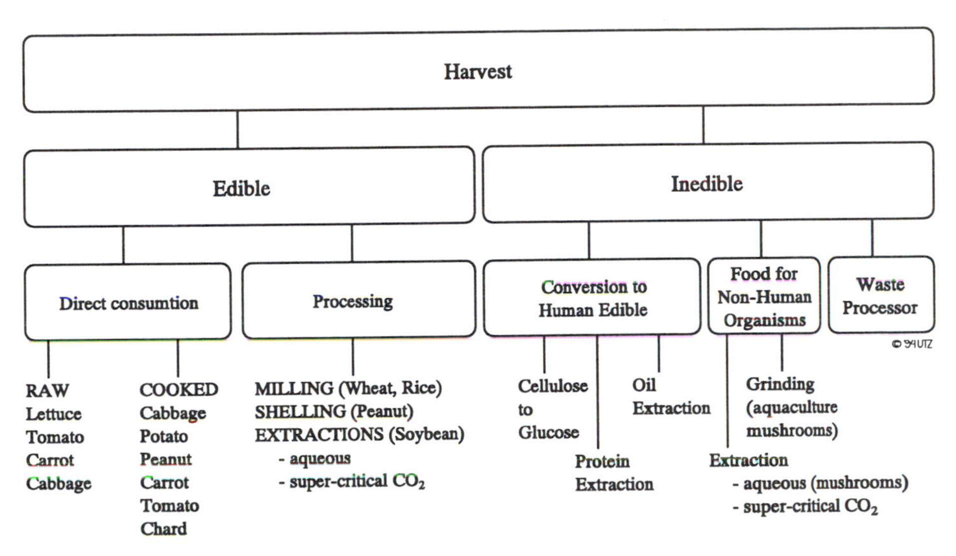

Figure VI.31: **Functional Flow of Materials in Food Processing Subsystem [58]**

It will probably be most important to develop food preparation techniques to provide fresh-cooked food from either resupplied or CELSS-produced products. In this respect a major problem will be the constraints of the space environment that have to be taken into account when developing cooking devices. Especially, the several characteristics of a microgravity environment, as described below, are of importance:

- There is no convection mechanism for heat transfer. Steady flames are only possible if combustion products are removed by artificial means, e.g., a fan. It is not possible to boil liquids by application of localized heat to a vessel.
- Liquids and particulate materials must be kept in closed containers. In particular, closed cups, perhaps with a straw for drinking, are needed for beverages, and dry food such as cookies could not be served on conventional plates. However, experience in Skylab suggests that viscous and adhesive foods such as stews can be served in bowls and eaten with a spoon.
- Many ordinary food preparation operations may require special equipment. It may be difficult or impossible to pour liquids or dry ingredients such as salt or sugar from one container to another or to transfer them with a spoon.

- Ingredients cannot be measured by conventional scales. Mass could be measured by inertial devices, but there may be difficulties because of sloshing or other motion of the sample relative to its container.
- Restraints are required for all personnel at work stations.

Problems with food processing in space already start with the pre-preparation of foods. For example, the behavior of foods stored in microgravity is not well known, and thus, it may be difficult to open and empty a can of peas. Tasks such as cleaning, peeling, trimming, and capturing the waste and rinse water could also present significant challenges. An overview of food preparation requirements is given in figure VI.32. [18]

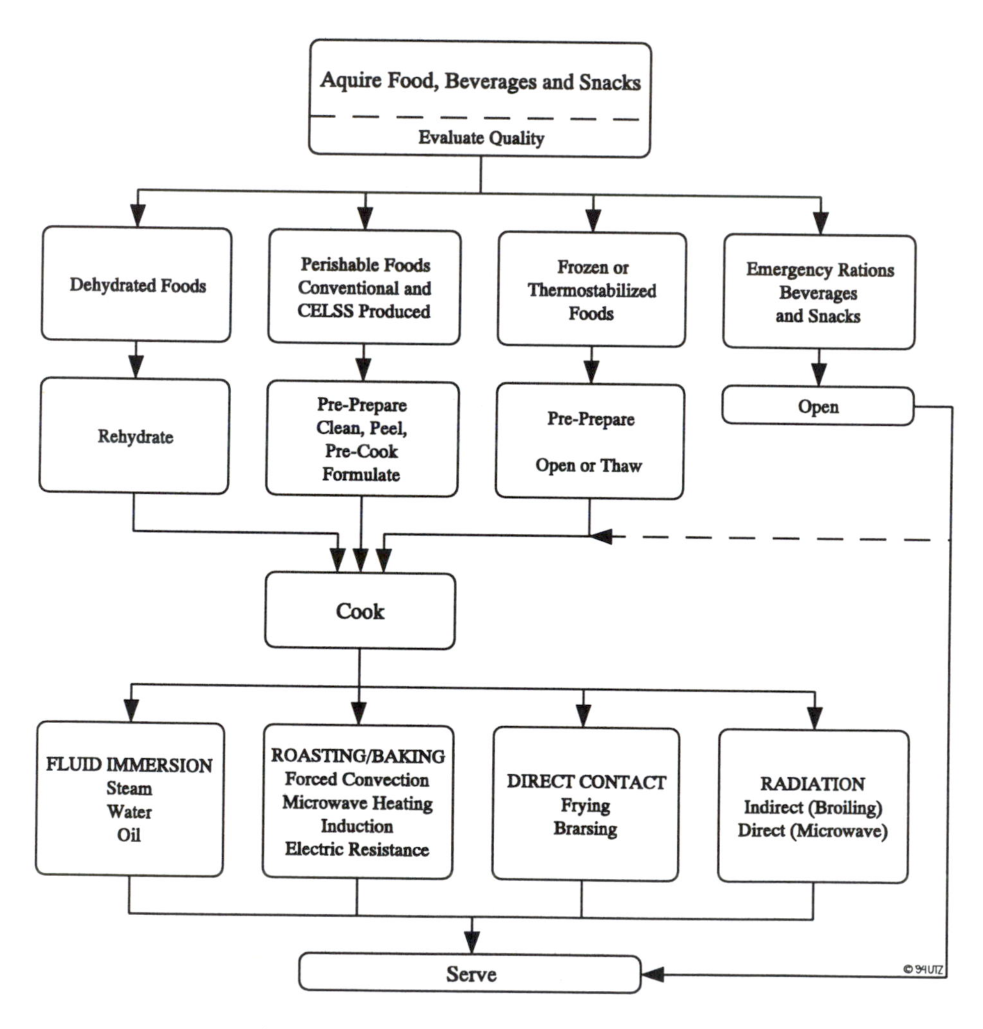

Figure VI.32: **Food Preparation Requirements [18]**

If, finally, a near microgravity environment is found to be unacceptable for food production, an artificial gravity environment could be provided by one or more of the following approaches (see section III.2.3):

- Small centrifuges to create a partial-g field to meet limited food preparation needs.

- Large low-g centrifuges inside a pressure vessel, big enough and rotating slowly enough so that personnel can work inside them for limited periods.

- Rotating sections of the habitat, either connected to the near zero-g areas by rotating pressure seals or attached by cables to the main habitat.

If any of these systems is required, the additional mass and cost would be charged to the CELSS, which would reduce its economic viability. The reasons to cook foods, i.e., expose them to the effects of heat, are to make them more attractive, more palatable, safer to eat, easier to chew, and in some cases to make their nutrient content more readily available. Heat may be transferred to foods by conduction, convection, and direct radiation or heat exchange via the oven walls. An oven is used to bake, roast, or broil meats, and back selected vegetables, grain and fruit products and combinations. Cooking processes cause both physical and chemical changes in foods. These changes result generally from high temperatures within the foods, which increase molecular activity and chemical interactions. The following parameters have to be considered for oven cooking procedures:

- *Time/Temperature Control*
  Is required to produce successful foods. It should be considered in conjunction with microprocessor and sensor technology.

- *Heat Transfer*
  Temperature gradients create the need for continued operator attention and manual operations such as turning meats to assure proper doneness and reduce sticking. Steam within an oven may serve as heat transfer medium.

- *Hydration*
  If water is included in a forced air system, a controlled-humidity atmosphere can be maintained. When dehydration is undesirable, this type of atmosphere will retard it in addition to accelerating heat transfer. Steam in an enclosed system can also function as a food process control and time saver.

- *Environmental consideration*
  The use of microwave heating in a CELSS would require meeting allowable levels for continuous human exposure to microwaves. The generation of magnetic fields may also result in, as yet, undefined environmental effects.

- *Energy Conservation*
  Although the potential for reducing the energy use for cooking may be small, the development of special cooking methods to reduce energy consumption is an important issue.

Several technologies developed for terrestrial cooking processes could be modified for use in a microgravity environment. The following concepts for cooking appliances indicate the approaches which could be considered to meet cooking process requirements:

- *Fluid Immersion*
  Figure VI.33 shows a concept based on the principle of a pressure cooker or fryer which could reach a broad range of cooking or rethermalization requirements by immersing the food product in water, steam, or oil to obtain the required heat transfer. A retaining basket positions the food product and assures immersion under reduced g conditions. The heat source could be electrical resistance or induction. An air lock would isolate the cooking process from the interior environment. A vent could be part of a heat recovery system to use the energy input to the cooking module efficiently. The cooker could be supplied with a safety valve to control the allowable pressures. Sensors and controls would regulate the time and temperatures required for processing specific food products.

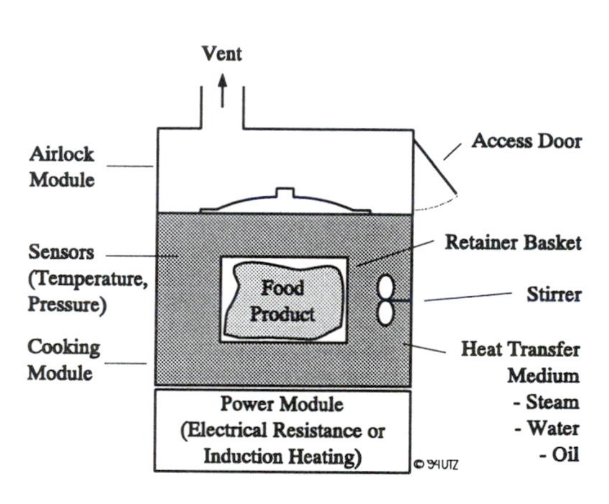

Figure VI.33: **Fluid Immersion Cooker [18]**

- *Roasting and Baking*
  In a reduced g environment, forced convection would be required for baking and roasting in an oven cavity. Figure VI.34 shows a combination microwave/forced-convection oven which could be designed to meet cooking process requirements beyond immersion heating, including browning, and duplicate the cooking processes used for a variety of food products heated

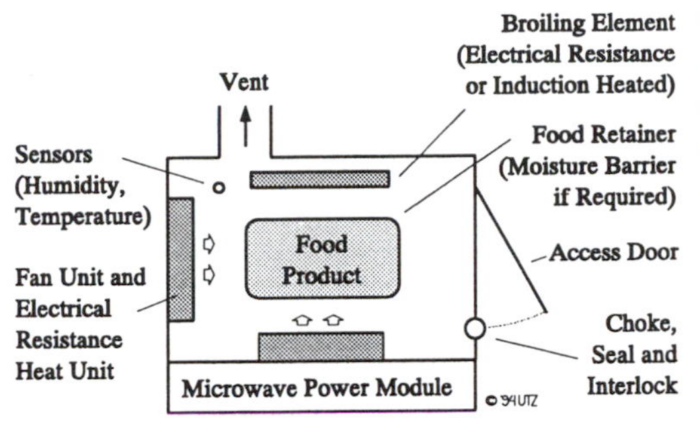

Figure VI.34: **Oven Cavity for Microwave/ Forced-Convection Heating and Broiling [18]**

in similar ovens on Earth. A moisture barrier could be included to heat food products with a high moisture content as option to immersion cooking. Broiling could be accomplished by positioning the food product in a fixture at the desired distance from a broiling element. The element would be heated either by electrical resistance or induction.

- *Direct Contact and/or Radiant Heating*

Grilling, frying and similar stovetop processes require heat to be transferred to the food product by direct conduction and /or radiant heating. Typically, the food is heated in a pan or on a hot surface with gravity maintaining adequate contact between the heated surface and the food. In a reduced g environment this would not be possible. Figure VI.35 shows the concept of a rotating grill which could create a near-one-g field to confine the food product and provide for conductive heat transfer. The contact surface could be electric-resistance or induction heated and a radiating element would be provided at the center. Induction heating of the contact surface would simplify the appliance, as rotating electrical brushes for power transfer would not be required. The cooking surface would be enclosed and access doors would be provided to perform the required cooking operations.

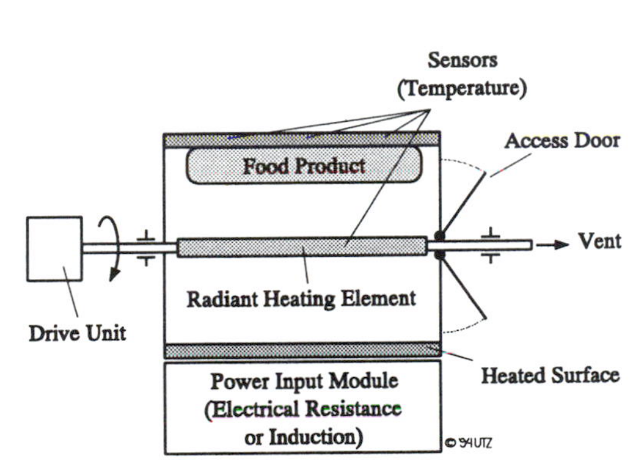

Figure VI.35: **Direct Contact and/or Radiant Heating Surface Cooking Processor [18]**

Menu planning and preparation will largely depend on crew size and mission duration. Dinnerware, utensils, and meal service equipment should be fabricated from easily cleanable and durable materials and should be of such construction as to minimize cleaning requirements. The design philosophy which may be appropriate for this equipment could be adapted from dining equipment developed for handicapped persons. Meal service sanitation, food storage, food preparation, and dining areas must be designed to be easily cleanable. The use of detergents and sanitizers should be kept to a minimum to improve the potential for recycling of the fluid waste streams. Innovative food preservation or packaging technique such as ice glazing of frozen foods and the coating of refrigerated perishable foods with edible glazes should be considered. Depending on the mission, it may be possible to recycle cellulose-based packaging materials for use either as a direct source of fiber in the diet or as medium for culturing foods.

VI References

[1] Alston J.
Continued Results of the SEEDS Space Experiment
LDEF - 69 Month in Space, NASA CP-3194, p. 1493-1495, 1992

[2] André M. et al
A Simplified Ecosystem Based on Higher Plants: ECOSIMP, a Model of the Carbon Cycle
Acta Astronautica, Vol. 27, p. 189-196, 1992

[3] Averner M.
Controlled Ecological Life Support System
Lunar Base Agriculture: Soils for Plant Growth, ASA, p. 145-153, 1989

[4] Averner M. et al
Problems Associated with the Utilization of Algae in Bioregenerative Life Support Systems
NASA CR-166615, 1984

[5] Berman B.; Jenkins D.
Space Bioscience
Space Science 1967, NASA SP-167, p.41-138, 1968

[6] Binot R.; Paul P.
BAF - An Advanced Ecological Concept for Air Quality Control
19th Intersociety Conference on Environmental Systems, SAE Technical Paper 891535, 1989

[7] Blüm V.; Kreuzberg K.
CEBAS-AQUARACK: Second Generation Hardware and Latest Scientific Results
Acta Astronautica, Vol. 27, p. 197-204, 1992

[8] Blüm V.; Kreuzberg K.
German CELSS Research with Emphasis on the CEBAS Project
Acta Astronautica, Vol. 23, p. 245-252, 1991

[9] Bork U. et al
Defective Embryogenesis of Arabidopsis Induced by Cosmic HZE-Particles
Life Science Research in Space, ESA SP-307, p.571-572, 1990

[10] Bugbee B.
Carbon Use Efficiency in Optimal Environments
19th Intersociety Conference on Environmental Systems, SAE Technical Paper 891572, 1989

[11] Caprara G.
The Complete Encyklopedia of Space Satellites
Portland House, New York, 1986

[12] David K.; Preiß H.
DEBLSS - Deutsche Biologische Lebenserhaltungssystemstudie
Dornier GmbH, Bericht TN-DEBLS-6000 DO/01, 1989

[13] David K. et al
Plant Production as Part of a Controlled Ecological Life Support System
Life Science Research in Space, ESA SP-307, p. 431-434, 1990

[14] Doll S.
Life Support Systems for Manned Space Exploration
Lecture Notes, International Space University, Kitakyushu, 1992

[15] Dutcher F.
Progress in Plant Research in Space
Advances in Space Research (to be published)

[16] El Bassam N.
Requirement for Biomass Production with Higher Plants in Artificial Ecosystems
Workshop on Artificial Ecosystems, DARA, p. 59-65, 1990

[17] Garland J.; Mackowiak C.
Utilization of the Water Soluble Fraction of Wheat Straw as a Plant Nutrient Source
NASA TM-103497, 1990

[18] Glaser P.; Mabel J.
Nutrition and Food Technology for a Controlled Ecological Life Support System
NASA CR-167392, 1981

[19] Grigsby D.
Final Results of Space Exposed Experiment Developed for Students
LDEF - 69 Month in Space, NASA CP-3194, p. 1479-1486, 1992

[20] Heath R. et al
A Generalized Photosynthetic Model for Plant Growth within a Closed Artificial Environ ment
20th Intersociety Conference on Environmental Systems, SAE Technical Paper 901331, 1990

[21] Hoff J. et al
Nutritional and Cultural Aspects of Plant Species Selection for a Controlled Ecological Life Support System
NASA CR-166324, 1982

[22] Karel M.
Problems of Food Technology in Space Habitats
Human Factors of Outer Space Production, AAAS Symposium 50, p. 147-157, 1980

[23] Kilgore M. et al
Microbiological Issues of Space Life Support Systems
Space Life Sciences, International Space University Textbook, Chapter 21, 1992

[24] Kliss M.; MacElroy R.
Salad Machine: A Vegetable Production Unit for Long Duration Space Missions
20th Intersociety Conference on Environmental Systems, SAE Technical Paper 901280, 1990

[25] Kohlmann K.
Biological-Based Systems for Waste Processing
SAE Paper 932251, 1993

[26] Kohlmann K. et al
Enzyme Conversion of Lignocellulosic Plant Materials for Resource Recovery in a Controlled Ecological Life Support System
Laboratory of Renewable Resources Engineering, Purdue University, 1995

[27] Kranz A.
Genetic Risc and Physiological Stress Induced by Heavy Ions
Life Sciences Research in Space, ESA SP-307, p.559-563, 1990

[28] Krikorian A.; Levine H.
Development and Growth in Space
Plant Physiology, Vol. X, Chapter 8, Academic Press, San Diego, 1991

[29] Leiseifer H. et al
Biological Life Support Systems
Environmental and Thermal Contol for Space Vehicles, ESA SP-200, p.289-298, 1983

[30] Lewis M.; Hughes-Fulford M.
Cellular Responses to Microgravity
Space Life Sciences, International Space University Textbook, Chapter 3, 1992

[31] Löser H.
Botanische Nutzlasten für Plattformen und Raumstationen
Zeitschrift für Flugwissenschaft und Weltraumforschung, No. 2, p. 116-121, 1988

[32] Löser H.
Concepts for the Life Support Subsystem in the EURECA Botany Facility
Zeitschrift für Flugwissenschaft und Weltraumforschung, No. 1, p.13-21, 1986

[33] Lork W.
Experiments and Appropriate Facilities for Plant Physiology Research in Space
Acta Astronautica, Vol. 17, No.2, p. 271-275, 1988

[34] Lovelock J.
The Ages of Gaia
Bantam Books, New York, 1990

[35] MacElroy R.; Averner M.
Space Ecosynthesis: An Approach to the Design of Closed Ecosystems for Use in Space
NASA TM-78491, 1978

[36] MacElroy R.
The Controlled Ecological Life Support Systems Research Program
AIAA Space Programs and Technologies Conference, AIAA-90-3730, 1990

[37] Mashinskiy A.; Nechitaylo G.
Birth of Space Plant Growing
Tekhnika-Molodezdhi, No. 4, p. 2-7, (Translation) NASA TM-77244, 1983

[38] Miller A.
Plant Cultivation in Space
Nauka i Tekhnika, No. 2, p. 17-23 (Translation: Joint Research Publication Services 56129), 1972

[39] Ming D., Barta D. et al
Development of Zeoponic Plant Growth Substrates for the Gravitational Biology Facility
Internal Report, NASA-JSC, 1994

[40] Ming D.
Manufactured Soils for Plant Growth at a Lunar Base
Lunar Base Agriculture: Soils for Plant Growth, ASA, p. 93-105, 1989

[41] Mitchell C.
CELSS Program Plan and Summary of the CELSS '93 PI Meeting
NSCORT Seminar Series Booklet Spring 1993, Purdue University, 1993

[42] Morgan P.
A Preliminary Research Plan for Development of a Photosyntetic Link in a Closed Ecological Life Support System
NASA CR-160399, 1979

[43] Morrow R. et al
The ASTROCULTURE™ Flight Experiment: Pressure Control of WCSAR Porous Tube Nutrient Delivery System
SAE Paper 932282, 1993

[44] Nelson M.
Bioregenerative Life Support for Space Habitation and Extended Planetary Missions
Space Life Sciences, International Space University Textbook, Chapter 22, 1992

[45] Nitta K.; Ohya H.
Lunar Base Extension Program and Closed Loop Life Support Systems
Acta Astronautica, Vol. 23, p. 253-262, 1991

[46] Nitta K.
Material Flow Estimation in CELSS
Acta Astronautica, Vol. 27, p. 205-210, 1992

[47] Olson R. et al
CELSS for Advanced Manned Mission
HortScience, Vol. 23(2), p. 275-286, 1988

[48] Oleson M.; Olson R.
Controlled Ecological Life Support Systems (CELSS) Conceptual Design Option Study
NASA CR-177421, 1986

[49] Oser H.
Space Research in Cell and Plant Biology
International Space University, Space Life Sciences Core Curriculum Notes, 1993

[50] Radmer R. et al
Algal Culture Studies Related to a Closed Ecological Life Support System (CELSS)
NASA CR-177322, 1982

[51] Raper C. et al
Use of Phytotrons in Assessing Environmental Requirements for Plants in Space Habitats
Human Factors of Outer Space Production, AAAS Symposium 50, p. 159-167, 1980

[52] Salisbury F.
Bioregenerative Life Support System: Farming on the Moon
Acta Astronautica, Vol. 23, p. 263-270, 1991

[53] Salisbury F.; Bugbee B.
Plant Productivity in Controlled Environments
HortScience, Vol. 23(2), p. 293-299, 1988

[54] Salisbury F.
Some Challenges in Designing a Lunar, Martian, or Microgravity CELSS
Acta Astronautica, Vol. 27, p.211-217, 1992

[55] Saugier B. et al
Modelling Dynamics of Simplified Ecological Systems Based on Higher Plants
Workshop on Artificial Ecosystems, DARA, p. 191-202, 1990

[56] Schwartzkopf S. et al
Conceptual Design of a Closed Loop Nutient Solution Delivery System for CELSS Imple mentation in a Microgravity Environment
19th Intersociety Conference on Environmental Systems, SAE Techn. Paper 891586, 1989

[57] Schwartzkopf S.
Design of a Controlled Ecological Life Support System
BioScience, Vol. 42, No. 7, p. 526-534, 1992

[58] Schwartzkopf S.
Lunar Base Controlled Ecological Life Support System (LCELSS)
Lockheed Missiles & Space Company, LMSC/F369717, 1990

[59] Shepelev Y.
Biological Life Support Systems
Translation of "Biologicheskiye Sistemy Zhizneobespecheniya", Academy of Sciences, Moscow, 1972

[60] Skoog I.
Lebenserhaltungssysteme
Handout of Presentation at the TU München, 1985

[61] Starikovich S.
A Month Alone with Chlorella
Khimiya i zhizn', No. 5, p. 58-63, (Translation) NASA TT F-16463, 1974

[62] Tamponnet C.
Higher Plants Compartment
Internal Note, ESTEC, 1992

[63] Tamponnet C. et al
Man in Space - A European Challenge in Biological Life Support
ESA Bulletin, No. 67, p. 38-49, 1991

[64] Tamponnet C.; Binot R.
Microbial and Higher Plant Biomass Selection for Closed Ecological Systems
Acta Astronautica, Vol. 27, p. 219-230, 1992

[65] Thompson B.
Controlled Ecological Life Support Systems (CELSS) in High Pressure Environments
Acta Astronautica, Vol. 19, No. 5, p. 463-465, 1989

[66] Westgate P.
Bioprocessing in Space
Enzyme Microbiology Technology, Vol. 14, p. 76-79, 1992

[67] Wieland P.
Designing for Human Presence in Space: An Introduction to Environmental Control and Life Support Systems (Draft)
NASA RP-1324, 1994

[68] Wilkins M.
Plant Biology
Life-Sciences Research in Space, ESA SP-1105, p. 37-47, 1989

[69] Wheeler R.; Tibbits T.
Controlled Ecological Life Support System Higher Plant Flight Experiments
NASA CR-177323, 1984

[70] BIORACK IML-2 (Handout)
Principal Investigator Meeting, ESTEC, 1990

[71] Continuous Hydroponic Wheat Production Using a Recirculating System
NASA TM-102784, 1989

[72] Controlled Ecological Life Support System - Use of Higher Plants
NASA CP-2231, 1982

[73] Controlled Ecological Life Support System - Research and Development Guidelines
NASA CP-2232, 1982

[74] Controlled Ecological Life Support System - Biological Problems
NASA CP-2233, 1982

[75] Greenhouse Design for a Martian Colony
University of Idaho

[76] Space Station Needs, Attributes, and Architectural Options Study, Volume 4
Boeing Aerospace Company, D180-27477-4, 1983

[77] Klabbers C.; Paul P.
Project Management of the BAF Project, Stork Comprimo B.V.
Personal Communications, 1995

VII BIOSPHERE 2 - LESSONS LEARNED FOR FUTURE CELSS RESEARCH

VII.1 THE RATIONALE OF BIOSPHERE 2

A biosphere may be defined as "a complex, stable, regenerating, adaptive, evolving system, containing life, composed of various ecosystems, operating in a synergetic equilibrium, essentially closed to material input, but open to energy and information exchange and capable of large scale and comparatively rapid cycles of transport and rearrangement of atoms and molecules. As a negentropic force a biosphere acts as an apparatus which utilizes life to generate and store the energy free to do work". [18]

Biosphere 2 is a materially closed, energetically open 1.28 hectare life support system in Arizona with 16000 m^2 of glass surface. It represents a seven biome, closed system, first approximation model of Earth's biosphere (Biosphere 1). Biosphere 2 started off with an estimated 20 tons of living biomass distributed in about 4000 species. Nominally, eight adult humans with a five-level cybernetic system connected to 1900 sensors are living in Biosphere 2 with (ideally) 100 % recycle of their water, food, waste, and air minus the air's leak rate. Further facts may be found in section VIII.2. [8, 18]

Basically, with a design goal of 100 years of operation, Biosphere 2 is intended to afford the opportunity for significant long-term studies of the maturation and dynamics of its ecosystems and to develop intelligent human management and technological practice. Designed as a private-venture endeavor, the development of environmental technologies and public education through eco-tourism are intended to be revenue generators and co-exist with scientific research at the Biosphere 2 project. In summary and according to its designers, the purpose of the Biosphere 2 is threefold:

1. To elucidate the laws of biospherics. Especially, to measure the biospheric hypothesis: How self-regulating is a biospheric system?

2. To create the corporate capacity to design, build, operate, and consult on the management of artificial biospheres, both on Earth and in space, and, hence, to investigate a broad spectrum of ecological interactions and to apply the knowledge gained to enrich human understanding of the life systems on Earth.

3. To assist in the ecological improvement of the human impacts on Earth's biosphere.

A closed life support system, separate from Earth's, frees the observer from constraints to the observation. Typical constraints present in the huge natural systems of Earth's biosphere are such things as the extremely long time periods necessary for gaseous cycles to close the loops in nature, e.g., all CO_2 in Biosphere 2's atmosphere passes into plant life at a rate 8000 times faster than in Earth's atmosphere. Natural cycles, therefore, become more accessible to investigation. A separate life system also creates a circumstance in which the observers can begin to compare the inside and outside systems and this comparison may form the basis for objective thought. Also, the response, systemic and biomic, of a range of biospheres to a range of conditions can be of great importance both in evaluating changes in Earth's biosphere and in designing planetary settlements. To accomplish this, Biosphere 2 has several specific experiments underway, to test and measure the response and adaptation of the total system and each of its biomes to the light and atmospheric conditions. These are further described in section VIII.4.

Also, at this point it may be ingenious for astronautics and biospherics to work out a partnership to make a conceptual design to carry out the mission to understand the planet Earth's biosphere, and to achieve human destinies in our larger home, the solar system. The constructors of Biosphere 2 propose that the bioregenerative technologies having been developed may find application in the initial life support systems for Moon and Mars exploration and settlement. These technologies may evolve from being a backup to physico-chemical life support systems and partially bioregenerative ones. Although Biosphere 2 has utilized soil-based systems, there might also be a place for hydroponic/aeroponic systems for food production in Moon and Mars habitations. Studies suggest that lunar regolith may only be a potential cropping medium with preceding biological treatment. Also, to supplement the plant nutrients already present in Martian soil may require amendments with organic material and microbial inoculations, e.g., it has been proposed to accomplish this by the composting and utilization of waste products from the flight and base crews. [9]

One aspect concerning Biosphere 2 not to be neglected is that the project has been criticized by some members of the scientific community for the overwhelming number of variables involved and the lack of adequate controls. In this respect, it may be noted that technologies developed for Biosphere 2 will probably not be applied in initial, but may be in advanced planetary bases. So while some of this criticism is justified, on the other hand, e.g., restoration ecologists understand the value of the process of ecosystem construction as a means of evaluating hypotheses and gaining new information about ecosystem function. Although such an approach obviously raises more questions than it answers, it may be recognized that new questions and new ways of thinking, here about ecosystems and the biosphere, are critical to the advancement of knowledge. [14]

VII.2 DESCRIPTION OF BIOSPHERE 2

Biosphere 2 is located in Oracle near Tucson, Arizona, at 32° 34' north latitude, 110° 51' west longitude, and 1190 m elevation. The facility is open to energy exchange, principally of solar and electrical energy and the removal of excess heat, and to a continual and substantial information exchange. Biosphere 2 contains seven biomes within its 180000 m^3 volume - rainforest, savannah, desert, ocean, and marsh, intensive agriculture and human habitat, plus an extensive mechanical system to assist in such functions as atmospheric and water circulation and heat exchange. Furthermore, Biosphere 2 includes the following specialized facilities:

- An analytical laboratory with gas chromatographs, mass spectrometer, flame ionization detector, ion chromatograph, and atomic absorption spectrophotometer.
- A medical facility designed to deal with all but extreme medical requirements. It includes stocks of many medicines and antibiotics, facilities for examination, X-ray equipment, equipment and supplies for minor surgery and for treatment of injuries including broken bones.
- A veterinary facility for care and treatment of the domesticated animals.
- Food processing equipment.
- A distributed computer system capable of monitoring sensors and controlling equipment within Biosphere 2. The system also provides a personal workstation for each crew member and is networked together with computers outside of Biosphere 2.
- A repair and maintenance shop equipped with lathe, milling machine, drill press, grinder, welder, and other tools and spare parts for repair and maintenance of the technical systems.
- An exercise room with exercise equipment.
- Video facilities enabling direct face-to-face meetings via video with persons outside, plus access to commercial Television programs.

More than 3000 species, excluding soil microbiota, introduced to Biosphere 2 were selected over a period of 5 years of research and development. A crew of eight people nominally resides in Biosphere 2, fully supported by and essential to the operation and research of the system. The crew's food is grown, and air, water, and waste are recycled, inside. In this chapter the basic features and design parameters of Biosphere 2 are summarized. [9, 18, 21]

Structure and Facilities

The structure of Biosphere 2 is a hermetically sealed container in order to eliminate physical material exchange with the outside (bottom: stainless steel, top: glass). The materials used, and the design parameters employed in Biosphere 2, were intended to provide for a useful lifetime of the apparatus of 100 years. The structure admits on average 40-50 % of ambient sunlight and the laminated glass excludes but 1 % of the normal UV. The "algal turf scrubber" systems, used to remove nutrients from marsh and ocean waters, are powered by artificial high intensity lights.

The Biomes

Biosphere 2 contains seven biomes within its 180000 m^3 volume. The footprint is divided into a habitat and an agricultural zone ("anthropogenic biomes"), and five "wilderness biomes" for bio diversity (rainforest, desert, savannah, marsh, ocean). The wilderness biomes are tropical in nature to provide the highest level possible of diversity. As many as possible of the almost 100 phyla, or divisions of life, are represented to simulate as closely as possible the conditions of life on Earth. Twenty-eight different soils have been created, tailored to specific biomes and special ecosystems within those biomes, such as rainforest foodplain (varzea), desert sand dune and salt playa, i.e., seasonally flooded savannah areas (billabongs). Soil horizons extend to a maximum depth of five meters, composed of varying strata. Biosphere 2 also includes an extensive mechanical system to assist in such functions as atmospheric and water circulation and heat exchange. A summary of the volumes, dimensions and areas of Biosphere 2 is given in tables VII.1 and VII.2. [9, 18]

	Volume [m^3]	**Soil [m^3]**	**Water [m^3]**	**Air [m^3]**
Intensive Agriculture	38000	2700	60	35200
Habitat	11000	2	1	11000
Rainforest	35000	6000	100	28900
Savannah / Ocean	49000	4000	3400	41600
Desert	22000	4000	400	17600
Lungs (at maximum)	50000	-	-	50000
Total	**205000**	**16700**	**4000**	**184300**

Table VII.1: **Volumes of Biosphere 2 [18]**

	North-South [m]	**East-West** [m]	**Area** [m²]	**Height** [m]
Intensive Agriculture	41	54	2200	24
Habitat	22	74	1000	23
Rainforest	44	44	2000	28
Savannah / Ocean	84	30	2500	27
Desert	37	37	1400	23
West Lung (airtight portion)	-	-	1800	-
South Lung (airtight portion)	-	-	1800	-
Total Airtight Footprint	-	-	**13000**	-
Glass surface	-	-	16000	-
Energy Center	-	-	2800	-
West Lung (weathercover dome)	48	48	2300	15
South Lung (weathercover dome)	48	48	2300	15
Ocean	37	19	700	7.6(max.)
Marsh	28	19	400	2

Table VII.2: **Areas and Dimensions of Biosphere 2 [18]**

Atmosphere and Leakage

The atmosphere in Biosphere 2 is not physically separated on the inside from area to area. It is variable in humidity and temperature. Two large pressure-volume expansion chambers, "lungs", regulate the air as it heats and cools. Each lung is a cylindrical tank, sealed on top by a flexible, impermeable, weighted membrane which rises and falls in response to the changes of air volume. The membrane does not stretch, but only changes shape. The pressure created by the membrane weight is essentially constant throughout the normal range of movement. One lung could serve the function. Two lungs allow for maintenance on one while temporarily using the other. In the system, the atmospheric content is nominally $5.7 \cdot 10^6$ moles of dry air. A one-meter movement of either membrane displaces 1115 m^3 of the system's atmospheric volume. Besides, all atmosphere is purified inside as well as water. The atmospheric components of Biosphere 2 at closure are given in table VI.3. [4, 9, 18]

Atmosphere Component	Pressure [kPa]	Total Mass [kg]
Oxygen	18.13	31800
Nitrogen	67.67	103775
Carbon dioxide	Fluctuates	
Water	Fluctuates	
Argon	0.83	1782

Table VII.3: **Atmospheric Components of Biosphere 2 at Closure [4]**

Temperature ranges (see table VII.4) in the internal ecosystems are maintained through a series of air handler units in the basement of Biosphere 2 which exchange heat with cooled or heated water passed through the facility. The air that the air handlers emit over the ecosystems can be varied depending on day/night or seasonal requirements, and the airflow of up to 4-5 km/h permits wind pollination and facilitates general air circulation within the enclosed structure. Air circulation rates are on the order of 5700 m³ per minute in each biome for summer cooling and about one-quarter as much for winter heating. Relative humidity is generally high, typically in the 60-90 %-range, and not infrequently up to 100 % when temperatures are falling. Condensation occurs within each biome at two different locations: on the cooling coils of the air handlers and on the airtight glazing of Biosphere 2. [5, 11]

Area	High [°C]	Low [°C]
Rainforest	35	13
Savannah	38	13
Desert	43	2
Intensive Agriculture	32	13

Table VII.4: **Temperature Ranges Inside Biosphere 2 [11]**

Average residence time of CO_2 in the facility atmosphere is about four days contrasted with 3-10 years in Earth's atmosphere. Diurnal flux of CO_2 can be as high as 600-800 ppm as opposed to annual variations of only 10-15 ppm in Earth's atmosphere. In order to control the CO_2 level inside Biosphere 2, a physico-chemical precipitator had been designed as a recycling system. It is able to take CO_2 from the incoming airflow, and precipitates calcium carbonate in a two-step process:

$$CO_2 + 2NaOH \rightarrow Na_2CO_3 + H_2O$$

$$Na_2CO_3 + CaO + H_2O \rightarrow CaCO_3 + 2NaOH$$

To return CO_2 into the atmosphere, limestone ($CaCO_3$) can be heated in an oven at 950° C until $CaCO_3$ dissociates, releasing CO_2, and regenerating CaO:

$$CaCO_3 + heat \rightarrow CaO + CO_2$$

This system was scaled such that it can lower CO_2 by about 100 ppm during an operating day. Other strategies which were employed to increase photosynthesis and decrease respiration during the low-light months included lowering night time temperatures, discontinuing composting, minimizing soil disturbance, prolonging active seasons of savannah and desert, pruning areas capable of rapid regrowth and dry-storing the cut biomass to slow decomposition. [10]

The maintenance of good air quality is among the prime challenges posed by the characteristics of a closed ecological system. Analogous problems were revealed with the advent of energy-efficient, tightly sealed buildings which made the "sick building syndrome" the object of considerable research. The outgassing compounds may be classified by source of their origin as "technogenic", i.e., from materials and equipment, "biogenic", i.e., from living plants, animals, soils, or "anthropogenic", i.e., from people, though some gases have multiple sources. Problems with the accumulation of trace gases in spacecraft have sought to reduce the problem by careful selection of materials. The conventional solutions to this problem include filtering methods using charcoal or catalytic oxidation which will require substantial investment of energy and/or expendable parts, such as filters. To solve this problem in Biosphere 2, research was conducted on the use of soil as a medium for the microbial metabolism, and consequent destruction, of trace gases. Concern about potential trace gas buildup in Biosphere 2 also motivated the design of a continuous air analysis system which monitors CO_2, O_2, and nine other trace gases, namely, CO, H_2S, SO_2, NH_3, NO, NO_x, O_3, CH_4, and total non-methane hydrocarbons. For air purification in Biosphere 2 the "soil bed reactor" technology (see section IX.3) has been developed. Soil bed reactors function by forcing air through the microbial communities of actively functioning soil. The entire volume of Biosphere 2 can be pumped through the soil bed reactor in less than a day's operation.

After closure in September 1991 for an initial two year experiment trial, there has been very little material exchange between Biosphere 2 and the environment. The air exchange, i.e., atmospheric leakrate, is estimated to be less than 10 % per year. This is far tighter than any experimental life support ever constructed. Previous Russian and U.S. facilities have had leak rates from 1-10 % per day. Leak rate is determined by two independent methods:

- By observation of the two expansion chambers (lungs)
- By periodic analysis of the decline in concentration of the inert gas SF_6 which was spiked into the Biosphere 2 atmosphere

Detection and sealing of underground leaks through the stainless steal liner is facilitated by a tunnel encircling the foundations from which trace gases can be detected in over 200 zones. [9]

Energetics

Electrical Peak Demands	**[kW]**
Biosphere 2 Airtight Enclosure	1500
Energy Center	1500
Total	**3000**
Generating Capacity	5250
Cooling Peak Demand	**[kJ/h]**
Intensive Agriculture	$11.4 \cdot 10^6$
Wilderness	$24 \cdot 10^6$
Total	**$35.5 \cdot 10^6$**
Heating Peak Demands	**[kJ/h]**
Intensive Agriculture	$3.6 \cdot 10^6$
Wilderness	$7.4 \cdot 10^6$
Total	**$11 \cdot 10^6$**
Solar Energy Entering Glass (Peak)	**[kJ/h]**
Intensive Agriculture	$8.7 \cdot 10^6$
Wilderness	$18.5 \cdot 10^6$
Total	**$27.1 \cdot 10^6$**

Table VII.5: **Energy Demands of Biosphere 2 [18]**

Biosphere 2 is energetically open to sunlight, electricity, and heat. Thus, plant life within can be photosynthetically driven without artificial light. An energy center external to Biosphere 2 generates electricity and provides thermal control of the facility's tropical biomes through energy exchange using an isolated closed loop piping system.

Evaporative water towers outside Biosphere 2 dissipate rejected heat.

Air handlers can circulate air up to 600 m^3/s throughout Biosphere 2. Velocities range from 5 m/s at a few localized discharge ducts to nearly imperceptible in many areas. Sunlight is the main driver of Biosphere 2's photosynthesis. The peak energy requirements of Biosphere 2 are summarized in table VII.5. [10, 18]

Food Production in the Agricultural Biome

Biosphere 2's intensive agriculture biome (IAB) is divided into 18 crop plots, a series of paddies (soil and tank) where fish/rice/azolla are grown year-round, an orchard area, and planting containers in the balcony and basement where incident sunlight and some use of high intensity light permit plant and tree growth. In addition, disease-resistant and pest-resistant crop varieties were selected, and individual cultivars can be rotated so that the same target is not presented (see figure VII.1). Overall, including herbs, some 86 varieties of crops are grown in Biosphere 2.

The agricultural zone was designed to provide all the food for the eight biospherians and their domestic animals. At closure, there was an initial supply of some three month's food previously grown in Biosphere 2. The goal was to supply the crew's nutrition during this initial two-year closure and leave a similar amount for the next crew. Light is a major factor limiting crop production and dictates use of crop varieties which do well with limited light. Despite this fact, there were no accommodations for supplemental growing lighting. Also, no pesticides or chemical fertilizers are used inside. The criterion for a non-polluting agriculture was required because in a small, tightly sealed environment the use of chemicals which might cause toxicity in air or water poses extreme and immediate hazards. Even in the 180000 m^3 volume of Biosphere 2, soil, water, and air buffering capacities are so small that there is no way of introducing pesticides and herbicides without serious health hazards. Thus, no conventional biocides are employed. Integrated pest management techniques in Biosphere 2 include use of resistant cultivars, introduction of beneficial insects, environmental manipulations, manual intervention, and use of non-toxic sprays.

Air-handler units inside the IAB regulate temperature and humidity. There are two main growing seasons. During the winter season from November to April temperatures are kept between 19-27° C. Temperatures during the rest of the year are kept in the 19-30° C range. Relative humidity is kept as low as possible, and generally averages below 45 % to control insects. Wheat and a winter variety of rice are grown during the shorter, winter day-length period. Sorghum, peanuts, and sweet potatoes predominate during the summer months. In addition, Biosphere 2 marks the first inclusion of animals in a closed system. There were three species of domestic animals originally included in the IAB: African Pigmy Goat, Ossabal Feral Swine, and Asian Jungle Fowl/Japanese Silky Hen crossbred chicken. At closure there were 4 does, 1 buck, 1 sow, 1 boar, 20 hens, and 3 roosters.

The domestic animal system was designed so that animals would be fed mainly crop residues that could not be eaten by humans. On a typical day, each goat doe is fed 4 kg of roughage, such as elephant grass. Worm beds insure a supply of worms for the soil as well as feed for the chickens. *Azolla*, a water fern, in the rice paddies is a source of high-protein feed (30 % with its associated bacteria) for the *Tilapia* fish growing there and a source of nutrients for the rice plants. The *Azolla* is also harvested and fed to the chickens and goats. A standard vitamin mix is given to all the animals daily as a dietary supplement. The domestic animals perform three functions besides that of eating crop residues. They produce milk, eggs, and meat, which to date have provided approximately 50 % of fat intake of the human diet. Also, they aid in the recycling of nutrients as well, consuming and thereby starting the recycling process for crop residues such as peanut and sweet potato greens, grain stalks, and banana leaves. Animal wastes go into the composting machine. Urine and animal pen wash-down water is processed along with human wastes and domestic waste water in the marsh recycling system. Finally, they provide companionship for the crew of Biosphere 2. Compact food processing equipment is employed to minimize human labor, e.g., grain thresher, seed cleaner, rice dehuller, hammer mill, and sugarcane press. Electric roto-tillers are used to turn plots. The goal is an agriculture that requires no more than 2-3 hours per person per day. [9, 10]

Water and Waste Management

In Biosphere 2 a complete hydrological cycle is included to cycle water from the salt water ocean, to the fresh water "rain" system, to the streams and marsh, and back to the ocean. Internal water recycling is accomplished by condensation, mainly at cooling coils. In the intensive agriculture, this is in an order of 7500 liters per day. Artificial rain is provided in the wilderness areas. It is calculated that an additional 750000 liters of water will be incorporated in living biomass as Biosphere 2 progresses from a relatively juvenile to a mature system. It is estimated, e.g., that the rainforest at initial closure contains only some 5-10 % of the biomass it will contain when it reaches maturity. So a reservoir of water for that purpose is being held in a tank incorporated into the south lung.

A scheme of the overall freshwater system of Biosphere 2 is shown in figure VII.2. The critical recycling step is condensation from the atmosphere as represented at the top of the diagram. Water vapor enters the atmosphere by evaporation and transpiration throughout all biomes, both from planted areas and from the exposed surfaces of water bodies. Condensate forms on the air handler cooling coils and is collected in trays from which it is pumped to collection tanks. Also, in cold weather, the outside air imparts enough cooling to the glass of the glazing to cause condensate to form on the inside surface. The condensate drains to the lower edge of the glass, is collected in a series of plastic extrudes troughs, and subsequently delivered to condensate tanks. Once collected, the condensate

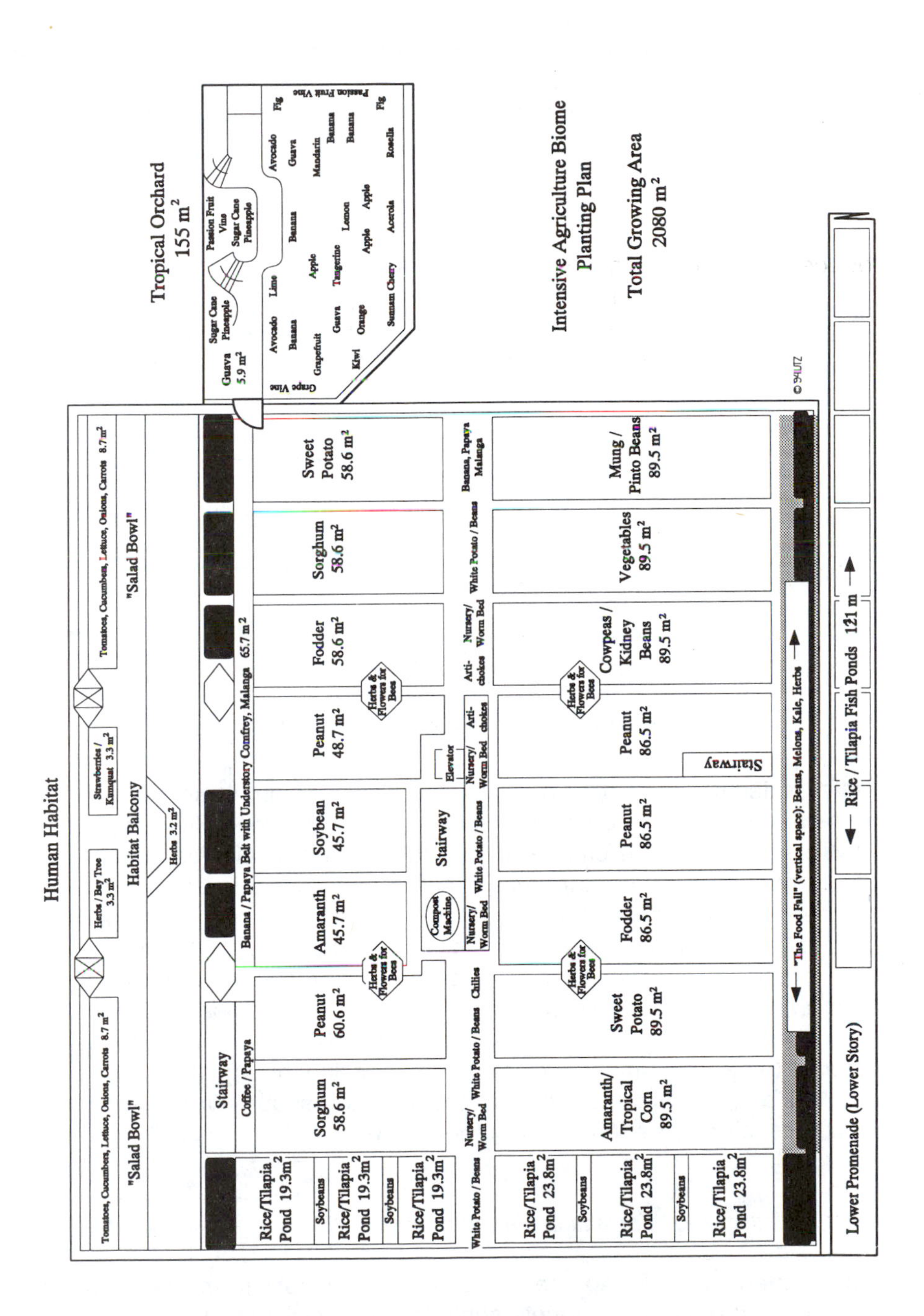

Figure VII.1: **Intensive Agriculture Biome Planting Plan [2]**

water is available for distribution. The dominant use by volume is for rainwater in the wilderness areas and irrigation of the agricultural systems. Another use is a misting system on the mountaintop in the rainforest for creating a fog. Another condensate use of major importance is for potable water. Therefore, condensate is passed through a two-stage filtration and UV-sterilization process to potable water tanks.

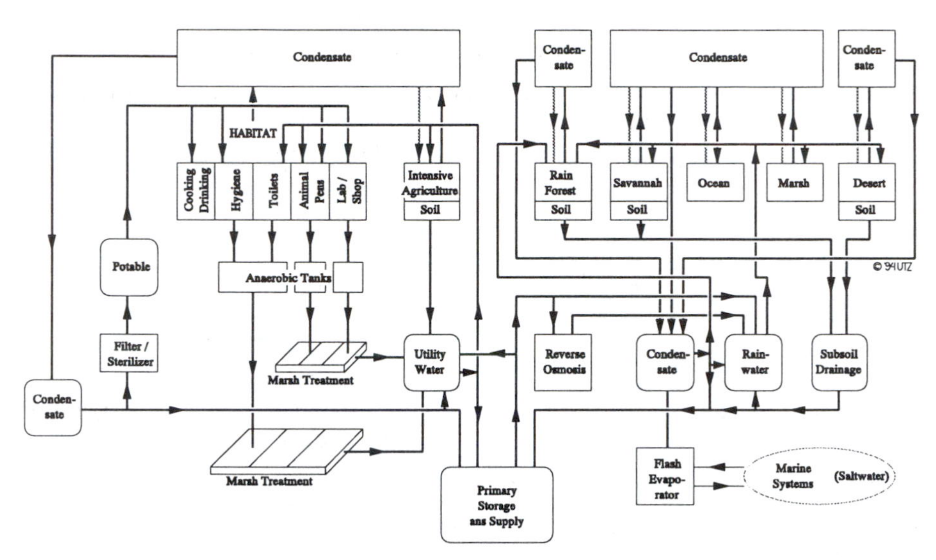

Figure VII.2: **Biosphere 2 Freshwater Systems [5]**

Waste products from both the humans and domestic animals are processed by means of anaerobic and aerobic microbial and plant systems. To maintain long-term soil fertility, inedible crop residues and domestic animal manure are composted and all human wastes from toilets and domestic waste water are processed in a two-step aquatic waste treatment system. Initial decomposition occurs in anaerobic holding tanks. Then batch treatment occurs in aerobic "marsh" lagoons which recirculate the water, exposing it to the aquatic plants, e.g., water hyacinth, canna, aquatic grasses and reeds, and their associated microbes which continue the regeneration process. The aquatic plants are fast growing and are periodically cut for fodder or used in decomposing. After passing through the marsh waste water system, the water is added to the irrigation supply for the agricultural crops, thus utilizing the remaining nutrients. [5, 10, 18]

Data Collection and Information Management

As mentioned earlier, Biosphere 2 is open to all kinds of information exchange. In addition to the natural cybernetic control feedback loops of the life systems in Biosphere 2, a five level "nerve" system using computers and over 2000 data collection points with sensors of various types is employed to monitor and control Biosphere 2. Furthermore, to study air, water, soil and organic samples, an analytic laboratory equipped with GC/MS, ion chromatograph and atomic absorption spectrometry is available to operate in a virtually non-polluting fashion inside. Airlocks facilitate the export of small amounts of soil, water, plant material and air samples as required for outside analysis and exchange with a minimum exchange of air.

The Biosphere 2 Nerve System is a state-of-the-art network, designed to have five levels as indicated in figure VII.3. The data flow and monitoring is indicated in figure VII.4. The first level is the raw data collection stage. Over 2000 sensors, each taking about 360 readings an hour, placed in strategic locations throughout the Biosphere, take readings of, e.g., air humidity, light levels, water flows, pH, carbon dioxide concentration, soil moisture, water salinity, and water, air, and soil temperature. Also, various sensors monitor the status and operation of equipment. The sensors then send electrical signals to the second level of the system, where they are converted. This level of the system also makes changes to some of the automatic control systems such as air pumps or the tide machine in the ocean.

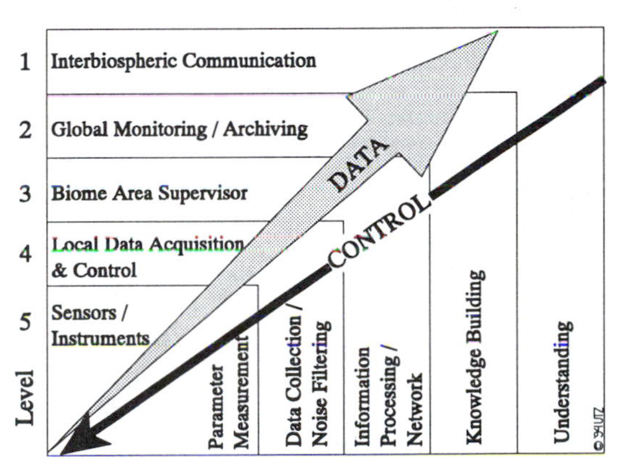

Figure VII.3: **The Five Levels of the Biosphere 2 Nerve System [20]**

If the sensors in the desert area, for example, detect a dangerous drop in temperature, the computers responsible for this area automatically pump in warm air. Without this kind of automatic care system, it would be very difficult to maintain the Biosphere. The data is further processed in level 3. Here, the data from every sensor is gathered for fifteen minutes. Then the highest value, the lowest value, and the average of the values are determined. This information, then, goes to the Mission Control of Biosphere 2 where technicians monitor the flow of data, and the historical data base. At level 3, each biome has its own computer, known as biome supervisor.

At level 4, the monitoring system will trigger an alarm to alert technicians to the problem, if measurement is not within the safe zone. The computers can automatically send a radio alarm to the people best able to deal with it. This way, the confusion of getting everyone excited over a problem that can be dealt with by one person, is eliminated. The fifth level of the Nerve System distributes information to other centers that are involved in similar research by fax, modem, or over computer networks. [9, 18, 20, 21]

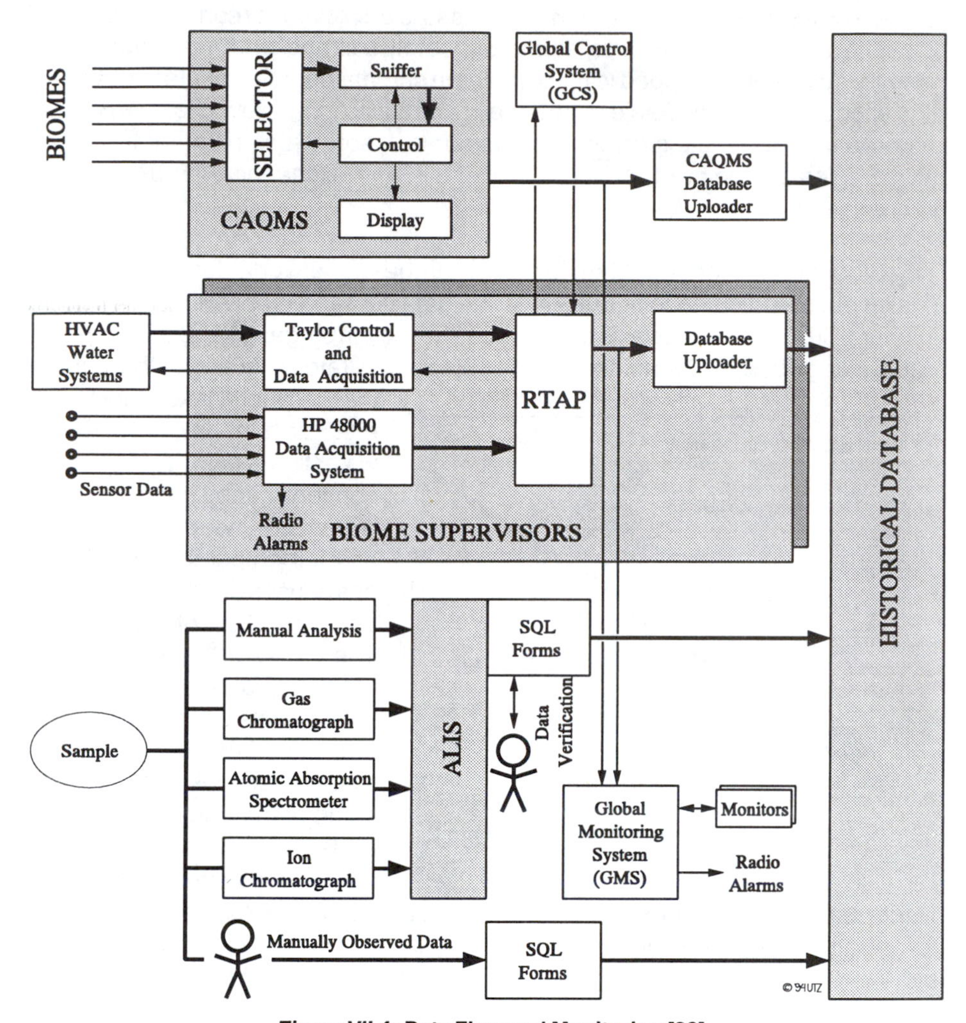

Figure VII.4: **Data Flow and Monitoring [20]**

VII.3 THE BIOSPHERE 2 TEST MODULE

The Biosphere 2 Test Module was the scientific-engineering prototype for Biosphere 2 (see figure VII.5). Before building a large scale apparatus, it was necessary to build a small-scale unit to test, measure, and learn how to operate the different variables. Like Biosphere 2, the test module admits ambient sunlight through a laminated glass and steel spaceframe superstructure (an average of 65 % of ambient Photosynthetically Active Radiation), and is sealed underground with a stainless steel liner. The size of the test module was determined primarily by the need to give life support to one human being. In addition, the test module was scaled to test not only ecological systems, but also experiments were conducted with the 480 m³ Biosphere 2 Test Module to test sealing technologies, as well as many other component technologies developed for Biosphere 2.

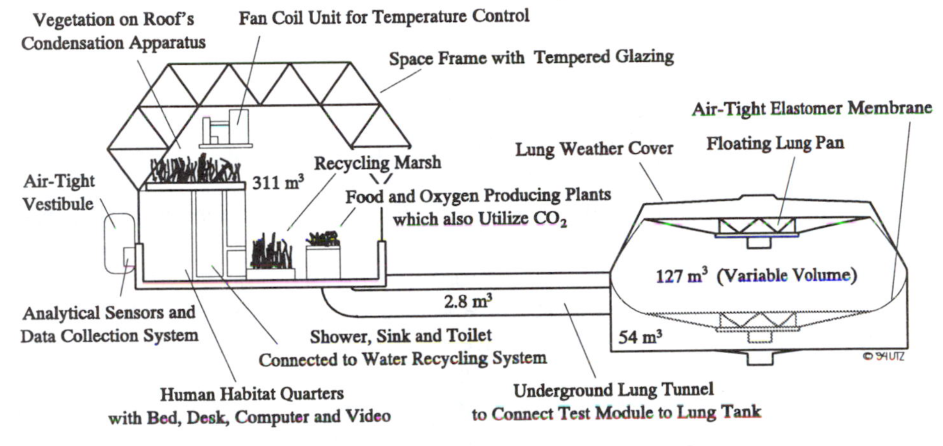

Figure VII.5: **Biosphere 2 Test Module [3]**

The human living quarter of the test module is comparable to a small apartment plus a workstation. Within an area of about 9 m² the habitat includes a kitchen, a bed which may be fold up, water-conserving toilet and shower, telecommunication systems, basic human physiological monitoring apparatus, and a workstation. A floor plan of the test module is given in figure VII.6.

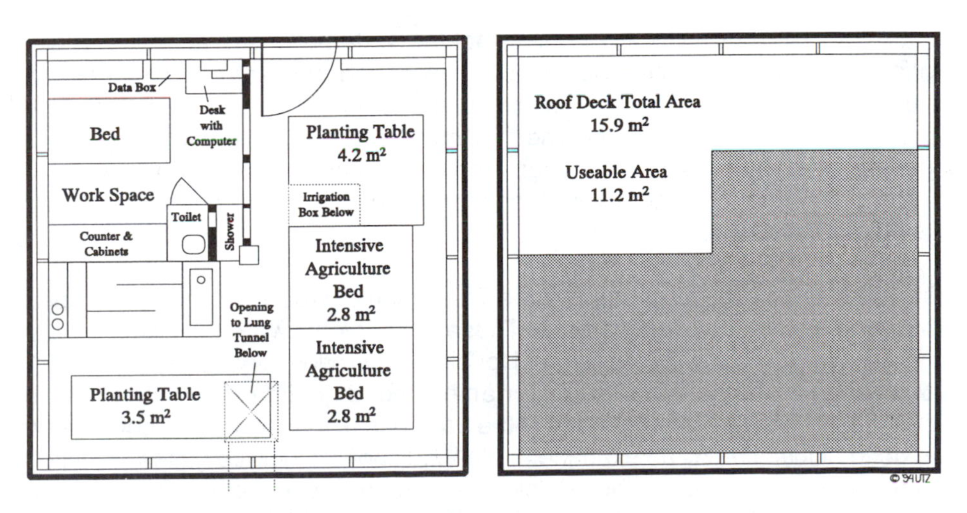

Figure VII.6: **Biosphere 2 Test Module Floor Plan (Excluding Lung) [3]**

In January 1987, the first unmanned experiments began in the test module. The series of experiments, which ran for a total of four months, marked the first time that soil-based systems were used in a closed ecological system. The questions investigated during these closures included:

- *Do plant species reproduce in a high humidity environment?*
- *Do plants and soil microbial filters manage to remove trace gases from the atmosphere?*
- *What effect do the reduced light, and especially UV, levels have?*
- *Are all of the functions of microbes within the soil ecology present?*

For the test module, plant species were chosen with a high growth rate, high photosynthetic rates, and selected at a young growth phase to maximize the amount of CO_2 which is used. Among the selected plants are savannah grasses, rainforest shrubs, small trees, desert cacti, and a sampling of agricultural crops. The soils are composed with low organic carbon and a high nutrient mixture of pumice, natural soil, and bat guano, to decrease the amount of soil respiration. The water recycling system consists of three subsystems for potable water, waste water from the habitat, and plant irrigation water. The sewage, kitchen, and domestic water is purified by the action of microbes and plants and then used to irrigate plants. The 2.6 m^2 system is designed to effectively and without malodor clean 19-57 liters of effluent per day. Potable water is distilled from the atmosphere by two dehumidifiers and sterilized with UV sterilizer systems, for both kitchen use and the 0.9 l/min shower.

After the first re-opening it was clear that the system worked: no massive die-offs had occurred, no microbe, insect, or plant had run destructively rampant, no green slime climbed the glass walls, the major functional microbes were present and in good quantity. The test module achieved air exchange rates (leakage) as low as 24 % per year. Also, the series of experiments provided information for doing detailed modeling of the basic parameters of closed ecological systems.

In 1988 and 1989, there were three test module experiments that included humans, lasting 3, 5, and 21 days. These experiments were the first to seal a human being into a system designed for bioregeneration of water, wastes, food, and air driven by natural sunlight for an extended period of time. In all of the closures involving a human, none of the trace gases reached levels considered toxic to human life. As an example, table VII.6 shows the trace gases identified in the test module during the first closure. During the third closure involving a human, an analytical system to continuously monitor CH_4, total non-methane hydrocarbons, NO_X, O_3, NO, CO_2, O_2, H_2S, SO_2, NO_2, and NH_3 was used. As an

Compound	Number of Isomers Found	Probable Origin
A: Identified by Gas Chromatograph, Mass Spectrometer		
Alkyl Substituted Cyclopentane	1	c
2-Butanone	1	c
Carbon Disulfide	1	b
Cyclohexane	1	c
Decahydonaphthalene (Decalin)	1	a
Decamethylcyclopentasiloxane	1	a
Decane	1	c
Dimethylbenzene	2	a
Dimethylcyclohexane	3	c
Dimethylcyclopentane	4	b
Dimethylhexane	2	c
Dimethyloctadienol Acetate	2	b
Dimethyloctane	2	c
Dimethyloctarine	1	b
Dimethylpentane	1	b
Ethylmethylcyclopentane	1	c
Ethylbenzene	1	c
Ethylcyclohexane	1	c
Heptane	1	c
Hexamethylcyclotrisiloxane	1	a
Hexane	1	c
Isopropyl Substituted Cyclopentane	1	b
Methyl (Methylethenyl) Cyclohexane	1	b
Methylbenzene	1	a
Methylbicyclohexene	1	b
Methylcyclohexane	1	c
Methylcyclohexene	1	c
Methylcyclopentane	1	c
Methylheptane	1	a
Methylhexane	2	c
Octamethylcyclotetrasiloxane	1	a
Substituted Cyclohexane	3	b
Substituded Cyclohexene	1	b
Tetrachlorethene	1	a
Tetrahydrofuran	1	a
1,1,1 Trichlorethane	1	a
Trichlormethane	1	a
Trimethylbicyclohepene	1	b
Trimethylcyclohexane	2	c
Trimethylcyclopentane	3	b
Trimethylpentane	1	c
Trimethylsilanol	1	a
B: Identified by Gas Chromatograph/Flame Ionizer Detector		
Ethane	1	c
Ethylene	1	c
Methane	1	c
Propane	1	a
C: Monitored with continuous sensors		
Ammonia	n/a	b
Carbon Monoxide	n/a	b
Formaldehyde	n/a	a
Hydrogen Sulfide	n/a	b
Nitrogen Dioxide	n/a	b
Ozone	not detectable	
Sulfur Dioxide	n/a	b

Probable Origin: a = Technogenic; b = Biogenic; c = a + b

Table VII.6: **Trace Organic Gases in the Test Module First Closure Involving a Human [3]**

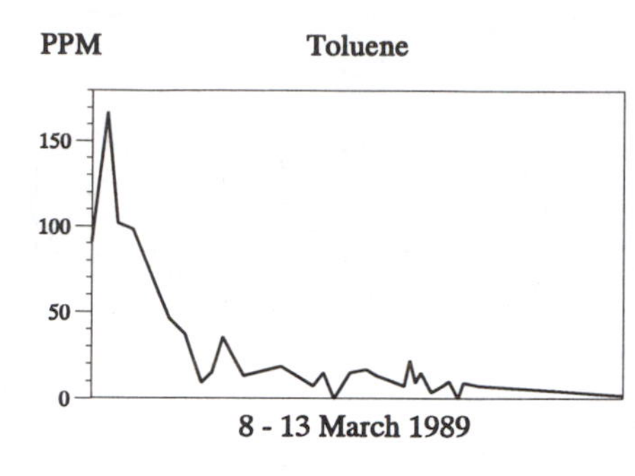

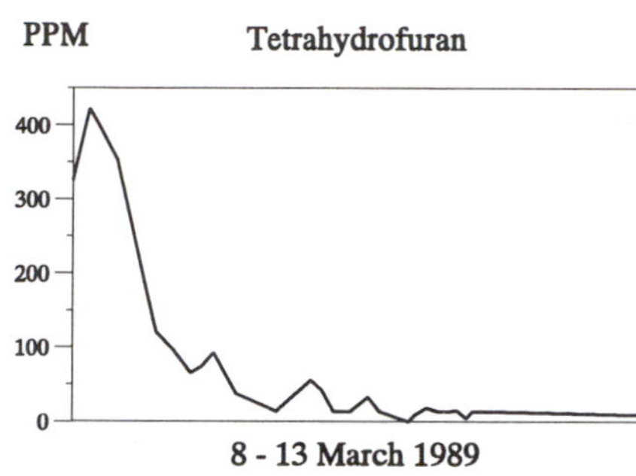

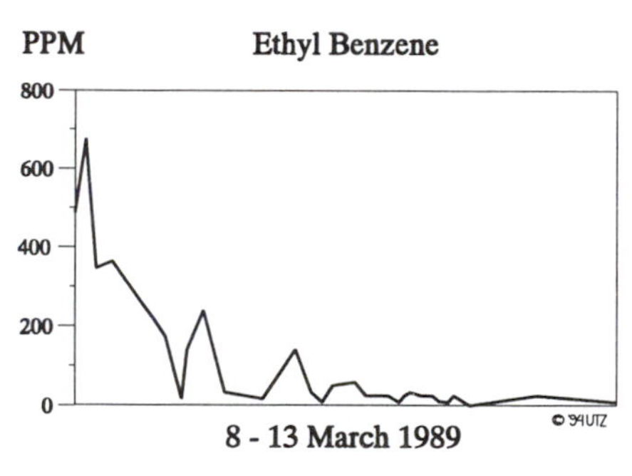

Figure VII.7: **Technogenic Gases in the Test Module Air in February / March 1989 [3]**

example, the data from this experiment for NO_x, O_3, SO_2, and CH_4 is illustrated in figure VII.8. For every compound the values stayed well below critical values. NO_x concentrations ranged from 0.15 to about 3 ppm (cautionary levels are in the order of 30 ppm), O_3 highs were at about 0.021 ppm (0.1 ppm), and SO_2 highs at 0.005 ppm (2 ppm). CH_4 levels were rising after human enclosure to about 150 ppm, which is still far below critical levels. The data for the post human enclosure phase suggested the hypothesis that it takes some time before methane-metabolizing microbes build up their populations to bring down atmospheric concentrations, then forming a classical negative feedback loop.

As another example, typical data for the behavior of technogenic gases, obtained before, during, and after the second human closure experiment (March 8-13, 1989), is shown in figure VII.7. All the graphs show an initial rise after closure and are then quickly brought down to extremely low levels by the action of the soil bed reactor and other biological metabolizers. Typical origins of the technogenic gases in this example are paints (toluene), glues (tetrahydrofuran), and resins (ethylbenzene). [1, 3, 12, 21]

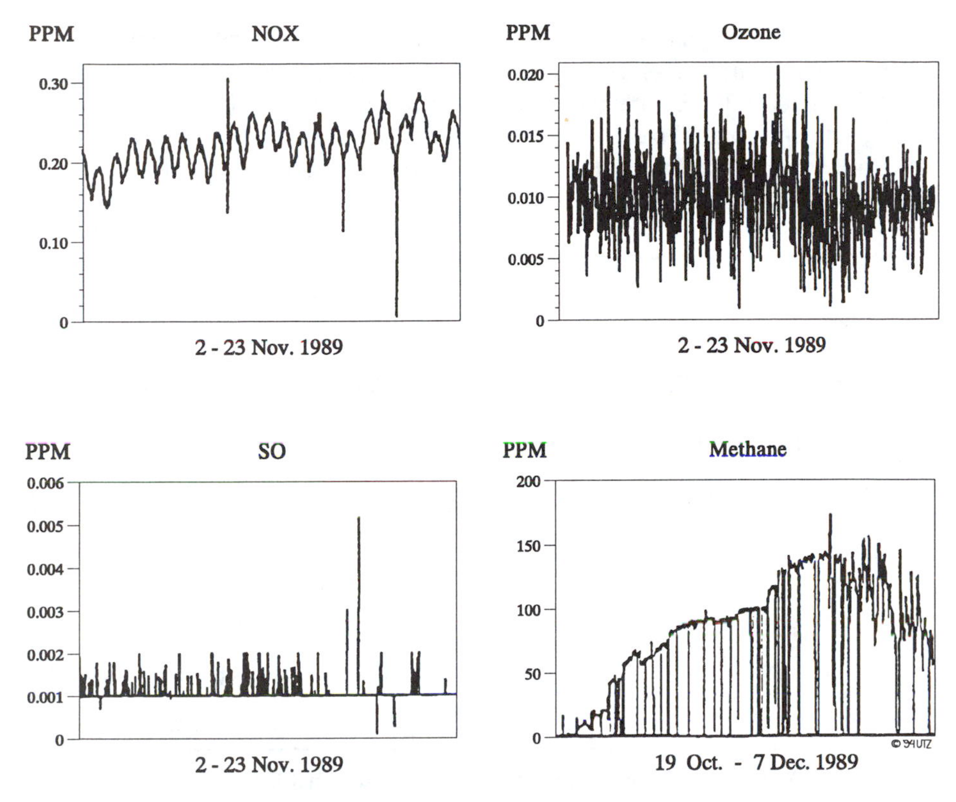

Figure VII.8: **NO_X, O_3, SO_2, CH_4 Levels During Test Module Human Closure (2 - 23 Nov.1989 - CH_4 including unmanned closure pre- and post-human habitation data) [3]**

VII.4 FIELDS OF RESEARCH

Biosphere 2 represents a symbiosis of ecology and technology. It provides a laboratory for studying how ecological systems self-organize, testing the relationship between biodiversity and stability. It also offers a chance to further our capacity to develop technologies that support rather than degrade the environment. In order to understand the research conducted in Biosphere 2 one has to know about the special characteristics of this artificial ecosphere. For example, the concentration of living biomass inside Biosphere 2 is some 100 times greater in relation to the atmosphere than in the global environment.

Soil organic carbon in proportion to the atmosphere is thousands of times greater. This reduces mean residence time of CO_2 in the atmosphere from an average of about three years for Earth's biosphere to about four days in Biosphere 2. The diurnal and seasonal fluxes of CO_2 are also much greater. Some differences between Biosphere 2 and Earth's biosphere are summarized in table VII.7.

Parameter	Earth	Biosphere 2
Ocean Surface Area	71 %	15 %
Maximum Ocean Depth	11 km	7.6 m
Maximum Atmosphere Height	17 km	23 m
Biomass Carbon : Atmospheric Carbon	1 : 1	9 : 1
Residence Time for Atmospheric Carbon	8-10 years	2-4 days
UV Radiation	Some at surface	Glass filters out
Community Size	Relatively large	Relatively small
Species Density	Relatively low	Relatively high

Table VII.7: **Comparison of Biosphere 2 and Planet Earth [21]**

Potential problems such as pollution from agrochemicals are tried to be eliminated in Biosphere 2 and others such as trace gas emissions are tried to be minimized by selection of natural or less toxic materials whenever possible. The quantity of paper and other potential solid wastes is restricted. Fire and open flames are also forbidden by mission rules to minimize air pollution problems. The humans are restricted from expanding into the wilderness biomes. Apart from an occasional bowl of fruit or some specialty items like coffee beans from rainforest trees, human food will come from agricultural crops. Should a fish cull from the ocean go to the kitchen or a grass cutting from the savannah go to domestic animals, a restoration of the biomass will be required by returning nutrients to the wilderness in appropriate fashion. The crew will regularly make soil and water tests in the biomes to determine where to distribute nutrients removed from marsh and ocean water by the algal scrubber systems.

Overall, the requirements of Biosphere 2's intensive agricultural biome are to remain indefinitely sustainable and highly productive (complete nutrition for eight people year round) without the use of chemical products. This requires an agricultural system that maintains soil fertility by returning nutrients to the soil and one which can cope with a wide variety of potential pests and diseases through methods such as beneficial insects, crop rotation, intercropping, etc.

Since the internal climate of Biosphere 2 is semi-tropical, such a system may be of greatest applicability to developing countries in the tropics, which at present are food-impoverished. Beyond this, and according to the constructors, the research program of Biosphere 2 reflects two principal goals:

- The study of ecosystems and biospheric interaction within materially closed systems as a tool for understanding basic global ecological processes.
- The advancement of the ability to create materially closed life support systems for space habitation.

Because of the given conditions this leads to the formulation of several experiments that are listed below:

1. Measuring and monitoring how the system responds to receiving less than 50% of the ambient radiation.
2. At the restricted light level, the carbon dioxide would be at a higher level than in Biosphere 1. It should be measured whether this level of CO_2 poses any problem for humans, animals, plants, or the coral reef system.
3. Investigate the atmospheric CO_2's influence on the acidity of the ocean.
4. Investigate the agricultural system, which is especially designed as a synthesis of tropical and subtropical food and is operating under predation by a variety of pests with no chemically toxic pesticides available and with 100 % waste recycle. If it is producing a complete diet for eight adults on half an acre this experiment may hold interesting implications for the "Hunger Belt" of planet Earth, as well as for designing small area agricultural systems for space settlements.
5. Monitor temporal changes in genetic diversity among selected Biosphere 2 species, including testing models of the extinction process against actual extinction levels.

Besides the experiments conducted by Space Biospheres Ventures, Biosphere 2 is also a host for research projects of other institutions as indicated in table VII.8. Especially, Biosphere 2 is a laboratory where the biological impact on air parameters can be closely examined. The high ratio of biomass to atmosphere and small reservoirs insure a most rapid biological impact on air composition. During a calendar year, in Biosphere 2 there will be almost a hundred cycles of CO_2 from the atmosphere into living or water systems and back into the air again. Similar studies are planned for other major nutrient cycles, such as nitrogen, phosphorus, etc., along with studies examining specific mechanisms and dynamics. This ability to track cycles which have an atmospheric component is a fundamental difference between studies in natural ecosystems, which are atmospherically open, and those in materially-closed ecological systems.

Response of varying ecosystems and individual species to heightened levels of such gases will be important in predicting possible feedback mechanisms. The crew actively manages any health-threatening buildups of atmospheric components.

Research Project	Investigator
Carbon Modeling	D. Botkin, University of California
Oxygen Dynamics	W. Broecker, Columbia's Lamont-Dougherty Lab
Coral Reef Vitality	P. Dustan, College of Charleston
Systems Modeling	H.T. Odum, University of Florida J. Corliss, NASA Goddard Spaceflight Center
Ocean Chemistry	B. Howard, Cornell University

Table VII.8: **Research Projects at Biosphere 2 [21]**

Furthermore, Biosphere 2 affords a laboratory for studying the interrelationships of biomes, and how their interplay affects overall system balances. In addition, the influence of eight human beings, including their technology and agriculture, can be studied in their impacts on individual ecosystems and overall system dynamics. The link between the health of the crew and the health of their enclosed environment will be much more immediate and quantifiable in Biosphere 2 than in the global environment. For example, studies are underway of the crew's blood chemistry and toxicology using pre-closure and post-closure samples. Since there is a careful monitoring of water and air quality the crew is exposed to inside Biosphere 2, it will be possible to see the responses of their blood systems on chemicals they have been previously exposed to as well as trace compounds presently found in the facility. Also, Biosphere 2 may yield interesting data on the importance of diversity in ecosystem operations. Studies of species loss, expansion and phenology will be relevant to determinations of the role of ecological self-organization in shaping what the Biosphere 2 system will become over time. Biosphere 2 offers a unique opportunity to observe to what extent "island" effects lead to ecosystems and species simplification and loss in terrestrial and aquatic ecosystems. [9, 18, 21]

VII.5 INITIAL RESULTS

This chapter summarizes some results of the Biosphere 2 experiment. During the first closure experiment a crew of eight was inside Biopshere 2 from September 26, 1991 to September 26, 1993. In March 1994 a second crew began a closure experiment, after a five and a half months transition phase, which ended in September 1994. In the future, Biosphere 2 will no longer operate with long-term crews. Using the airlocks, which minimizes air exchange, scientists and engineers will be able to enter the facility for shorter periods (from days to months) depending on the work that needs to be done. The results of the first years of Biosphere 2 operating, including the resident crews, are summarized below. [22]

Nutrition and Crew Health

In Biosphere 2, the first crew, four men and four women (ages 25 to 67 years), have been sealed for two years, living on food crops grown inside. Average ages of males (42 years) and females (36 years) did not differ significantly. All were in good health and non-smokers. During the study all sustained physical activity judged equivalent to 3-4 h of manual farming daily. 2-4 months before closure all subjects received a complete physical examination, urinalysis, respiratory spirometry, etc. During closure, every 2 weeks one subject of each sex received a symptom review, physical examination, urinalysis, and blood analysis.

The diet of the crew was essentially vegetarian, with intake of six varieties of fruits (banana, fig, guava, lemon, papaya, and kumquat), five cereal grains (oats, rice, sorghum, wheat, and corn), split peas, peanuts, three varieties of beans, 19 vegetables and greens, white and sweet potato, and small quantities of goat milk and yogurt (average, 84 g/day), goat meat, pork, chicken, fish, and eggs (average, 2.5, 6.0, 3.6, 2.0, and 3.0 g/day, respectively). Sufficient amounts of all essential nutrients were supplied by the diet, except for vitamins D (the enclosure's glass transmits only a trace of UV radiation), B_{12} (deficient in the low animal-product diet), and calcium (average, 500 mg/day). The diet contained adequate protein (63 g/day), was high in fiber (52 g/day), and low in cholesterol (36 mg/day) and fat (about 10 % of energy intake). Individuals also received daily vitamin/mineral supplements consisting of about 50 % of the recommended daily allowance or "safe and adequate" amounts of known essential vitamins and minerals.

The diet, low in calories (average Sep.1991 - Sep.1993: 7448 kJ/day), and nutrient-dense, conforms to that which in numerous animal experiments has promoted health, retarding aging, and extended maximum life span. Medical data on the eight subjects showed significant changes, comparing preclosure data with data through 6 months of closure, as shown in table VII.10. These values of the study provided evidence that radical and possibly beneficial changes in risk factors can be produced in normal affluent individuals in Western countries quickly and reproducibly by reduction in cholesterol and blood pressure, when applying a carefully chosen diet. [17]

	Amount	R.D.A *	% of R.D.A
Calories [kJ]	8925	7140 - 10500	125 - 85
Protein (grams)	69	58	119
Fat (grams)	29	28	104

** Recommended Daily Allowance*

Total Agricultural Production [kg]

Vegetables		Grains		Fruit	
Green Beans	8	Rice	196	Apple	1
Beet Greens	273	Sorghum	131	Banana	1024
Beet Roots	308	Wheat	113	Fig	39
Bell Pepper	13			Guava	41
Carrots	88			Kumquat	4
Chili	63	**Starchy Vegetables**		Lemon	10
Cabbage	83	White Potato	198	Lime	4
Cucumber	17	Sweet Potato	1335	Orange	6
Eggplant	155	Malanga	84	Papaya	639
Kale	11	Yam	20		
Lettuce	90				
Onion	107			**Animal Products**	
Bok Choy	12	**High Fat Legume**		Goat Milk	407
Snow Pea	1	Peanut	24	Goat Meat	8
Squash Seed	8	Soy Bean	14	Pork	35
Summer Squash	187			Fish	10
Swiss Chard	58			Eggs	6
Swt. Pot. Greens	64	**Low Fat Legume**		Chicken Meat	8
Tomato	288	Lab Lab Bean	63		
Winter Squash	261	Pea	15	**Grand Total**	**6530**

Table VII.9: **Human Consumption and Agricultural Production in Biosphere 2 [10]**

Parameter	Pre-Closure	6 Months after Closure
Mean Weight	74 kg (men) 62 kg (women)	62 kg (men) 54 kg (women)
Mean Systolic / Dia- stolic Blood Pressure	109/74 mmHg	89/58 mmHg
Total Serum Cholesterol	191 ± 11 mg/dl	123 ± 9 mg/dl
High Density Lipoprotein	62 ± 8	38 ± 5
Triglyceride	139 mg/dl (men) 78 mg/dl (women)	96 mg/dl (men) 114 mg/dl (women)
Fasting Glucose	92 mg/dl	74 mg/dl
Leukocyte Count	$6.7 \cdot 10^9$ cells per liter	$4.7 \cdot 10^9$ cells per liter

Table VII.10: **Changes of Physiological Values of the Biospherians [17]**

Food Production

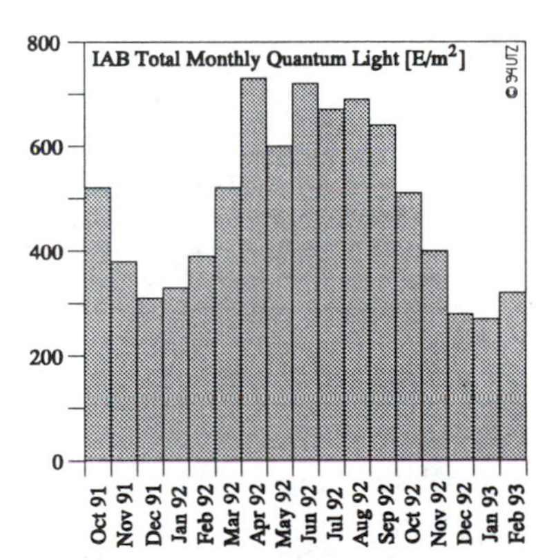

Figure VII.9: **Light Received in the Intensive Agricultural Biome [10]**

Production for the first year is summarized in table VII.9. Environmental conditions, e.g., temperature, moisture, and humidity, were manipulated to minimize conditions pests prefer. In the first autumn, the fall of 1991, three plots of peas were lost due to fusarium fungus infection of the roots. This was probably caused by excessive ground moisture. Over the first 18 months, control of ground moisture improved through experience operating the environmental technologies to adjust temperature and humidity in correlation with changing outside weather conditions. The light received in the agricultural biome is shown in figure VII.9. [10]

Carbon Dioxide Level

CO_2 is of primary concern because of its role as a plant nutrient. At lowered concentrations it is capable of slowing plant growth and at sufficiently elevated concentrations it becomes toxic to both plants and animals. The carbon dioxide concentration in the atmosphere showed both seasonal and diurnal characteristics.

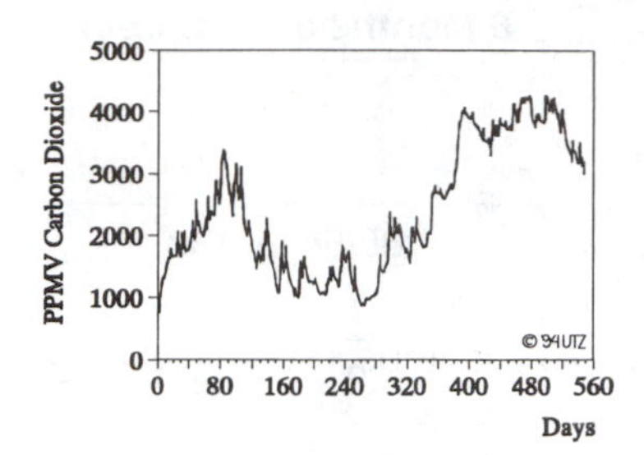

Figure VII.10: **Carbon Dioxide Levels (26 Sept '91 to 30 Mar '93) [10]**

Typical day/night variation has been on the order of 600 ppm CO_2. During the two years of closure, CO_2 levels ranged from a low of about 1000 ppm to over 4000 ppm, the peaks coinciding with unusually dark cloudy periods (see figure VII.10).

There is a strong fluctuation between daylight hours when CO_2 is strongly drawn down, because of intensive photosynthesis, and night hours when respiration is unchecked, resulting in a rapid rise in CO_2 concentrations. Data from the first two years of closure has also shown a strong seasonal variance in CO_2 levels. During the winters when ambient light fell to its lowest level, average CO_2 was in the range of 2500 - 4000 ppm. By contrast, during the summers when days were significantly longer and total light input greatest, carbon dioxide in the Biosphere 2 atmosphere averaged 1000 - 2500 ppm. Outside ambient PPF averaged 16.8 moles/m²/day during December 1991, and 53.7 moles/m²/day during June 1992. On average, 40-50 % of this is received inside Biosphere 2 because of structural shading and glass interception of sunlight. [9, 21]

Oxygen Level

The decline in atmospheric oxygen which had occurred after closure was an unanticipated development. Since closure in September 1991, oxygen had declined from Earth ambient level of 20.94 %. Much of the decline occurred during the first four months after closure. By the end of January 1992, it had reached 18 %. Since the end of April 1992, the decline in oxygen has been fairly linear at a rate of approximately 0.25 % per months, to 16.95 % (June 1992), 16.04 % (September 1992 - corresponds to an altitude of about 3300m), and to slightly lower than 14.5 % (middle of January 1993). On medical recommendation, when this level was reached, pure oxygen was injected into Biosphere 2 over a period of several weeks to increase its atmospheric concentration to 19 %. In June 1993 the oxygen was down to 17.8 % again (see figure VII.11).

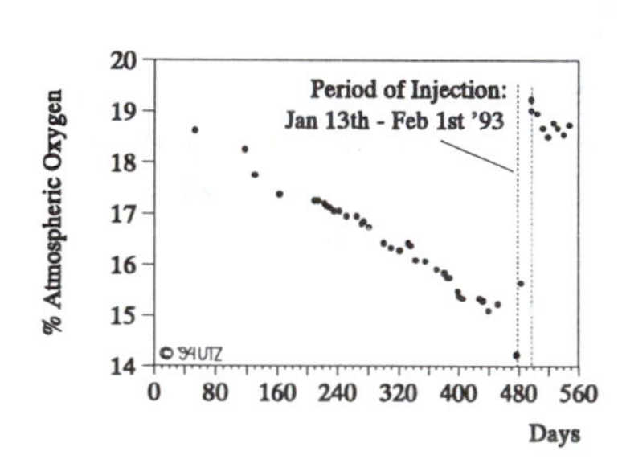

Figure VII.11: **Oxygen Concentration (26 Sep '91 to 30 Mar '93) [10]**

O_2 loss was concurrent with an apparent loss of active carbon from the system probably as CO_2. This was determined by mass balance of isotropic abundance's in the soil and plant tissue, etc. This and the rate of O_2 loss varying inversely with the sun light flux on Biosphere 2, supports the hypothesis of oxygen loss through oxidation of soil organic material and subsequent deposition of the CO_2 as calcium carbonate. About 0.9 % oxygen were attributable to oxygen sequestered in the calcium carbonate produced by the CO_2 precipitator. The conclusions were established by measuring C_{12}/C_{13} isotope ratios in several materials and applying mass balance equations. Several soil reactions may account for oxygen sequestering, including oxidation of reduced iron, sulfur or nitrogen soil components, oxidation of soil organic materials and subsequent formation of $CaCO_3$. [6, 21]

Leakage

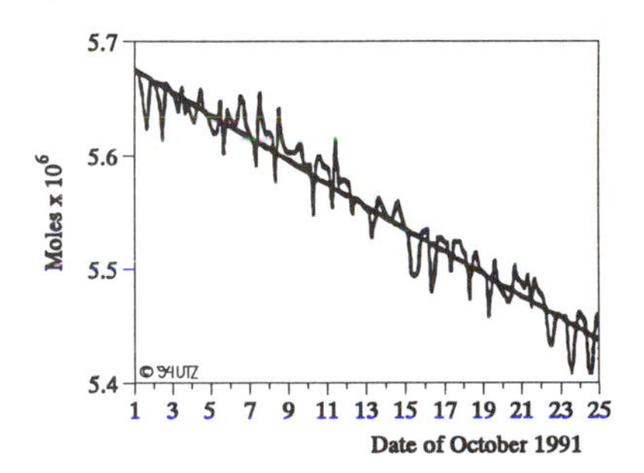

Figure VII.12: **Progressive Dilution of SF_6 and He [4]**

In Biosphere 2, there have been important atmospheric changes which would have been difficult or impossible to observe had it not been effectively sealed. Experimentation with various pressures during September 1991 - January 1992 established a relation between leak rate and pressure that extrapolated to an estimated leak rate of 6 % per year at the maintained operating pressure. In addition, measurement of the progressive dilution of a marker trace gas (SF_6) confirmed that the leak rate is not larger than 10 % per year (see figure VII.12).

Leakage through even very small holes is driven by the differential pressure between inside and outside. These differential pressures are cyclical in nature, typically following diurnal photoperiod cycles or weather patterns. In an Earth-based CELSS, both negative differential pressures of atmosphere are created as the resultant of three influences:

- Thermal expansion / contraction
- Transition of water between liquid and vapor phases
- External barometric pressure variations

The resultant may be typically in the order of 5000 Pa. By providing a flexible expansion chamber, the differential pressure range can be reduced by two, or even three orders of magnitude, which correspondingly reduces the leakage.

Also, permeation may occur through the lung membranes or elastomeric seals. Permeation rate is generally dependent on the gas species and differential partial pressure, and will tend to shift the atmospheric composition. [4, 12]

Trace Gases

Contaminants in the atmosphere of Biosphere 2 have not been a problem. Several gases, attributable to construction materials, have been detected in low part per billion levels and posed no health threats to the crew. Other trace gases, CO, NO_2, NO, SO_2, H_2S, NH_3, and O_3 were also all in part per billion levels. Methane was at about 60 ppm, and N_2O at about 30 ppm. Air analysis from the first ten months of Biosphere 2 closure showed the presence of 130 trace gases, both of biogenic and technogenic origin. But the trace gases have stayed at safe enough levels that use of the soil bed reactors has not been required. This same mechanism of soil metabolism occurs through natural mixing and diffusion of air through the agricultural and wilderness ecosystem soils of Biosphere 2, though at a slower rate. Toxic compounds in air and water were at an all time low in June 1993, probably due at least in part to a decline in the rate of off-gassing from construction materials as they age. [9, 21]

Coral Reef Research

The creation of Biosphere 2's ocean and coral reef was amongst the most challenging part of the project. The 3400 m^3 ocean with a reef is populated by some 40 species of hard and soft coral, and over five dozen types of tropical fish, snails, shrimp, and crabs. Among the environmental variables measured inside Biosphere 2's ocean were pH, light, alkalinity, dissolved nutrients, dissolved oxygen and temperature. The coral reef not only survived its installation and the first two-year closure without large losses, but also considerable reproduction of corals was documented during the transition research. Only one species of corals was lost of the original ones. The outbreak of disease and bleaching of corals and their apparent slowdown and control has been witnessed. This problem, though a serious and widespread phenomenon in natural reefs, has thus far been inadequtely studied. Currently, research applications are being made to support continued long-term coral studies and comparisons between the Florida Keys, a degraded and unhealthy reef, Belize and Bahama reefs, and Biosphere 2. [13]

Materials Cycles

The management of shifting equilibria among carbon, oxygen, the several species of nitrogen, and other elements within and between biomes is one of the more critical and intriguing scientific challenges of Biosphere 2. The intensive agriculture biome (IAB) appeared to play a central role in those dynamics. Its many balances are the subject of ongoing research and will be dealt with in future publications. [10]

Plant Diversity

During closure, the Biosphere 2 facility has sustained a diverse assemblage of life. Most plant species have persisted and canopy biomass has increased since closure. Some species dominance shifts have occurred in the vegetation, particularly in the desert ecocommunity. [8]

VII References

[1] Allen J.
Biosphere 2 - The Human Experiment
Penguin Books, New York, 1991

[2] Allen J.
Historical Overview of the Biosphere 2 Project
NASA, Conference Publication, 1990

[3] Alling A. et al
Biosphere 2 Test Module Experimentation Program
NASA, Conference Publication, 1990

[4] Dempster W.
Methods for Measurement and Control of Leakage in CELSS and their Application and Performance in the Biosphere 2 Facility
(not published)

[5] Dempster W.
Water Systems of Biosphere 2
International Conference on Life Support and Biospherics, Proceedings, p. 331-356, 1992

[6] MacCallum, T. - Analytical Systems Manager of Biosphere 2
Personal Communication, 1992-1993

[7] Nelson M. et al
Atmospheric Dynamics and Bioregenerative Technologies in a Soil-Based Ecological Life Support System: Initial Results from Biosphere 2
(not published)

[8] Nelson M.; Dempster W.
Biosphere 2 - A New Approach to Experimental Ecology
Journal of Environmental Conservation
(to be published)

[9] Nelson M. et al
Biosphere 2 and the Study of Human/Ecosystems Dynamics
Humans as Components of Ecosystems (Proceedings), 1991

[10] Nelson M.; Silverstone S.; Poynter J.
Biosphere 2 Agriculture: Testbed for Intensive, Sustainable, Non-polluting Farming Systems
Outlook on Agriculture
(to be published)

[11] Nelson M.
Bioregenerative Life Support for Space Habitation and Extended Planetary Missions
Space Life Sciences, International Space University Textbook, 1992

[12] Nelson M. et al
Using a Closed Ecological System to Study Earth's Biosphere
BioScience, Vol. 43, No. 4, p. 225-236, 1993

[13] Nelson M.
Commencement of Second Closure Experiment in Biosphere 2
Life Support & Biospheric Science, Vol. 1, No. 2, p. 103-104, 1994

[14] Petersen J. et al
The Making of Biosphere 2
Restoration and Management Notes, 10:2, p. 158-168, 1992

[15] Poynter, J. - Intensive Agriculture Systems Manager of Biosphere 2
Personal Communication, 1993

[16] Redor J. et al
CES-HABLAB: A Closed Ecological System and Habitability Test Facility
ESA, Preparing for the Future, Vol. 2, No. 2, p. 1-3, 1992

[17] Walford R. et al
The Calorically Restricted Low-Fat Nutrient-Dense Diet in Biosphere 2
Proceedings of the National Academy of Sciences (USA), Vol. 89, p. 11533-11537, 1992

[18] Zabel B.
Biosphere 2 (Lecture notes)
International Space University, Kitakyushu, 1992

[19] Zabel B. - General Manager of Construction of Biosphere 2
Personal Communication, 1992-1993

[20] Biosphere 2's Monitoring and Control System
The Biosphere Press, 1993

[21] Biosphere 2 Newsletter, Vol. 2, No. 3
Space Biospheres Ventures, 1992

[22] The Biosphere 2 Magazine, Vol. 2, No. 1
Space Biospheres Venture, 1995

VIII FUTURE LIFE SUPPORT IN SPACE

VIII.1 FUTURE MANNED SPACE EXPLORATION

There are many reasons to continue the exploration of space. If mankind decides to return to the Moon and proceed to Mars, the reasons may be to:

- Satisfy the need to explore, to strive, to seek, to find
- Increase the pool of scientific knowledge
- Enhance the understanding of life in the universe and find out if life once existed on Mars
- Ignite the human spirit
- Potentially utilize the resources of Moon and even Mars

One of the major challenges of today, concerning the space exploration of tomorrow, is to develop an evolutionary approach for space development. This is especially true for future manned space exploration. In this respect, three different strategies may be distinguished:

1. *Human expeditions*, emphasizing on highly visible, near-term aims by sending humans to Mars or its moons. These expeditions would be similar in scope and objective to the Apollo program, with infrastructure development only done to the degree necessary to support short-duration trips.

2. *Science outpost*, emphasizing scientific exploration as well as the investigation of technologies and operations needed for permanent habitation.

3. *Evolutionary expansion*, focusing on the exploration and settlement of the inner solar system in a series of steps, with continued development of technologies, experience, and infrastructure. A roadmap for such a human exploration process is given in figure VIII.1.

In any case, the key to future manned space exploration will lie in the furnishing of technologies enabling the "umbilical links" to Earth to be cut. The practical impossibility of relying upon frequent resupply of consumables from Earth will dictate extensive adoption for planetary bases of closed-loop, regenerative life support systems, likely based on integrated physico-chemical and biological components.

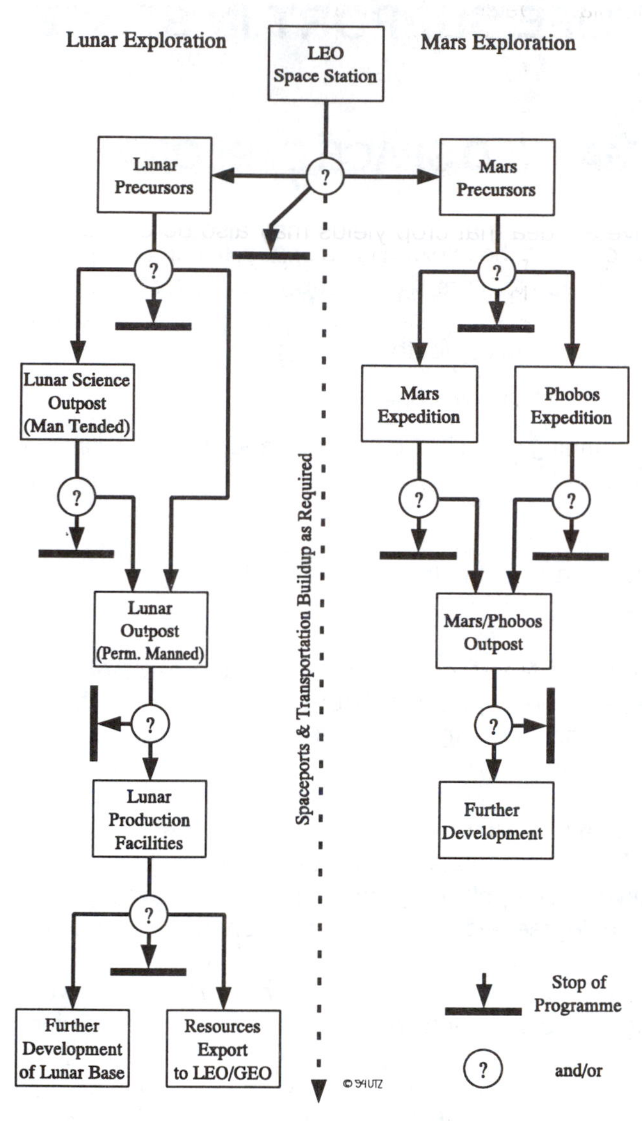

Figure VIII.1: **Roadmap of Human Space Exploration Options [5]**

The harshness of the planetary environments requires that adequate life support has to be provided by technological means for biological processes, like plant and animal growth and reproduction. In the solar system a variety of environments exist that are potential sites for scientific installation, work bases and colonies. These include high pressure environments like the surface of Venus (>100 bar) or the deep ocean zones of Earth (>>100 bar). The high pressures could be dealt with either by maintaining the ambient pressure of the human habitat at 1 bar or less, or maintaining the pressure of these environments at levels close to that of ambient environmental pressures. Each of these approaches has advantages and disadvantages. Habitats maintained at a pressure of 1 bar in regions of high external pressure would have high structural strength requirements, but the inhabitants would live under normal physiological conditions. Habitats designed to operate with internal pressures at levels near the high ambient pressure would have lower structural strength requirements, but the inhabitants would be exposed to abnormal living conditions. Humans can live and work, at least for short periods of time, at elevated pressures up to 66 bar. Bacteria are capable of long term growth at high pressures. In experiments (at 10 bar) on the effects of high pressure on the metabolic functions of land plants it was found that the seed outgrowth varies with the plant species. Combination germination-outgrowth experiments also showed that both of these portions of the life-cycle work under elevated pressure. However, the effects were also varying depending upon plant species. Combination germination-outgrowth-photosynthesis experiments demonstrated that for the plant species tested, photosynthesis and the resulting biomass incorporation occurs at elevated pressures, but also to varying degrees.

These results already give an idea that crop yields may also be dramatically increased compared to those obtained on Earth by deliberately manipulating the environmental parameters. At a planetary base, plants could be grown inside dedicated modules ("greenhouses"). Either conventional soil-type agriculture, or hydroponics/aeroponics techniques could be applied. The big difference between agriculture on Earth and in a planetary base is that in the terrestrial ecosphere, the stability of bioregenerative processes is guaranteed by huge buffers of gases, water, and biomass, i.e., atmospheres, oceans, forests, etc., while in a small-scale CELSS, the overall ecology needs to be actively controlled or "engineered", to reduce the sizes of the various reservoirs and still be a able to maintain stability. Also, there are physical limits to the extent of the regenerative processes that are feasible, imposed by:

- Loss of water through vacuum venting during the reclamation process
- Loss of air and water vapor via atmospheric leakage and in airlock operations
- Loss of carbon and nitrogen in oxidation-process residues

Therefore, self-sufficiency will have to be based not just on indefinite regeneration of consumables, but also local planetary resources have to be utilized to make up for losses such as these in the CELSS and, thereby, minimize the need for resupplying raw materials from Earth. Local resources to be exploited include:

- *Gravity* can be extremely useful at a planetary base for operating gravity-sensitive hardware, such as phase separators or dust/particle collectors. It also allows heat transfer by natural convection and it helps the growth of plants if they can sense the "up" direction. Also the collection of hydroponics/aeroponics liquid would be easier than in weightlessness.

- *Sunlight* may be useful for habitat and greenhouse illumination. Glazing or light piping (through fiber optics) may be used, with adequate radiation protection (see section VI.4.3.4).

- *Soil* may be used as a source of useful minerals, as construction material, plant-growth substrate (as glass- or rock-wool), as heat sink, or adsorbing material in water filtration.

- *Atmosphere* - Atmospheric mining, i.e., concentration of atmospheric water, nitrogen, carbon dioxide, etc., can be conducted from a fixed site, and thus, may be an option for Mars compared with soil mining, which requires transportation, excavation, in-situ operators, and some other logistics.

At this point it should be noted that for the use of resources on Moon and Mars many similar technologies may be applied, but that there are two major differences:

- Since Moon has virtually no atmosphere, no atmosphere compression is possible
- Teleoperation from Earth of robotics equipment on Mars is basically impossible, due to the long transmission times. [5, 11, 19]

VIII.2 SPACE STATION

The future International Space Station will, just like the already existing MIR station, serve as a research laboratory in space and potentially as a staging base for missions to Moon and Mars. Being a completely closed system, the space station life support system will be more complicated than any of the past, most of which were basically open loop systems. Operation over the long term is a requirement that greatly increases the life support system complexity. Especially, long-term operation must accommodate simple on-orbit maintenance requiring little crew time. Anyway, still mainly physico-chemical systems will be applied, e.g., carbon dioxide will be reduced to recover the useful oxygen in the form of water, oxygen will be produced by water electrolysis, eliminating the need for oxygen resupply, hygiene waste water, urine, condensate, and CO_2 reduction product water will all be recycled to keep water resupply to a minimum, etc. Candidate technologies for the International Space Stations' life support system are summarized in table VIII.1. In addition, the Space Station may also serve as a research facility for the development of bioregenerative life support elements, i.e., CELSS technology, which will be required for permanently manned bases on Moon and Mars.

Functions and Functional Subsystems	Potentially Used Technologies
Air Revitalization System	
Carbon Dioxide Concentration	Electrochemical CO_2 Concentrator, Steam Desorbed Amine
Carbon Dioxide Reduction	Sabatier (Methane) Reactor, Bosch (Carbon) Reactor
Oxygen Generation	Static Feed H_2O Electrolysis, Acid Electrolyte, H_2O Electroly.
Trace Contamination Control	High Temperature Catalytic Oxidizer, Regenerated Carbon, Expendable Carbon
Atmosphere Monitoring	Mass Spectrometer
Atmosphere Pressure and Composition Control	
Oxygen Storage	High Pressure Gas, Cryogenic
Nitrogen Storage	High Pressure Gas, Cryogenic
Nitrogen Generation	Catalytic N_2H_4 Decomposition, Cryogenic
Composition Control	Shuttle Technology
Pressure Control	Shuttle Technology
Temperature and Humidity Control	
Temperature Control	Stainless Steel Plate Fin
Humidity Control	Stainless Steel Plate Fin, Slurper
Ventilation Circulation	Ventilation Fans, Anemostats
Water Reclamation System	
Pretreatment	Oxone with H_2SO_4, Biopal, Antifoam & H_2SO_4
Water Recovery, Urine	Vapor Compression Distillation, Thermoelectric Membrane
Water Recov., Condens. & Hyg.	Ultrafiltration, Reverse Osmosis
Post Treatment	Activated Charcoal, UV Enhanced Ozone Oxidation
Water Quality Monitoring	Electrochem. Organic Content, Total Organic C, pH, Conduct.
Biocide Addition & Monitoring	I_2 Injection, Steam Sterilization
Microorganism Monitoring	I_2 (Biocide) Spectrophometrie Monitor., Microorgan. Monitor
Water Storage	Stainless Steel, Metal Bellows Tanks
Personal Hygiene	
Cold	Stainless Steel Cooler
Hot	Cartridge Type Electric Heater
Handwash	Covered Spray & Air Transport
Full Body Shower	Enclosed Spray & Handheld Sheet
Laundry (Washer / Dryer)	Spin, Tumble Wash / Tumble Air Dry
Waste Management	
Toilet	Modified Shuttle Commode
Urinal	Shuttle Technology
Solids Collection	Stainless Steel Receptacles
Trash Compaction	Mechanical Shredding / Gridding
Compacted Solids Storage	Stainless Steel, Super Critical Wet Oxidation
Concentr. Waste Liquid Storage	Stainless Steel, Metal Ballots, Super Critical Wet Oxidation

Table VIII.1: **Space Station Life Support Subsystem Candidate Technologies**

VIII.3 LUNAR BASE

The establishment of a manned lunar base can be discussed in terms of three distinct functions:

1. The scientific investigation of Moon and its environment and the application of special properties of Moon to research problems.
2. The capability to use the materials of Moon for beneficial purposes throughout the Earth-Moon system.
3. To conduct research and development, leading to a self-sufficient, self-supporting lunar base.

Analogous to the task of a base in Antarctica, a lunar base may provide logistical and supporting laboratory capability to rapidly expand knowledge of lunar geology, geophysics, environmental science, and resource potential through wide-ranging field investigations, sampling, and placement of instrumentation. Access to large, free-vacuum volumes may enable new experimental facilities such as macroparticle accelerators. The fact of having a fixed platform may enable new astronomical interferometric measurements to be obtained. The challenge of long-term, self-sufficient operations on Moon can spur scientific and technological advances in material sciences, bioprocessing, physics, and chemistry based on lunar materials, and reprocessing systems. Nevertheless, the prospect of returning to the Moon to establish a growing lunar base leads to two basic questions to be answered:

- How can man stay on the Moon long enough to begin serious exploration?
- How can a lunar base evolve from a small, occasionally occupied outpost to a continuously inhabited, self-sufficient lunar base?

In this context, life support is probably the main critical factor. It is very likely that a lunar base life support system will initially use recycling technology developed for the space station. Eventually, however, the lunar base will require a higher degree of life support system closure than provided for a space station in an Earth orbit. This is because of the more permanent nature of a lunar base and the greater distance from Earth and, therefore, increased transportation costs of providing expendables from Earth. Thus, it is very likely that at least a CELSS will have to be applied at an advanced lunar base, also, because Moon offers a gravity gradient that will make the engineering of systems involving fluid flow much easier. As an example, some candidate configurations for lunar life support, as identified by a Lockheed study, are outlined in figure VIII.2. Each of the conceptual design candidates was based on a generic system structure consisting of five subsystems (atmosphere regeneration, water purification, waste processing, food production, and biomass production), along with three other interfacing systems (in-situ resource utilization, extrahabitat activity, and system monitoring and maintenance). The designs reflect the requirement to provide life support for a nominal crew of 30 persons, with the capability to accommodate a range from 4 to 100. A breakeven analysis between concepts 1 and 5 was yielding a breakeven time for concept 5 of about 2 years.

Concept	Configu-ration	Resupply Mass [kg]	Self Suf-ficiency [%] *	System Mass [kg]	System Volume [m³]	System Power [kW] **
1	Physico - chemical with food resupply	35	-	28850	230	115
2	Physico - chemical with carbohydrate synthesis	20	43	31000	255	150
3	Hybrid with animal food production	30	14	93250	1050	165
4	Hybrid with plant food production	2	92	211200	2075	685
5	Hybrid with plant and animal food production	< 0.1	> 99	222700	2320	595

* Relative to Concept 1 ** Assuming a plant production system with artificial light

Table VIII.2: **Engineering Estimates for Candidate Lunar Base Life Support Configurations [16]**

In option 1, food is provided through resupply, and waste materials are stored. It provides the minimum initial launch cost, power consumption, crew time requirement, and system complexity, but it has the highest logistics costs and the lowest self-sufficiency. In summary, although this candidate makes sense as the first step in lunar life support and development, it is an interim option only, as the base must develop a capability for self-sufficiency as quickly as practical. The second option incorporates the same air and water recycling technologies as used in option 1, but adds the capability for producing carbohydrates for human consumption. Over 90 % of human's energy needs are from carbohydrates, which are therefore of high importance for life support. By adding this capability, the resupply mass requirement for the base is significantly reduced, and self-sufficiency is increased. Option 3 was considered as another potential means for closing the food loop by using a wide variety of animal species as potential food sources. It was found that the system complexity increases substantially, along with a small increase in system self-sufficiency. The methods and technologies which could be employed for implementing this design are extremely uncertain, however. In addition, the system mass increased significantly, and the power requirements increased by about 45 %. The fourth option is the usually discussed CELSS concept. It provides an extremely high degree of self-sufficiency by almost totally closing the food loop. It also provides a number of potential psychological benefits, many of which have been described by cosmonauts during long stays in space. Option 5, which is outlined in figure VIII.2, simply adds the animal production capability to concept 4.

When comparing several options for life support it is necessary to carry out trade studies (see section IV.3.5) in order to make an optimal choice. In this specific case important topics for trade studies for lunar base life support systems include:

- Lighting for plant photosynthesis
- Waste processing technology selection
- Animals as human food in a CELSS
- Aquaculture system feasibility
- Food processing technology review
- Dietary/nutritional evaluation
- Feasibility of using membranes for gas separation
- Crew time requirements for CELSS implementation
- In-situ resource utilization

Another important aspect is a detailed evaluation of the technology research and development requirements. In general, research and technology needs fall into four areas. First of all, the performance of existing, applicable life support technologies must be more precisely characterized with respect to several basic measures, including mass flows, power requirements, potential for mass closure, and interface requirements. Secondly, system- and interface-definition studies must be conducted to verify operational interaction of different life support system designs. Then, although many of the required technologies are in commercial use on Earth, this hardware is sized to support very large numbers of people. Accordingly, research and development efforts must also be directed at miniaturizing existing hardware for use in space. Finally, the suite of research and development efforts described will require the design and construction of hardware testbeds. [5, 8, 9, 16]

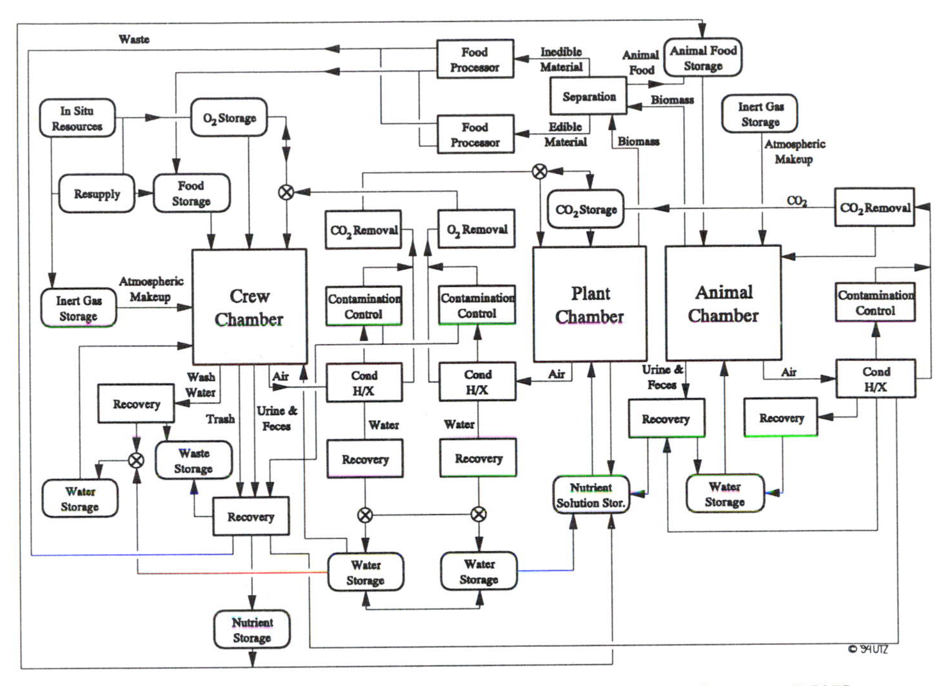

Figure VIII.2: **Block Diagram of Lunar Base Life Support Concept 5 [15]**

Lunar Resources

As mentioned above, another important aspect concerning the establishment of a permanently manned lunar base is the potential use of in-situ resources (see section III.5.1.4). Processing and utilization of local lunar resources might be an economic means for reducing future space transportation demands. Especially, the in-situ production of indigenous oxygen seems to be an interesting possibility, because the element is of vital importance for propulsion and life support. Lunar rocks and soil contain about 45 % of chemically bound oxygen. Extraction concepts are based upon the treatment of lunar soils, rocks, or their components by physical, chemical, or electrochemical means. All those concepts need high amounts of energy for oxygen release and/or reagent recycling. Corresponding average power requirements are in the order of 100 to 1000 kW. Moreover, the lunar soil contains considerable portions of silicone, iron, calcium, aluminum, magnesium, and titanium, which can be extracted as metals, possibly as coproducts of the same process which extracts oxygen. The range of products, e.g., structural beams and plates, solar cells, or pipes that can be generated through the processing of lunar regolith is limited only by the ability to develop the necessary technology. [7, 19]

Lunar Habitat Design

In order to discuss the design of a lunar habitat, with or without utilization of in-situ resources, it may be helpful to recall the specific conditions on the Moon. As mentioned in chapter III, the lunar environment is characterized by low gravity (16.5 % Earth-g), no atmosphere, and a day that lasts 29.53 Earth days with the Sun shining intensely for half of that time (extremely high temperatures) and darkness reigning for the other half (extremely low temperatures). There is apparently no water and virtually no carbon, i.e., no CO_2. Especially, the vacuum of space is a serious problem, since there will always be leaks from structures on Moon. Thus, it will be necessary to construct a structure which is as tight as possible, in order to minimize the amount of atmospheric gases that have to be resupplied. To ease the engineering problems of building such structures in space that contain an artificial atmosphere, pressure within the structures can be reduced, particularly the N_2 partial pressure. Also, temperature control on Moon, but also on Mars, will be extremely challenging, because it will be not as easy as in free space to radiate excess heat. Additionally, the long lunar night probably requires that the power storage devices brought from Earth would have to be quite massive. This is because crops might be grown in underground growth units to avoid harmful ionizing radiation and micrometeorites, and therefore, artificial light might be necessary, although some sunlight may be brought in by fiber optics during the lunar day (see section VI.4.3.4).

Other important environmental aspects are the radiation, but also meteorites that hit the lunar surface. During the first flights to Moon, the Apollo astronauts only stayed for a few days, so they did not need massive shielding from radiation.

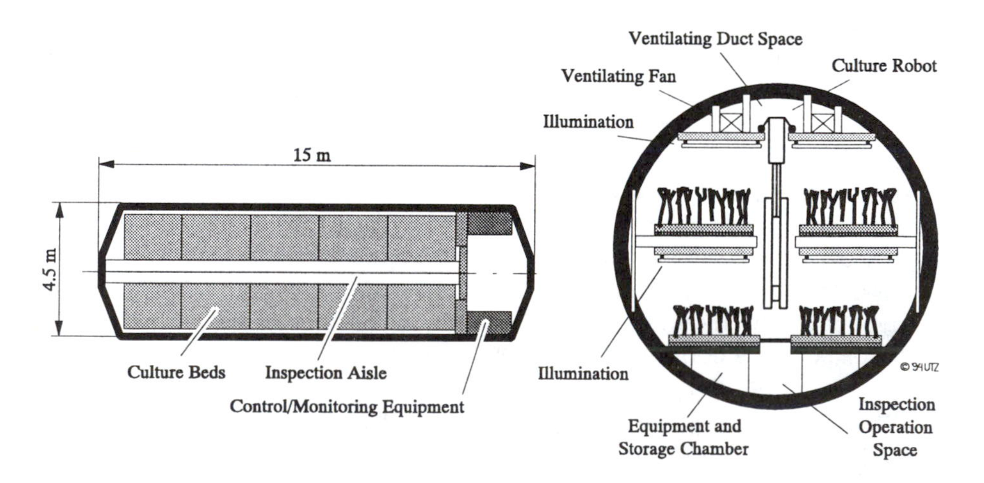

Figure VIII.3: **Lunar Base Plantation Module [9]**

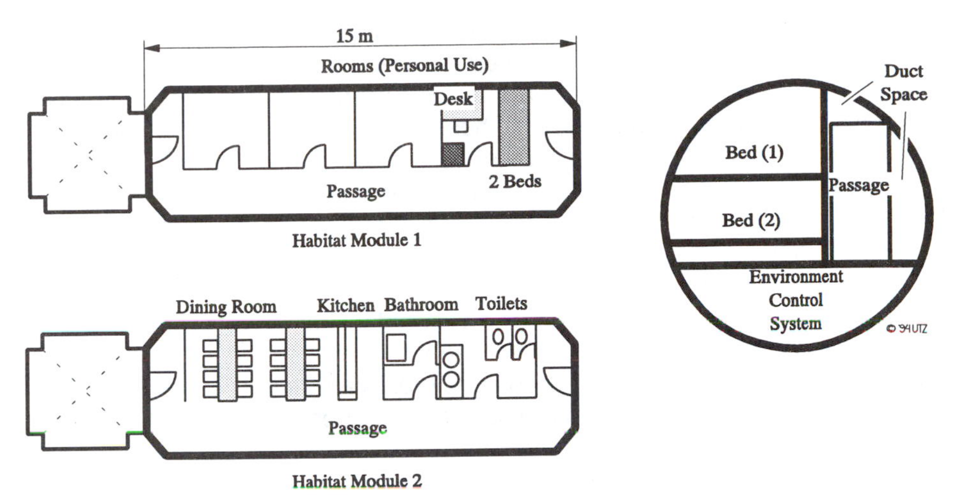

Figure VIII.4: **Lunar Base Habitat Modules [9]**

People staying for any length of time must be protected, however, not only from solar flares but also from galactic cosmic radiation. Advanced lunar habitats could be constructed underground in tunnels, giving natural protection, e.g., it has been proposed that lunar lava tubes could be ideal initial sites for lunar outposts because they provide a constant temperature environment, a natural shield from radiation, and meteorite protection. Early habitats, on the other hand, may be prefabricated space station-type modules or inflatable structures brought from Earth. One way to provide shielding for these habitats is to simply pile a large amount of regolith on them. This might be the first use of lunar material at an outpost. A thickness of several meters would be required for long stay times. Burying the modules under a pile of soil has some draw-backs, however. It places constraints on the design of habitat modules and airlocks and may limit operations near the habitat. The soil could slide off, so it would have to be stabilized with supports or retaining walls. A thick layer of soil would make it difficult to add rooms or to connect new

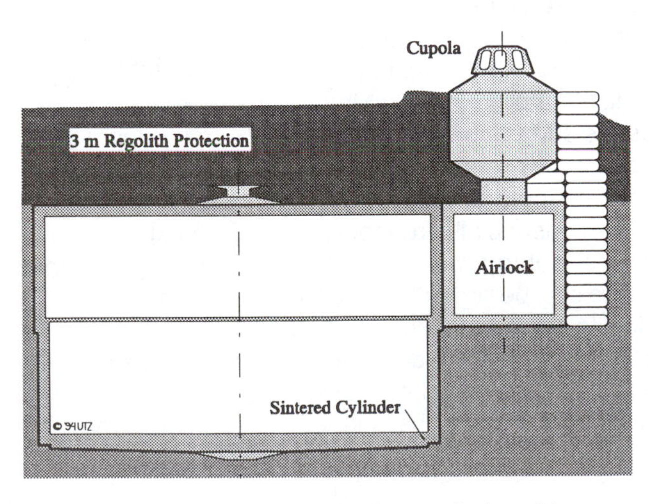

Figure VIII.5: **The Lunar Habitat as designed by the University of Houston [3]**

wires, antennae, or pipes as the outpost gets improved. Putting soil in sand bags would solve some of these problems but not all of them. An improvement over loose soil would be a radiation protection system constructed of cast basalt or sintered blocks made from regolith. These can provide much denser radiation protection compared to loose soil, therefore, the shield can be less thick. Just as examples, design proposal for lunar habitats as developed in Japan and at the University of Houston are given above. Figures VIII.3 and VIII.4 show tentative drafts of lunar base plantation and habitat modules developed in Japan.

The other concept for a lunar habitat, developed by the University of Houston, is given in figure VIII.5. The design uses a machine that simultaneously excavates and sinters the lunar regolith to create a cylindrical hole of 4-12 m depth. The hole is then enclosed with cast basalt slaps, allowing the volume to be pressurized.

A very comprehensive overview of all the design aspects of a lunar base is given in "The Lunar Base Handbook" [3] by the author of this volume.
[2, 4, 10, 13, 19]

VIII.4 MARTIAN BASE

When discussing the establishment of a permanently manned base on Mars the main difference that has to be taken into account, in comparison to discussions on a lunar base, is, of course, the much bigger distance of Mars from Earth. Even a fast mission from Earth to Mars would take in the order of 200 days. Therefore, the self-sufficiency of a permanently Martian base is essential. For cost, time, and safety reasons a Martian base could not rely on resupply from Earth. Under these conditions it is a prerequisite to first establish a lunar base in order to gain experiences in planetary base development and to verify the developed technologies. This is especially true for life support equipment, but also resource utilization technology. In many respects the engineering requirements for a Martian base may be similar to those for a lunar base. Therefore, the principal conceptions for life support on Moon and Mars are about the same and one may the refer to the conclusions made in the previous chapter. But still there remain five main differences between the lunar and Martian environment: the higher gravity, the distance from the Sun, the length of the days, the existence of water on the surface, and the fact that Mars has an atmosphere.

The Martian environment is characterized by 38 % Earth-g. Mars' greater distance from the Sun means lower irradiance levels, but light levels also vary with the Martian seasons, especially because of the high ecliptical orbit, but also because of the tilted axis (25°).Thus, sunlight energy is only at 37-52 % of that on Moon. The Martian year is 687 Earth days long and the Martian days last 24.7 hours. The thin atmosphere of Mars contains 20 times as much carbon dioxide as Earth's atmosphere (see section III.5.2). Thus, at least principally, carbon may not have to be carried to Mars. Also, the Martian atmosphere can be processed to release oxygen for life support or propellant use. Carbon monoxide, which could be a moderate performance rocket fuel, would be the coproduct. By combining this oxygen with a small amount of hydrogen, water could be produced. Since life support technologies routinely deal with the conversion of CO_2 to other compounds, including methane, a direct application of this technology to the Martian atmosphere would allow for the production of oxygen, methane, and water by bringing only small amounts of hydrogen. Thus, also large quantities of propellant could be leveraged from minimal import mass. Another good aspect of atmosphere utilization is that no mining would be involved but comparatively simple gas handling equipment can be used. Also, the existing atmosphere provides some shielding from radiation. Planetary scientists agree that water is available at the poles of Mars in the form of ice. It is likely, but not certain that water is available elsewhere on the planet, perhaps as a permafrost layer or bound as mineral hydrate. The two moons of Mars, Phobos and Deimos, may also be rich in water. According to studies 10-100 tons per month of water could be extracted from dry, but hydrated soil using a 50 kW thermal source, while the dehumidification of the atmosphere could liberate about 3 tons of water in one month using a 50 kW source. Furthermore, oxygen can be obtained from silicate rocks as on Moon. Similar as for Moon, the use of in-situ raw materials for the construction of Martian base structures has been suggested, including:

- Simple rock fracturing and soil moving
- Production of Mars cement
- Brick production using microwave and laser technology

The facts mentioned above yield the following main criteria for an initial landing and outpost site on Mars:

- Accessibility from orbit
- Presence of useful raw materials
- Low elevation for landing ease and radiation protection
- Earth communication availability
- Solar energy potential
- Interesting science prospects

An additional criterion that might facilitate base site selection is the base evolution potential for long-term exploration and development of Mars. Criteria concerning this evolution potential are:

- Amount and degree of importance of scientific data available
- Relationship to other planets or existing bases
- Transportation
- Communication
- Power
- Maintenance and logistics capability
- Construction factors
- Role in interplanetary missions
- Commercial potential

In this respect it should be noted that, as on Moon, lava tubes usable for initial base sites appear to be located in sites of interest for initial Mars bases. This is especially interesting for thermal control reasons, since the temperature fluctuates tremendously on Mars: -75° C at night, to a maximum of 20° C at noon in the summer on the Martian equator. [2, 10, 13, 19]

VIII References

[1] Chicarro A.; Scoon G.; Coradini M.
Mission to Mars
ESA SP-1117, 1990

[2] Cordell B.
Human Operations, Resources, and Bases on Mars
Engineering, Construction, and Operations in Space II, American Society of Civil Engineers, p. 759-768, 1990

[3] Eckart P.
Back to the Moon - The Lunar Base Handbook
To be published, 1996

[4] Fahey M. et al
Indigenous Resource Utilization in Design of Advanced Lunar Facility
Journal of Aerospace Engineering, Vol. 5, No. 2, p. 230-247, 1992

[5] Fairchild K.; Roberts B.
Options for Human Settlement of the Moon and Mars
Lunar Base Agriculture: Soils for Plant Growth, ASA, p. 1-3, 1989

[6] Henninger D.
Life Support Systems Research at the Johnson Space Center
Lunar Base Agriculture: Soils for Plant Growth, ASA, p. 173-191, 1989

[7] Lingner S. et al
Lunar Oxygen Production by Soil Fluorination - Concepts and Laboratory Simulation
Zeitschrift für Flugwissenschaft und Weltraumforschung, Vol. 17, No. 4, p. 245-252, 1993

[8] MacElroy R.; Klein H.
The Evolution of CELSS for Lunar Bases
Lunar Bases and Space Activities of the 21st Century, W. Mendell, (ed.), Lunar Planetary Institute, p. 623-633, 1985

[9] Mendell W. et al
Strategies for a Permanent Lunar Base
Lunar Base Agriculture: Soils for Plant Growth, ASA, p. 23-35, 1989

[10] Nitta K.; Ohya H.
Lunar Base Extension Programme and Closed Loop Life Support Systems
Acta Astronautica, Vol. 23, p. 253-262, 1991

[11] Novara M.
Life Support on Moon and Mars
ESA Bulletin 57, p. 32-39, 1989

[12] Salisbury F.
Bioregenerative Life Support System: Farming on the Moon
Acta Astronautica, Vol. 23, p. 263-270, 1991

[13] Salisbury F.
Some Challenges in Designing a Lunar, Martian, or Microgravity CELSS
Acta Astronautica, Vol. 27, p. 211-217, 1992

[14] Salisbury F.; Bugbee B.
Wheat Farming in a Lunar Base
Lunar Bases and Space Activities of the 21st Century, W. Mendell, (ed.), Lunar Planetary Institute, p. 635-645, 1985

[15] Sauer R.
Metabolic Support for a Lunar Base
Lunar Bases and Space Activities of the 21st Century, W. Mendell, (ed.), Lunar Planetary Institute, p. 647-651, 1985

[16] Schwartzkopf S.
Lunar Base Controlled Ecological Life Support System (LCELSS)
Lockheed Missiles & Space Company, Contract NAS 9-18069, 1990

[17] Seboldt W. et al
Sauerstoffgewinnung aus Mondgestein - Schlüsseltechnologie für die bemannte Weltraumerkundung?
DLR-Nachrichten, Heft 71, p. 15-20, 1993

[18] Silberberg R. et al
Radiation Transport of Cosmic Ray Nuclei in Lunar Material and Radiation Doses
Lunar Bases and Space Activities of the 21st Century, W. Mendell, (ed.), Lunar Planetary Institute, p. 663-669, 1985

[19] Sullivan T.; McKay D.
Using Space Resources
NASA Johnson Space Center, 1991

[20] Tamponnet C. et al
Implementation of Biological Elements in Life Support Systems: Rationale and Development Milestones
ESA Bulletin 74, p. 71-82, 1993

[21] Regenerative Stoffwirtschaft in der Raumfahrt
DARA, Bericht 50 RS 8902, 1992

IX POTENTIAL TERRESTRIAL APPLICATIONS

IX.1 INTRODUCTION

Life support technologies have a wide spectrum of potential terrestrial applications in addition to their role in space exploration and habitation, since some of the problems that they address have strong counterparts in the terrestrial environment. The problem to date is that not many of these potential applications have been defined. Especially from the research concerning bioregenerative life support in space, many spin-off technologies may be derived. Since there is rising criticism on most investments in space research it seems time to do so and demonstrate that the so often quoted spin-offs may really be developed. Tentatively, the several potential spin-offs from space-related research on closed ecological systems and life support contributing to the solution of actual terrestrial problems may be divided into the following categories:

- Ecological research and environmental protection
- Development of methods for atmosphere, water, and waste bioregeneration
- High efficient cultivation of plants and microorganisms
- Design and support of habitats and research facilities

This chapter summarizes some developments of terrestrial applications that have already been derived from life support research, and also lists promising projects in this respect that are basically waiting to be implemented.

IX.2 BASIC ECOLOGICAL RESEARCH

Earth has always been subject to change from what is known of its geological and life history. What is different about the present time is that humans now are beginning to add their own impacts to the natural changes, especially because of the increasing number of humans. Currently, some 5.5 billion people live on Earth. Projections show that this number may increase to some 10 billion by 2025. Everybody is familiar with the dramatic effects of the actions of these humans and their so-called technosphere, especially air and water pollution and exploitation of natural resources, on the terrestrial environment. But although mankind is aware of its destruction work there is no comprehensive plan how to preserve the endangered biosphere of our planet, especially with respect to the

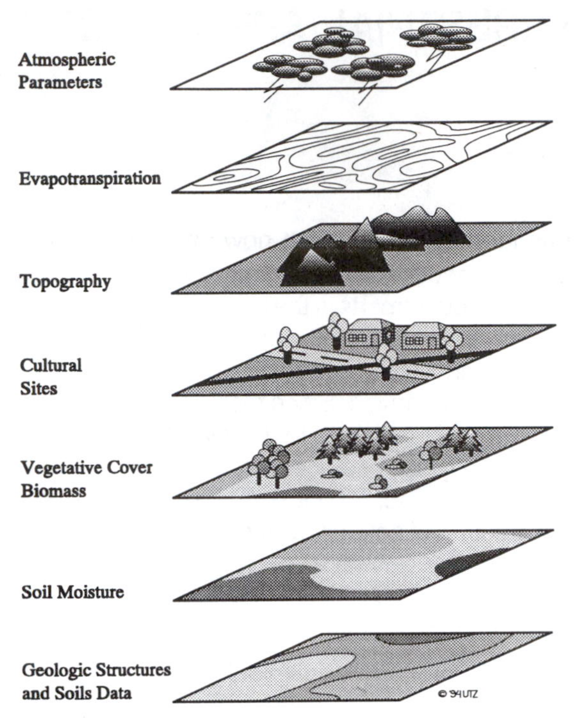

Figure IX.1: **Parameters for a Geobased Information System [16]**

increasing number of humans inhabiting it. In the future, mankind may have to be the captain of spaceship Earth, may have to try to achieve control over some of its natural processes. But therefore scientists have to start putting together models to try to understand the dynamic, ecological processes of the biosphere. It has to be understood where the Sun plays a role, where clouds play a role, what the oceans do, what land does, what the water does. Figure IX.1 gives an illustration of some of the parameters that have to be fit together in such a "geobased information system". [16]

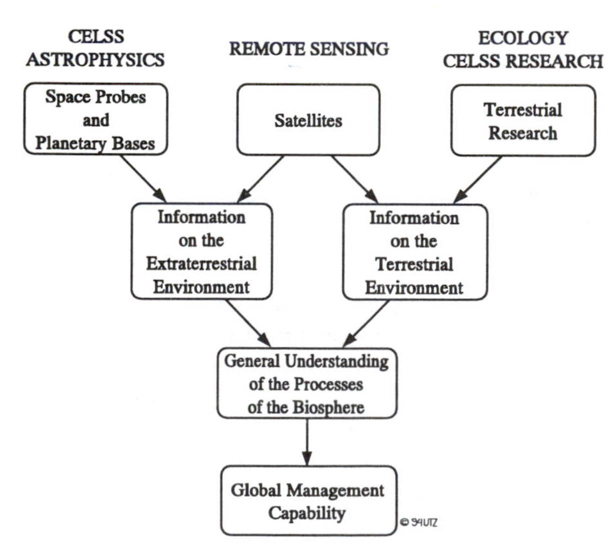

Figure IX.2: **Synergistic Approach towards the Understanding of the Biosphere**

This goal of really understanding the processes of planet Earth can only be achieved by a new interdisciplinary program that has to face the challenge, but also the chance, of creating unique synergistic effects. Therefore, especially the fields of ecology, remote sensing, and life support research will have to closely work together. Moreover, such a collaboration concept, as symbolized in figure IX.2, may yield new impulses and public awareness for the work of all of the disciplines involved.

The contribution of life support research, and in this respect especially closed ecological systems research, to such a program will to a large extent consist of an enhanced understanding and quantitative estimates of terrestrial buffering capacities. In closed systems, buffers would be specifically defined, either as naturally occurring absorbers or storage depots, or as mechanical or physico-chemical devices. In exploring buffers, the true energy requirements of species would become apparent, thus enlightening the relationship between organisms and inanimate world, as well as identifying what are now obscure energetic relationships among organisms. Other aspects of investigation will include better estimates of the significance of non-biological energy trapping, e.g., the role of lightning in the atmosphere in fixing nitrogen for biological purposes, or the role of atmospheric photochemistry in eliminating organic toxins. Also, small, materially closed systems have short cycle times for nutrients and gases and allow very intensive monitoring, thus facilitating detailed examination of mechanisms of interest to scientists studying Earth's geosphere and biosphere.

In this respect it may be interesting to consider the buffering capacity available to one square meter of land on Earth. In addition to the contents and the dynamics of 1 m^2 of the soil to a depth of about 1 m, there exist 2.42 m^2 of ocean surface. The average depth of the ocean is about 3400 m, so that the volume of ocean water corresponding to 1 m^2 of land is about 8300 m^3. The atmosphere above 3.42 m^2 of land and ocean extends to 60 km or more, but if the atmospheric gases were compressed to standard pressure, they would occupy about 1260 m^3. In addition to the large volume buffering, the whole Earth ecosystem utilizes the physical solubility of CO_2 and O_2 in the oceans as well as chemical equilibria to maintain average concentrations of CO_2 and O_2 all over the Earth. The chemical equilibria governing the capacity of the oceans to absorb CO_2 are complex. Physically dissolved CO_2 is in equilibrium with bicarbonate and carbonate ions. The soluble ions are, in turn, in equilibrium with insoluble Ca or Mg carbonates, and all equilibria are greatly influenced by, and affect, the pH level. Both the oceans and the atmosphere perform the additional function of distribution. Some areas of Earth, i.e., the deserts and polar regions, produce little O_2, while tropical rainforests produce large amounts. The atmosphere, acting as a transport medium, homogenizes the noncondensing gases so that only very slight variations in O_2, N_2, and CO_2 concentrations exist worldwide. Condensable gases, such as water vapor, are less homogeneously distributed. Nevertheless, water distribution is vital for the survival of the terrestrial ecosystem.

A closed ecological system, i.e., a system closed to mass exit or entry, but open to energy input and with controlled energy loss, may mimic a test tube environment and will permit study of such parameters and processes of ecological systems, as described in the preceding paragraph. The fact that the enclosed system will not be a natural one may even allow new ways of evaluating interactions between species, of studying the true metabolic and energy requirements of groups of organisms, of investigating environmental triggers of metabolic and behavioral changes, and of reexamining classical predator-prey and competition interactions. The advantages of closed system studies lie primarily in the realm of easily established control experiments, and of permitting specific and regulated variables to affect the system. [12]

In parallel to this work, the modeling of the terrestrial, environmental processes is only possible if these processes themselves, i.e., the relevant parameters, are continuously monitored, both on Earth and from space. Therefore, it will not only be necessary to place the required instruments on the ground and satellites in space, but also to establish centers with sufficient computer capacity for data processing and evaluation all over the world. Moreover, the information acquired should be accessible for the public at large and used for ecological diagnosis and creation of an alarm system. The main parameters to be determined, once such a system has been built up, are summarized in the listing below:

- The global distribution of energy input to and energy output from Earth
- The structure, state variables, composition, and dynamics of the atmosphere from the ground to the mesopause
- The physical and biological structure, state, composition and dynamics of the land surface, including terrestrial and inland water ecosystems
- The rates, important sources and sinks, key components and processes of Earth's biogeochemical cycles
- The circulation, surface temperature, wind stress, sea state and the biological activity of the oceans
- The extent, type, state, elevation, roughness, and dynamics of glaciers, ice sheets, snow, sea ice, and the liquid equivalent of snow in the global cryosphere
- The global rates, amounts and distribution of precipitation
- The dynamic motions of Earth, i.e., geophysics, as a whole, including both rotational dynamics and the kinematic motions of the tectonic plates [16]

As a prerequisite for such an ambitious program, but maybe even more if the results of such a program yield specific measures that have to be taken in order to preserve the terrestrial biosphere, an ecological morality and ethics will have to established. At the latest, the new generations, who are to live and act in the 21st century have to realize that the entire life on Earth needs moral protection from the boundless aggression of the technosphere. It will be necessary to "ecologize the technology".

I. Gitelson, who came up with these ideas, even claims that the "survival of humanity as it is and the biosphere make it necessary to mold the ecological morality, ethics and restrictive rules of behavior for the relations of man with the biosphere. The technologies, destructive for the turnover of the matter in the biosphere, must be declared antiecological, to be prohibited and substituted by the ecologically acceptable. The human society is to work out and adopt for the generations to come solid moral principals and ethical principals for relations of humanity with the living nature". Gitelson has also pointed out that man-made biospheres can be of exceptional educational importance, because they may clearly demonstrate the principle of organization of the life on Earth and its mechanisms, and they contribute to the cultivation of an ecological morality. [7]

IX.3 ATMOSPHERE, WATER, AND WASTE REGENERATION

As mentioned in the previous chapter, there are several parallels between the processes in artificial closed ecological systems and those of the terrestrial biosphere. Also, some of the developments in conjunction with bioregenerative life support systems research may find application in terrestrial areas. Some of these potentials, parallels and the potential terrestrial applications are summarized below:

- The problem of clean air regeneration in spacecraft cabins or sealed space outposts is analogous to the so-called "sick building syndrome" often associated with relatively tightly sealed, energy efficient buildings. A static atmosphere, without a means of removing trace gases from outgassing of materials and from people themselves, eventually leads to accumulation of trace gases and possible health problems.
- The challenge to recycle human waste and regenerate domestic/hygiene water in space habitats is similar to the problem of developing recycling systems in the global arena to prevent the degradation of water caused by urban sewage disposal.
- The goal of developing bioregenerative life support systems which completely sustain humans in a clean and healthy atmosphere, and which do not produce pollution as a byproduct, has parallels with humanity's need to create technologies that will permit development without eroding the habitability or life support capacity of our home planet.

An example for a spin-off technology from closed ecological system research is the use of biological reactors for the scrubbing of trace atmospheric organic contaminants that may be used in both open and closed environments.

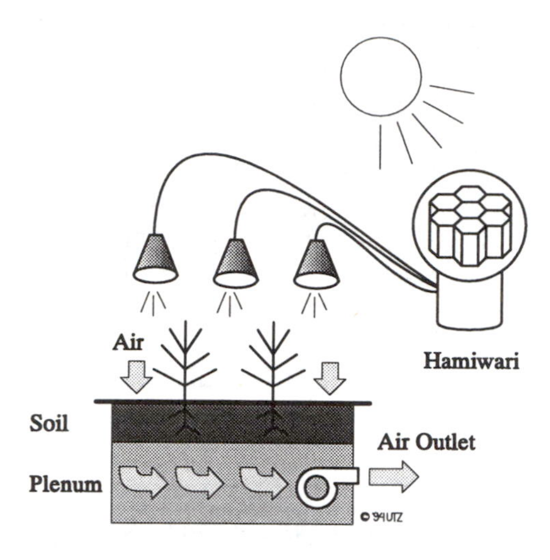

Figure IX.3: **Soil Bed Reactor [6]**

The basic configuration of such a so-called soil bed reactor (SBR) is that air is moved through a living soil that supports a population of plants (see figure IX.3). Research results have indicated that a SBR has no impact on plant productivity or phenology. That means that functioning soils could be used for both intensive biomass production and air purification.

Upon exposure to the soil, contaminants are either passively adsorbed onto the surface of soil particles, chemically transformed in the soil to usable compounds that are taken up by the plants or microbes, or the compounds are directly used by the microbes as a metabolic energy source and converted to CO_2 and water. The number and type of compounds degradable by soils is large. A compilation of compounds known or suspected to be degradable in soils is given in table IX.1.

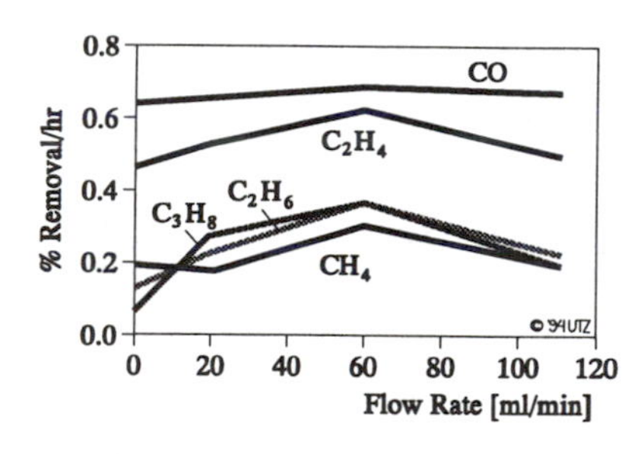

Figure IX.4: **Reducing the Atmospheric Concentration of Organic Compounds by a SBR [6]**

It has been found that SBRs are highly variable in their behavior. Nevertheless, a SBR substantially reduces the concentration of organic compounds within a closed system. Figure IX.4 shows the results of an experiment conducted by the Environmental Research Laboratory (ERL) of the University of Arizona. While the methane (CH_4), ethane (C_2H_6), and propane (C_3H_8) levels are reduced, carbon monoxide (CO) is relatively unaffected by the operation of the SBR, and the ethylene (C_2H_4) concentrations were even higher when the SBR was operating than when it was not. Despite this fact, the atmospheric concentrations of C_2H_4 and CO were reduced to less than 20 % of their original levels.

The ERL has also investigated the use of small SBRs for use in office and home environments. The result was that a SBR is also effective in minimizing airborne biological particulates. In addition, the ERL is working on a methodology of using SBRs and agricultural production for simultaneously reducing CO_2, CH_4,

Compound	Reference	Compound	Reference
Acetaldehyde	Fuller W.F. et al 1983	Ethylcyclohexane	Stirling L.A. et al. 1977
Acetic Acid	Zavarzin G.A. et al. 1977	Ethylene	DeBont J.A.M. 1976
Acetoin	Bohn H.L. 1977	Fluoro-4-nitrobenzoate (2-)	Horvath R.S. et al. 1972
Acetylene	Smith K.A. et al. 1973	Fluorobenzoate (o-)	Horvath R.S. et al. 1972
Acrolein	Fuller W.F. et al 1983	Fluoride	Bohn H.L. 1977
Alkyl Benzene Sulfonate	Horvath R.S. et al. 1972	Formaldehyde	Grundig M.W. et al. 1987
Aldehydes	Fuller W.F. et al 1983	Formate	Hou C.T. 1980
Ammonia	Hutton W.E. et Al. 1953	Heptadecycylohexane	Beam H.W. et al. 1974
Anthracene	Dalton H. et al. 1982	Hexadecane	Beam H.W. et al. 1974
Benzene	Dalton H. et al. 1982	Hydrogen Sulfide	Smith K.A. et al. 1973
Benzonate	Dalton H. et al. 1982	Hydrogen	Zavarzin G.A. et al. 1977
Bicyclohexyl	Higgins I.J. et al. 1979	Isoprene	Van Ginkel C.G. et al. 1987
Bromomethane	Dalton H. et al. 1982	Isopropyl Benzene	Higgins I.J. et al. 1979
But-2-ene	Higgins I.J. et al. 1979	Isopropylcyclohexane	Stirling L.A. et al. 1977
Butadiene (1,3-)	Van Ginkel C.G. et al. 1987	Isopropyltoluene (p-)	Horvath R.S. et al. 1972
Butane	Hou C.T. 1980	Lactic Acid	Bohn H.L. 1972
Butene (1-)	Dalton H. et al. 1982	Limonene	Dalton H. et al. 1982
Butene (cis-2-)	Dalton H. et al. 1982	Methane	Anthony C. 1982
Butene (trans-2-)	Dalton H. et al. 1982	Methanol	Dalton H. et al. 1982
Butylbenzene (n-)	Horvath R.S. et al. 1972	Methyl Mercaptans	Fuller W.F. et al 1983
Butylcyclohexane (n-)	Horvath R.S. et al. 1972	Methyl Sulfide	Smith K.A. et al. 1973
Butyric Acid	Bohn H.L. 1972	Methylcatechol (3-)	Horvath R.S. et al. 1972
Cadaverine	Bohn H.L. 1977	Methylcyclohexane	Stirling L.A. et al. 1977
Caprolactone	Stirling L.A. et al. 1977	Methylnaphtalene (1-)	Higgins I.J. et al. 1979
Carbon Monoxide	Bartholomew et al. 1982	Methylnaphtalene (2-)	Higgins I.J. et al. 1979
Chlorbenzoate (m-)	Dalton H. et al. 1982	Naphtalene	Dalton H. et al. 1982
Chlorofluoromethanes	Bohn H.L. 1977	Nitric Oxide	Bohn H.L. 1972
Chloromethane	Dalton H. et al. 1982	Nitrous Oxide	Goyke N. et al. 1989
Chlorophenol (m-)	Higgins I.J. et al. 1979	Ozone	Turner N.C. 1973
Chlorotoluene (m-)	Higgins I.J. et al. 1979	Octadecane	Perry J.J. 1979
Cinerone	Horvath R.S. et al. 1972	Organophosphorus	Bohn H.L. 1977
Cresol (m-)	Higgins I.J. et al. 1979	Pentachlorophenol	Lagas P. 1988
Cresol (o-)	Higgins I.J. et al. 1979	Pentanol (n-)	Higgins I.J. et al. 1979
Cyanides	Bohn H.L. 1977	Phenol	Schmidt S.K. et al. 1985
Cycloheptane	Beam H.W. et al. 1974	Phenyldecane (1-)	Higgins I.J. et al. 1979
Cycloheptanone	Beam H.W. et al. 1974	Phenylnonane (1-)	Higgins I.J. et al. 1979
Cyclohexanediol (1,2)	Beam H.W. et al. 1974	Phosgene	Turner N.C. 1973
Cyclohexanediol (1,3)	Stirling L.A. et al. 1977	Propane	Bohn H.L. et al. 1988
Cyclohexanediol (1,4)	Stirling L.A. et al. 1977	Propene	Dalton H. et al. 1982
Cyclohexandione (1,2)	Stirling L.A. et al. 1977	Propylbenzene (n-)	Horvath R.S. et al. 1972
Cyclohexane	Stirling L.A. et al. 1977	Propylene	Hou C.T. 1980
Cyclohexanol	Beam H.W. et al. 1974	Putrescine	Bohn H.L. 1977
Cyclohexanone	Beam H.W. et al. 1974	Pyridine	Dalton H. et al. 1982
Cyclohexene	Stirling L.A. et al. 1977	Pyrrolidone	Horvath R.S. et al. 1972
Cyclohexene Oxide	Stirling L.A. et al. 1977	Skatole	Bohn H.L. 1972
Cyclooctane	Beam H.W. et al. 1974	Styrene	Higgins I.J. et al. 1979
Cyclopentanone	Beam H.W. et al. 1974	Sulfur Dioxide	Smith K.A. et al. 1973
Cymene (p-)	Dalton H. et al. 1982	Terpenes	Rasmussen R.A. 1972
Decane (n-)	Higgins I.J. et al. 1979	Tetrachloromethane	Galli R. et al. 1989
Dialkyl Sulfides	Fuller W.F. et al 1983	Tetradecane	Perry J.J. 1979
Dichloroatechol (3,5-)	Horvath R.S. et al. 1972	Toluene	Dalton H. et al. 1982
Dichlorodiphenylmethane (p,p-)	Horvath R.S. et al. 1972	Toluidine (p-)	Higgins I.J. et al. 1979
Diethylether	Dalton H. et al. 1982	Tridecane (n-)	Perry J.J. 1979
Dimethyl Disulfide	Oremland R.S. et al. 1989	Triethylamine	Fuller W.F. et al 1983
Dimethylether	Dalton H. et al. 1982	Trichlorobenzoate (2,3,6-)	Horvath R.S. et al. 1972
Diphenyl-2,2,2-trichlorethane (1,1-)	Horvath R.S. et al. 1972	Trichloroethane (1,1,1-)	Galli R. et al. 1989
Dodecane (n-)	Perry J.J. 1979	Trichloromethane	Galli R. et al. 1989
Dodecyclohexane	Beam H.W. et al. 1974	Trichlorophenoxy Acetic Acid 2,4,5	Horvath R.S. et al. 1972
Ethane	Dalton H. et al. 1982	Xylene (m-)	Higgins I.J. et al. 1979
Ethanol	Zavarzin G.A. et al. 1977	Xylene (o-)	Horvath R.S. et al. 1972
Ethylbenzene	Dalton H. et al. 1982	Xylene (p-)	Horvath R.S. et al. 1972

Table IX.1: **Compounds Known or Suspect to be Decomposable by a Soil Bed Reactor [6]**

SO_2 and other emissions from power plants and increasing agricultural productivity. Currently, the first indoor Air Purifiers called AIRTRON, are being sold by Space Biospheres Ventures (Biosphere 2). [6]

Another promising contribution to the solution of Earth's environmental problems is the development of a means to utilize both air and water pollution as a nutrient source for growing green plants, i.e., using nature to clean our environment. Sewage may then be used as a nutrient solution for growing plants while the plan roots and associated microorganisms convert sewage to clean water. This may be a very economical means of treating sewage, especially for rural areas and small cities. Research in this direction has been conducted by Wolverton. While microorganisms have always been used by engineers to treat sewage and industrial waste water, the use of higher pants in completing nature's cycle was a new contribution to that process. Although microorganisms are a vital part of waste water treatment, it is important to have vascular plants growing in these treatment filters to feed off the metabolic by-products of microorganisms and to prevent slime layer formation from dead microorganisms. Aquatic plant roots can also add trace levels of oxygen to help maintain aerobic conditions in plant-microbial wastewater treatment filters. An example for an artificial marsh waste water treatment system as developed by Wolverton is given in figure IX.5.

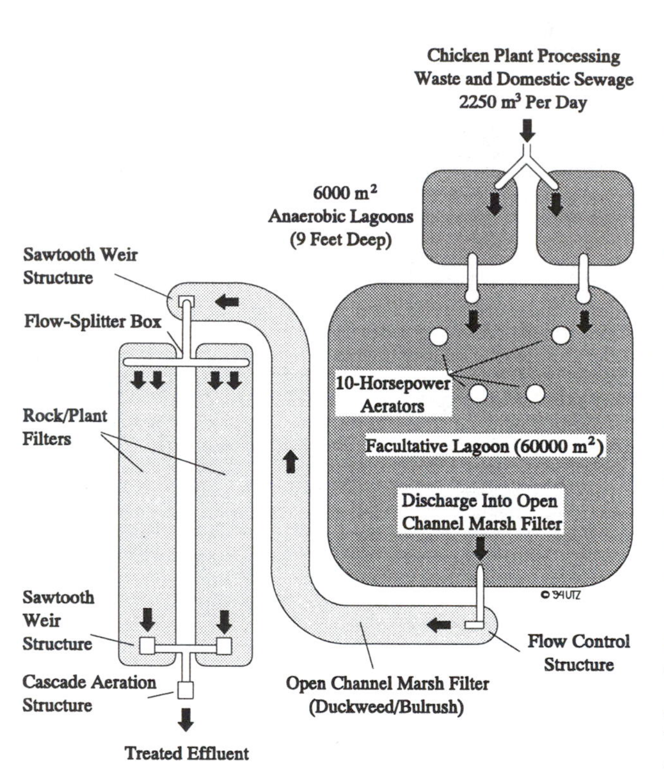

Figure IX.5: **Example of an Artificial Waste Water Treatment System [20]**

The largest aquatic plant rock filter system to date is installed at Denham Springs, Louisiana. This system is treating approximately 11000m³ of domestic sewage per day. Also, the chemical manufacturing industry, paper mills, the textile industry, and animal processing plants are beginning to use the aquatic plant wastewater treatment process as an economical and environmental safe method of dealing with their wastewater. Some numbers concerning the effectiveness of such a system are given in table IX.2. Air emission problems from point sources like smoke stacks, incinerators, etc., may be solved by converting the air pollutants into water pollution and purifying the polluted water using aquatic plant microbial marsh filters.

A further step in the development of plant and microbial filter biotechnology is to incorporate the complex waste water treatment / indoor air purification concept into a real home environment. An example of such a concept developed by Wolverton is given in figure IX.6. The bioregenerative system basically adds a room, filled with houseplants that purify the air while feeding off the waste water to the house. [19]

Chemicals [mg/l]	Marsh Plants in Rock Filter	Influent	Effluent *
Trichlorethylene	Torpedo Grass Southern Bulrush	3.60 9.90	0.0009 0.05
Benzene	Torpedo Grass Southern Bulrush Reed	7.04 12.00 9.33	1.52 5.10 0.05
Toluene	Torpedo Grass Southern Bulrush Reed	5.62 11.47 6.60	1.37 4.50 0.005
Chlorobenzene	Torpedo Grass Southern Bulrush	4.85 10.65	1.54 4.90
Phenol	Caltail Reed	101.00 104.00	17.00 7.00
P-Xylene	Reed	4.07	0.14
Pentachlorophenpl (PCP)	Torpedo Grass	0.85	0.04
Potassium Cyanide	Torpedo Grass	3.00	< 0.20
Potassium Ferric Cyanide	Torpedo Grass	12.60	< 0.20

*24-hour retention time

Table IX.2: **Effectiveness of Artificial Marshes for Treating Industrial Waste Water [19]**

Another concept is the BioHome, developed at the NASA Stinnes Space Center. This is a 60 m² habitat with the purpose to evaluate the efficiency of bioregenerative technology in a closed system. The structure is 14 m long and 4.9 m wide with a 30 cm thick fiberglass insulation. The BioHome is divided into

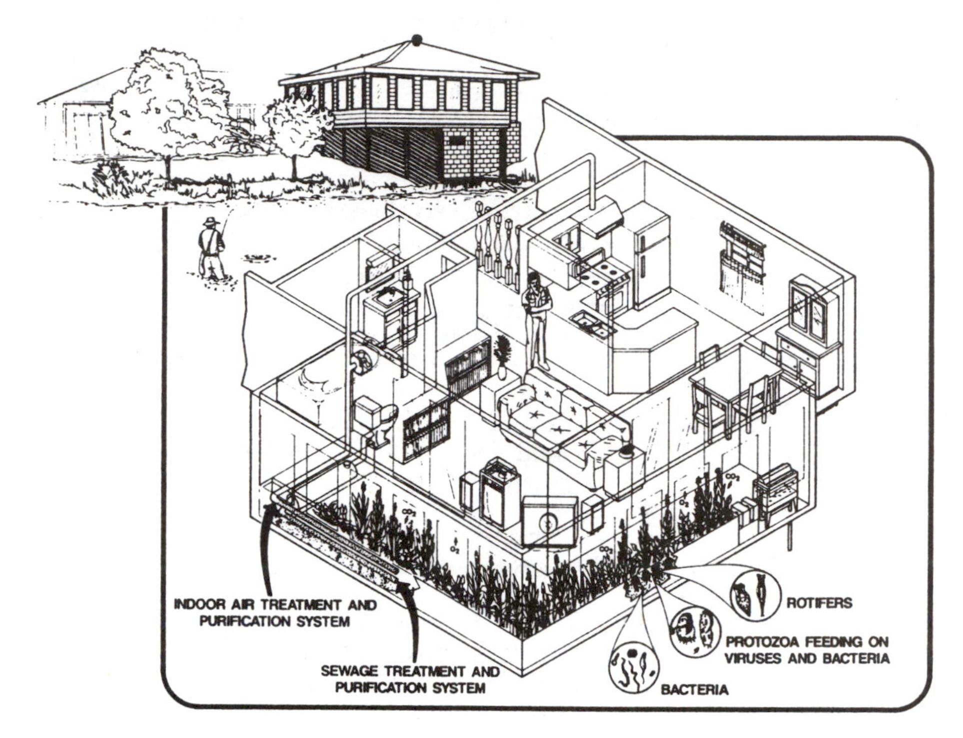

Figure IX.6: **A Real Home Environment Featuring Bioregenerative Technology [19]**

two areas: the living area and the waste treatment area. In the living area plant filters have been included to dissipate off-gassing products. The wastewater facility is essentially a small artificial wetland system adapted for inclusion in BioHome. Waste water flows from the exterior septic tank to a series of 20 cm diameter PVC pipes and finally in a 375 liter aquarium. Segment 1 of the pipe is empty in order to facilitate further settling of solids. Pipes 2 and 3 are approximately 50 % full of lava rock which functions to promote development of a biofilm. Segments 4 and 5 also have plants but the substrate inside is granular activated carbon, while segment 6 includes carbon in about the first meter of pipe, followed by zeolite. The zeolite works to remove the ammonia from the system. The source of drinking water in the BioHome is water vapor obtained from plants. The water quality was found to be well within the necessary requirements. Although there are so many plants in the treatment facility that it almost resembles a jungle, only about 11 liters of water were produced per day. This is not enough to supply one person with sufficient amounts of drinking, food preparation, and hygiene water. [9, 10]

IX.4 BIOMASS PRODUCTION AND RESEARCH

A wide field where closed ecological systems research may find some terrestrial applications is the area of biomass production and research. Here, the aspects of an efficient and optimal use of resources and area is of major interest, especially taking into account the progressive destruction of the environment and the growing world population. A list of specific, prospective research subjects in this context is given below:

- Optimization of the yield of certain plant cultures
- Optimization of plant production in devastated and extreme climatic areas
- Optimization of composting methods
- Development of chemical and toxin free plant protection
- Development of genetic engineered plants by, e.g., concerning plant resistance and use of N_2
- Development of optimized glass house cultures, e.g., with respect to substrates, control systems, saving of energy
- Development of new plant species
- Development of intensive methods of pisci-culturing
- Development of biological methods of soil-cleaning
- Development of new methods for plant culturing
- Development of toxin-free plant production methods
- Production of single cell food

Zeoponics (see section VI.4.3.2.4) also hold the promise of being a future commercial spin-off technology. Zeoponic substrates may potentially be used in a variety of commerical applications, including:

- substrates for plants in greenhouses,
- substrates for potting mixes for house plants,
- fertilizers on golf greens,
- field applications as a slow-release fertilizer, e.g., for iron,
- slow-release fertilizers in areas where environmental issues may be of concern, e.g., sandy soils near surface water or aquifers. [3]

Another important aspect is the potential use of closed ecological systems as biotechnology testbeds. In a such a complex but closed and controlled ecosystem, genetically engineered microorganisms and plants could be tested before their release into the outside world. Similarly, the full impact of chemicals, such as weed killers and insecticides, on representative ecosystems could be investigated without the risk to the terrestrial biosphere inherent in existing procedures. Of course, there are many more possible applications in this field that may be added to this tentative summary. [5, 13]

IX.5 SUPPORT OF TERRESTRIAL ANALOGS

Any development in the field of space life support research may, of course, also find application in the construction and improvement of terrestrial analogs or similar systems. These analogs include:

- Submarines and submarine habitats
- Underground habitats
- Antarctic Bases

Like space habitats submarines are relatively small vehicles largely isolated from biosphere. Therefore, in many respects basically the same systems as onboard a space habitat may be used for temperature and humidity control (CH_X), CO_2 removal LiOH or regenerable sorbents), or fire detection and suppression.

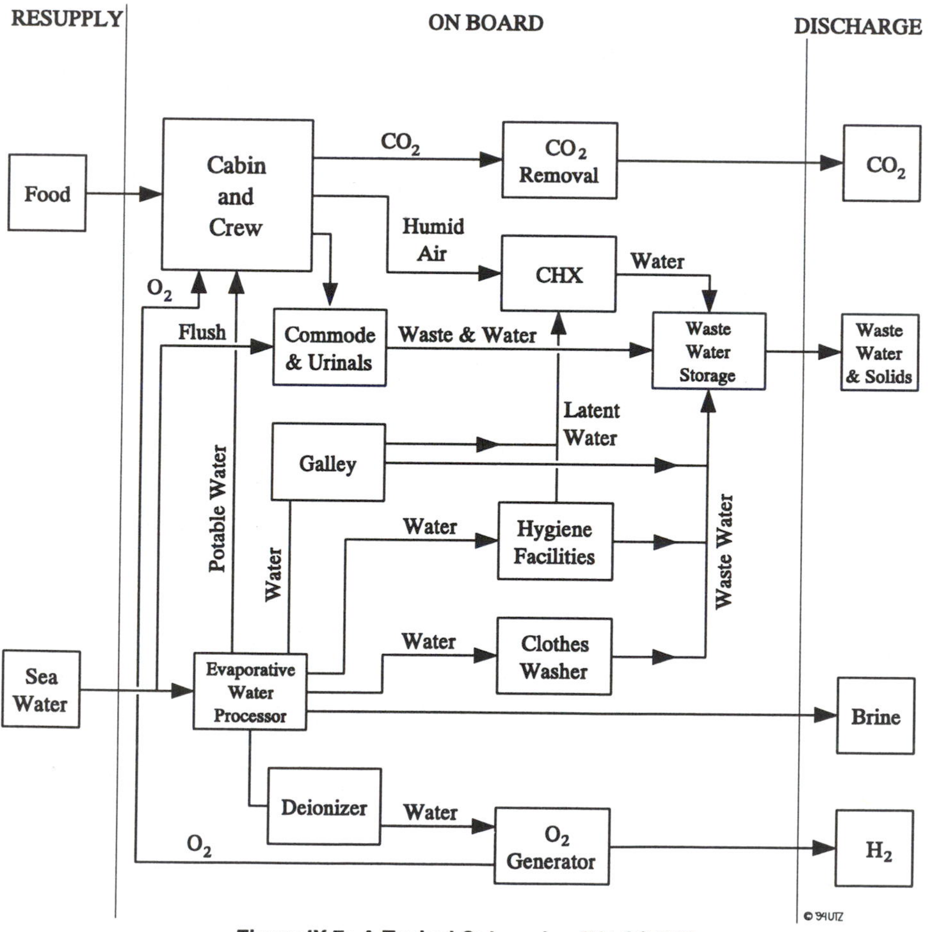

Figure IX.7: **A Typical Submarine ECLSS [18]**

This is also indicated in the scheme of a typical submarine ECLSS in figure IX.7. But although basically the same life support functions have to be fulfilled, it is easier to support life in a submarine than in a space habitat. For example, on nuclear powered submarines (which may remain isolated from the surface for 90 days or more), power is not in such short supply as in space habitats. Also, gravitational effects provide natural convection and gravity-assisted separation of liquids and gases, and gas leckage is not a problem, since the surrounding hydrostatic pressure is much higher than the internal atmospheric pressure. Above all, unlimited amounts of surrounding seawater are available as in-situ resources for water supply, water electrolysis for oxygen generation, and waste handling. Moreover, the safety requirements do not have to be as strict as for a space habitat, because a submarine still has the possibility to rise to the surface.

Also, present and proposed life support systems may be considered for use in underground habitats. Nevertheless, the trade criteria for an underground habitat are quite different from the requirements for space station and submarine application due to the difficulty in rejecting heat. Spacecraft can use radiation, submarines convection, whereas an underground habitat must reject heat via conduction to the surrounding rock or to a phase change device.

The analog of an Antarctic base and a planetary base is seen in the similarities with the types of human activity, the remoteness, the extreme environments, the isolation, and the problems associated with the resupply needed to support humans both in space and on the Antarctic continent. In detail this leads, for example, to the following similarities: lack of fresh food, cramped conditions, lack of usual social and family relations, artificial lighting, artificially purified air, and sensory deprivation, i.e., no smell, darkness, no humidity, no colors. In this respect, an Antarctic base may provide an excellent testbed for the simulation of lunar and Martian bases, especially of the psychological aspects of living under such extreme conditions for longer periods. The major "drawbacks" of Antarctic bases for space habitat simulation are that, like onboard submarines, gravity is still present and that atmospheric gases are, of course, not in short supply.

Presently, the largest base on the Antarctic continent is the McMurdo Station - it may have a summer population of up to 1000 people. Thus, concerning Antarctic bases it also may be attractive to find ways to reduce the resupply mass. For example the application of recycling systems may reduce the waste production, especially, since the dumping of waste into nature by the crews of the bases in Antarctica has already lead to some criticism in the public. [18]

IX References

[1] Allen J.
Business and Life in Space
NASA, Conference Publication, 1990

[2] Backhaus R. et al
Technikfolgenabschätzung zur Erdbeobachtung unter umweltstrategischen Aspekten
DLR Nachrichten, Heft 72, p. 22-27, 1993

[3] Barta D. et al
Development of Zeoponic Plant Growth Substrates for the Gravitational Biology Facility
Internal Report, NASA-JSC, 1994

[4] Dams R. et al
Air Purification Systems for Submarines
International Conference on Life Support and Biospherics, Proceedings, p. 263-271, 1992

[5] David K.; Preiß H.
DEBLSS - Deutsche Biologische Lebenserhaltungssystemstudie
Dornier GmbH, Bericht TN-DEBLSS-6000 DO/01, 1989

[6] Frye R.; Hodges C.
Soil Bed Reactor Work of the Environmental Research Lab of the University of Arizona in Support of the Research and Development Biosphere 2
NASA, Conference Publication, 1990

[7] Gitelson I.
Biotechnological Life Support Systems - Their Role in Moulding Ecological Morality and Ethics
International Conference on Life Support and Biospherics, Proceedings, p. 445-447, 1992

[8] Glenn E.; Frye R.
Soil Bed Reactors as Endogenous Control Systems for CELSS
Workshop on Artificial Ecosystems, DARA, p. 41-56, 1990

[9] Johnson A. et al
Assessment of Internal Contamination Problems Associated with Bioregenerative Air/Water Purfication Systems
20th Intersociety Conference on Environmental Systems, SAE Technical Paper 901379, 1990

[10] Johnson A.
The Biohome: A Spinoff of Space Technology
NASA, Conference Publication, 1990

[11] Knott B.
The CELSS Breadboard Project
NASA, Conference Publication, 1990

[12] MacElroy R.; Averner M.
Space Ecosynthesis: An Approach to the Design of Closed Ecosystems for Use in Space
NASA TM-78491, 1978

[13] Redor J. et al
CES-HABLAB: A Closed Ecological System and Habitability Test Facility
Preparing for the Future, Vol. 2, No. 2, ESA, 1992

[14] Ruppe H.O.
Die grenzenlose Dimension Raumfahrt, Vol. 1 & 2
ECON Verlag, Düsseldorf, 1980

[15] Scott C. et al
Life Support Research and Development, a Department of Energy Program for the Space Exploration Initiative
International Conference on Life Support and Biospherics, Proceedings, p. 83-91, 1992

[16] Soffen G.
Earth Observing Satellite: Understanding the Earth as a System
NASA, Conference Publication, 1990

[17] Sribnik F.
Life Support System for an Underground Habitat - Relation to Submarines and Space Station LSS
International Conference on Life Support and Biospherics, Proceedings, p. 509-525, 1992

[18] Wieland P.
Designing for Human Presence in Space
NASA RP-1324, 1994

[19] Wolverton B.; Wolverton J.
Bioregenerative Life Support Systems for Energy-Efficient Buildings
International Conference on Life Support and Biospherics, Proceedings, p. 117-126, 1992

[20] Wolverton B.
Plants and their Microbial Assistants: Natures Answer to Earth's Environmental Pollution Problems
NASA, Conference Publication, 1990

Index

Symbols

A

D

E

G

H

M

N

O

P

Q

R